Tensile Testing

Second Edition

Edited by
J.R. Davis
Davis & Associates

ASM International
Materials Park, Ohio 44073-0002
www.asminternational.org

First printing, December 2004
Digital printing, July 2009
Third printing, August 2020

Comments, criticisms, and suggestions are invited, and should be forwarded to ASM International.

Prepared under the direction of the ASM International Technical Book Committee (2004–2005), Yip-Wah Chung, Chair (FASM).

ASM International staff who worked on this project include Scott Henry, Senior Manager of Product and Service Development; Bonnie Sanders, Manager of Production; Carol Polakowski, Production Supervisor; and Pattie Pace, Production Coordinator.

Library of Congress Cataloging-in-Publication Data

Tensile testing / edited by J.R. Davis.—2nd ed.
p. cm.
Includes bibliographical references and index.
ISBN 0-87170-806-X
1. Materials—Testing. 2. Brittleness. 3. Tensiometers. I. Davis, J. R. (Joseph R.)
TA418.16.T46 2004
620.1′126—dc22 2004057353

ISBN-13: 978-0-87170-623-2 (print)
ISBN-10: 0-87170-806-X (print)
ISBN: 978-1-61503-215-0 (pdf)

SAN: 204-7586

ASM International®
Materials Park, OH 44073-0002
www.asminternational.org

Printed in the United States of America

Contents

Preface

In the preface to the first edition of *Tensile Testing*, editor Patricia Han wrote "Our vision for this book was to provide a volume that could serve not only as an introduction for those who are just starting to perform tensile tests and use tensile data, but also as a source of more detailed information for those who are better acquainted with the subject. We have written this reference book to appeal to laboratory managers, technicians, students, designers, and materials engineers." This vision has been preserved in the current edition, with some very important new topics added.

As in the first edition, section one opens with an introduction that discusses the fundamentals and language of tensile testing. Subsequent chapters describe test methodology and equipment, the use of tensile testing for design, and the use of tensile testing for determining the formability of sheet metals.

The second section consists of five chapters that deal with tensile testing of the major classes of engineering materials—metals, plastics, elastomers, ceramics, and composites. New material on testing of adhesively bonded joints, welded joints, and threaded fasteners has been added.

The third section contains chapters that review testing at elevated and low temperatures and special tests carried out at very high strain rates. Although these subjects were introduced in the first edition, they have been substantially expanded in this book.

In the fourth and final section, a glossary of terms related to tensile testing and properties has been compiled. Comprehensive tables provide tensile yield strengths of various materials and compare the elastic modulus of engineering materials.

In summary, this edition retains much of the flavor of the first edition while introducing readers to a number of additional topics that will extend their knowledge and appreciation of the tensile test.

Joseph R. Davis
Davis & Associates
Chagrin Falls, Ohio

CHAPTER 1

Introduction to Tensile Testing

TENSILE TESTS are performed for several reasons. The results of tensile tests are used in selecting materials for engineering applications. Tensile properties frequently are included in material specifications to ensure quality. Tensile properties often are measured during development of new materials and processes, so that different materials and processes can be compared. Finally, tensile properties often are used to predict the behavior of a material under forms of loading other than uniaxial tension.

The strength of a material often is the primary concern. The strength of interest may be measured in terms of either the stress necessary to cause appreciable plastic deformation or the maximum stress that the material can withstand. These measures of strength are used, with appropriate caution (in the form of safety factors), in engineering design. Also of interest is the material's ductility, which is a measure of how much it can be deformed before it fractures. Rarely is ductility incorporated directly in design; rather, it is included in material specifications to ensure quality and toughness. Low ductility in a tensile test often is accompanied by low resistance to fracture under other forms of loading. Elastic properties also may be of interest, but special techniques must be used to measure these properties during tensile testing, and more accurate measurements can be made by ultrasonic techniques.

This chapter provides a brief overview of some of the more important topics associated with tensile testing. These include:

- Tensile specimens and test machines
- Stress-strain curves, including discussions of elastic versus plastic deformation, yield points, and ductility
- True stress and strain
- Test methodology and data analysis

It should be noted that subsequent chapters contain more detailed information on these topics. Most notably, the following chapters should be referred to:

- Chapter 2, "Mechanical Behavior of Materials Under Tensile Loads"
- Chapter 3, "Uniaxial Tensile Testing"
- Chapter 4, "Tensile Testing Equipment and Strain Sensors"

Tensile Specimens and Testing Machines

Tensile Specimens. Consider the typical tensile specimen shown in Fig. 1. It has enlarged ends or shoulders for gripping. The important part of the specimen is the gage section. The cross-sectional area of the gage section is reduced relative to that of the remainder of the specimen so that deformation and failure will be

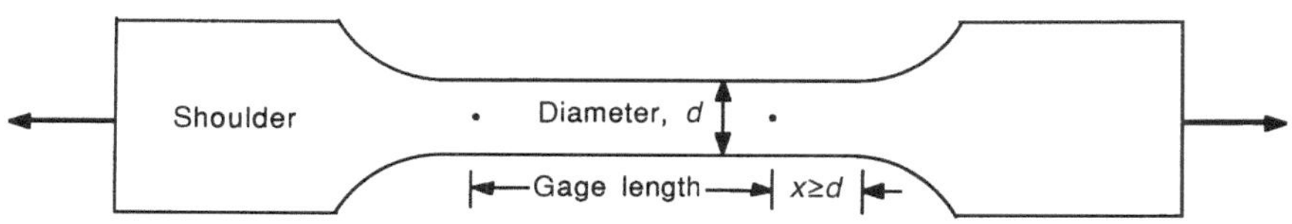

Fig. 1 Typical tensile specimen, showing a reduced gage section and enlarged shoulders. To avoid end effects from the shoulders, the length of the transition region should be at least as great as the diameter, and the total length of the reduced section should be at least four times the diameter.

localized in this region. The gage length is the region over which measurements are made and is centered within the reduced section. The distances between the ends of the gage section and the shoulders should be great enough so that the larger ends do not constrain deformation within the gage section, and the gage length should be great relative to its diameter. Otherwise, the stress state will be more complex than simple tension. Detailed descriptions of standard specimen shapes are given in Chapter 3 and in subsequent chapters on tensile testing of specific materials.

There are various ways of gripping the specimen, some of which are illustrated in Fig. 2. The end may be screwed into a threaded grip, or it may be pinned; butt ends may be used, or the grip section may be held between wedges. There are still other methods (see, for example, Fig. 24 in Chapter 3). The most important concern in the selection of a gripping method is to ensure that the specimen can be held at the maximum load without slippage or failure in the grip section. Bending should be minimized.

Testing Machines. The most common testing machines are universal testers, which test materials in tension, compression, or bending. Their primary function is to create the stress-strain curve described in the following section in this chapter.

Testing machines are either electromechanical or hydraulic. The principal difference is the method by which the load is applied.

Electromechanical machines are based on a variable-speed electric motor; a gear reduction system; and one, two, or four screws that move the crosshead up or down. This motion loads the specimen in tension or compression. Crosshead speeds can be changed by changing the speed of the motor. A microprocessor-based closed-loop servo system can be implemented to accurately control the speed of the crosshead.

Hydraulic testing machines (Fig. 3) are based on either a single or dual-acting piston that moves the crosshead up or down. However, most static hydraulic testing machines have a single acting piston or ram. In a manually operated machine, the operator adjusts the orifice of a pressure-compensated needle valve to control the rate of loading. In a closed-loop hydraulic servo system, the needle valve is replaced by an electrically operated servo valve for precise control.

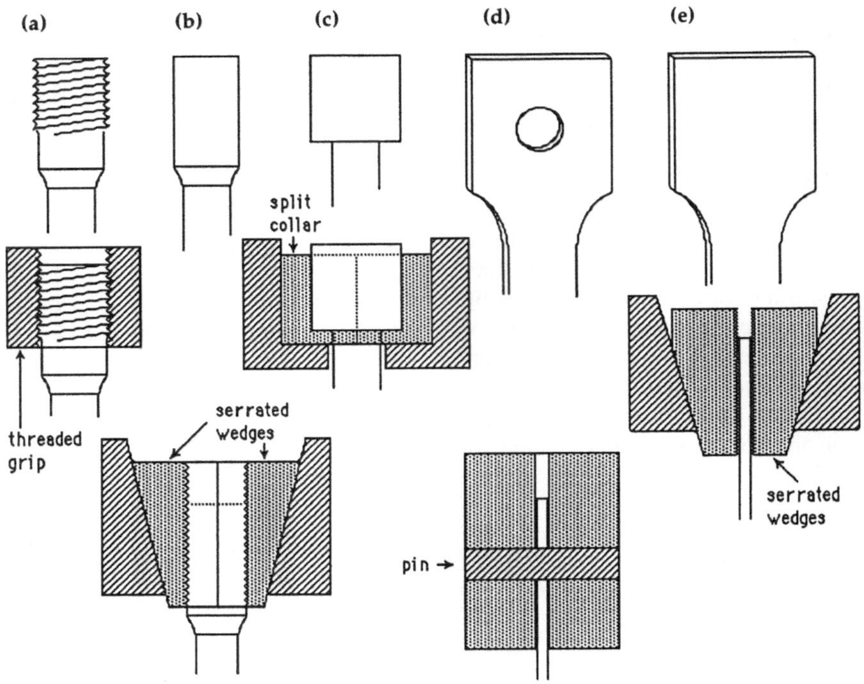

Fig. 2 Systems for gripping tensile specimens. For round specimens, these include threaded grips (a), serrated wedges (b), and, for butt end specimens, split collars constrained by a solid collar (c). Sheet specimens may be gripped with pins (d) or serrated wedges (e).

In general, electromechanical machines are capable of a wider range of test speeds and longer crosshead displacements, whereas hydraulic machines are more cost-effective for generating higher forces.

Stress-Strain Curves

A tensile test involves mounting the specimen in a machine, such as those described in the previous section, and subjecting it to tension. The tensile force is recorded as a function of the increase in gage length. Figure 4(a) shows a typical curve for a ductile material. Such plots of tensile force versus tensile elongation would be of little value if they were not normalized with respect to specimen dimensions.

Engineering stress, or nominal stress, s, is defined as

$$s = F/A_0 \quad \text{(Eq 1)}$$

where F is the tensile force and A_0 is the initial cross-sectional area of the gage section.

Engineering strain, or nominal strain, e, is defined as

$$e = \Delta L/L_0 \quad \text{(Eq 2)}$$

where L_0 is the initial gage length and ΔL is the change in gage length ($L - L_0$).

When force-elongation data are converted to engineering stress and strain, a stress-strain curve (Fig. 4b) that is identical in shape to the force-elongation curve can be plotted. The advantage of dealing with stress versus strain rather than load versus elongation is that the stress-strain curve is virtually independent of specimen dimensions.

Elastic versus Plastic Deformation. When a solid material is subjected to small stresses, the bonds between the atoms are stretched. When the stress is removed, the bonds relax and the material returns to its original shape. This re-

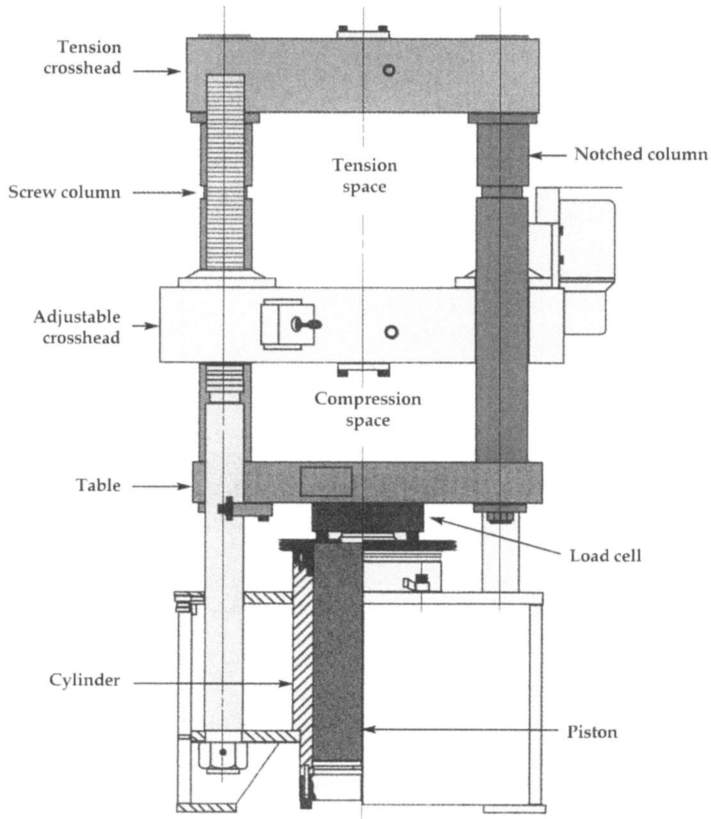

Fig. 3 Components of a hydraulic universal testing machine

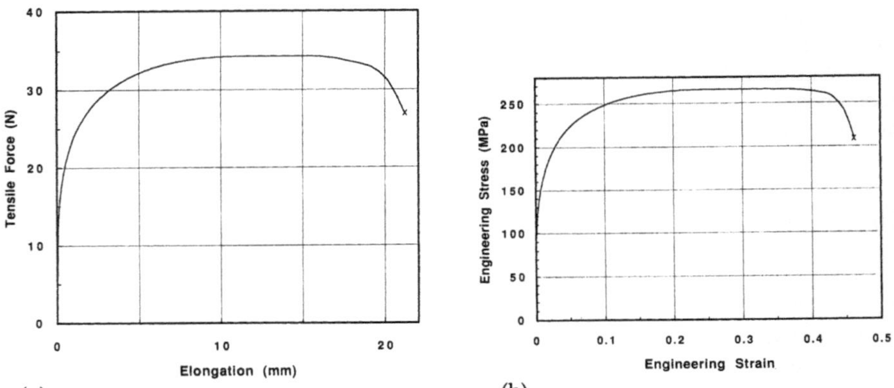

Fig. 4 (a) Load-elongation curve from a tensile test and (b) corresponding engineering stress-strain curve. Specimen diameter, 12.5 mm; gage length, 50 mm.

versible deformation is called *elastic deformation.* (The deformation of a rubber band is entirely elastic). At higher stresses, planes of atoms slide over one another. This deformation, which is not recovered when the stress is removed, is termed *plastic deformation.* Note that the term "plastic deformation" does not mean that the deformed material is a plastic (a polymeric material). Bending of a wire (such as paper-clip wire) with the fingers (Fig. 5) illustrates the difference. If the wire is bent a little bit, it will snap back when released (top). With larger bends, it will unbend elastically to some extent on release, but there will be a permanent bend because of the plastic deformation (bottom).

For most materials, the initial portion of the curve is linear. The slope of this linear region is called the *elastic modulus* or *Young's modulus:*

$$E = s/e \quad \text{(Eq 3)}$$

In the elastic range, the ratio, υ, of the magnitude of the lateral contraction strain to the axial strain is called *Poisson's ratio:*

$$\upsilon = -e_y/e_x \text{ (in an } x\text{-direction tensile test)} \quad \text{(Eq 4)}$$

Because elastic strains are usually very small, reasonably accurate measurement of Young's modulus and Poisson's ratio in a tensile test requires that strain be measured with a very sensitive extensometer. (Strain gages should be used for lateral strains.) Accurate results can also be obtained by velocity-of-sound measurements (unless the modulus is very low or the damping is high, as with polymers).

When the stress rises high enough, the stress-strain behavior will cease to be linear and the strain will not disappear completely on unloading. The strain that remains is called plastic strain. The first plastic strain usually corresponds to the first deviation from linearity. (For some materials, the elastic deformation may be nonlinear, and so there is not always this correspondence). Once plastic deformation has begun, there will be both elastic and plastic contributions to the total strain, e_T. This can be expressed as $e_T = e_e + e_p$, where e_p is the plas-

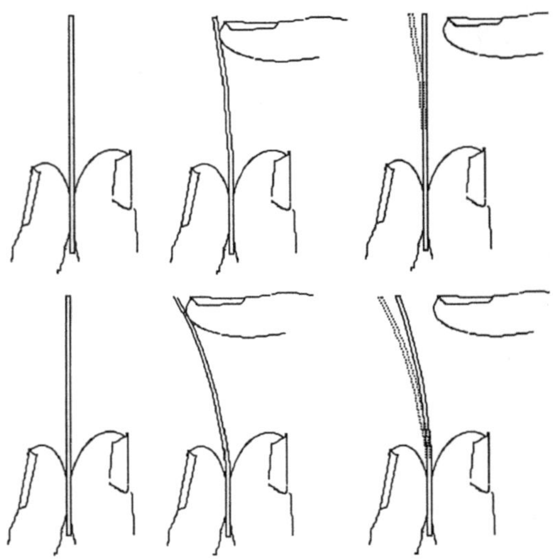

Fig. 5 Elastic and plastic deformation of a wire with the fingers. With small forces (top), all of the bending is elastic and disappears when the force is released. With greater forces (below), some of the bending is recoverable (elastic), but most of the bending is not recovered (is plastic) when the force is removed.

tic contribution and e_e is the elastic contribution (and still related to the stress by Eq 3).

It is tempting to define an *elastic limit* as the stress at which plastic deformation first occurs and a *proportional limit* as the stress at which the stress-strain curve first deviates from linearity. However, neither definition is very useful, because measurement of the stress at which plastic deformation first occurs or the first deviation from linearity is observed depends on how accurately strain can be measured. The smaller the plastic strains that can be sensed and the smaller the deviations from linearity can be detected, the smaller the elastic and proportional limits.

To avoid this problem, the onset of the plasticity is usually described by an offset *yield strength*, which can be measured with greater reproducibility. It can be found by constructing a straight line parallel to the initial linear portion of the stress-strain curve, but offset by e = 0.002 or 0.2%. The yield strength is the stress at which this line intersects the stress-strain curve (Fig. 6). The rationale is that if the material had been loaded to this stress and then unloaded, the unloading path would have been along this offset line and would have resulted in a plastic strain of e = 0.2%. Other offset strains are sometimes used. The advantage of defining yield strength in this way is that such a parameter is easily reproduced and does not depend heavily on the sensitivity of measurement.

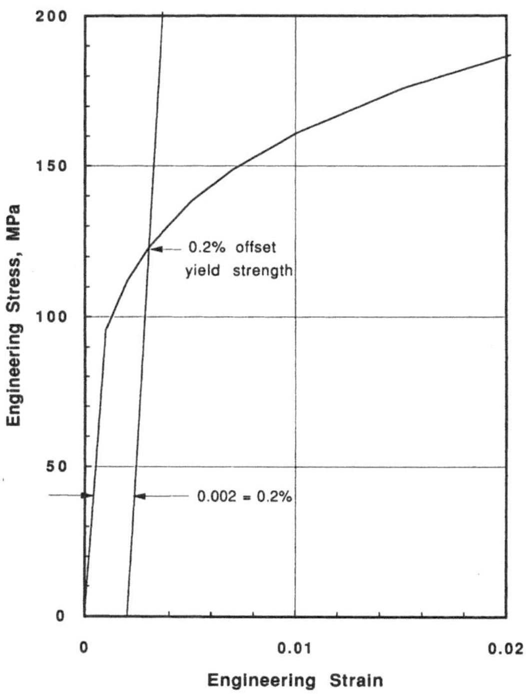

Fig. 6 The low-strain region of the stress-strain curve for a ductile material

Sometimes, for convenience, yielding in metals is defined by the stress required to achieve a specified total strain (e.g., e_T = 0.005 or 0.5% elongation) instead of a specified offset strain. In any case, the criterion should be made clear to the user of the data.

Yield Points. For some materials (e.g., low-carbon steels and many linear polymers), the stress-strain curves have initial maxima followed by lower stresses, as shown in Fig. 7(a) and (b). After the initial maximum, all the deformation at any instant is occurring within a relatively small region of the specimen. Continued elongation of the specimen occurs by propagation of the deforming region (Lüders band in the case of steels) along the gage section rather than by increased strain within the deforming region. Only after the entire gage section has been traversed by the band does the stress rise again. In the case of linear polymers, a yield strength is often defined as the initial maximum stress. For steels, the subsequent *lower* yield strength is used to describe yielding. This is because measurements of the initial maximum or *upper* yield strength are extremely sensitive to how axially the load is applied during the tensile test. Some laboratories cite the minimum, whereas others cite a mean stress during this discontinuous yielding.

The tensile strength (ultimate strength) is defined as the highest value of engineering stress* (Fig. 8). Up to the maximum load, the deformation should be uniform along the gage section. With ductile materials, the tensile strength corresponds to the point at which the deformation starts to localize, forming a neck (Fig. 8a). Less ductile materials fracture before they neck (Fig. 8b). In this case, the fracture strength is the tensile strength. Indeed, very brittle materials (e.g., glass at room temperature) do not yield before fracture (Fig. 8c). Such materials have tensile strengths but not yield strengths.

Ductility. There are two common measures used to describe the ductility of a material. One

*Sometimes the upper yield strength of low-carbon steel is higher than the subsequent maximum. In such cases, some prefer to define the tensile strength as the subsequent maximum instead of the initial maximum, which is higher. In such cases, the definition of tensile strength should be made clear to the user.

is the percent elongation, which is defined simply as

$$\%El = [(L_f - L_0)/L_0] \times 100 \quad \text{(Eq 5)}$$

where L_0 is the initial gage length and L_f is the length of the gage section at fracture. Measurements may be made on the broken pieces or under load. For most materials, the amount of elastic elongation is so small that the two are equivalent. When this is not so (as with brittle metals or rubber), the results should state whether or not the elongation includes an elastic contribution. The other common measure of ductility is percent reduction of area, which is defined as

$$\%RA = [(A_0 - A_f)/A_0] \times 100 \quad \text{(Eq 6)}$$

where A_0 and A_f are the initial cross-sectional area and the cross-sectional area at fracture, respectively. If failure occurs without necking, one can be calculated from the other:

$$\%El = \%RA/(100 - \%RA) \quad \text{(Eq 7)}$$

After a neck has developed, the two are no longer related. Percent elongation, as a measure of ductility, has the disadvantage that it is really composed of two parts: the uniform elongation that occurs before necking, and the localized elongation that occurs during necking. The second part is sensitive to the specimen shape. When a gage section that is very long (relative to its diameter), the necking elongation converted to percent is very small. In contrast, with a gage section that is short (relative to its di-

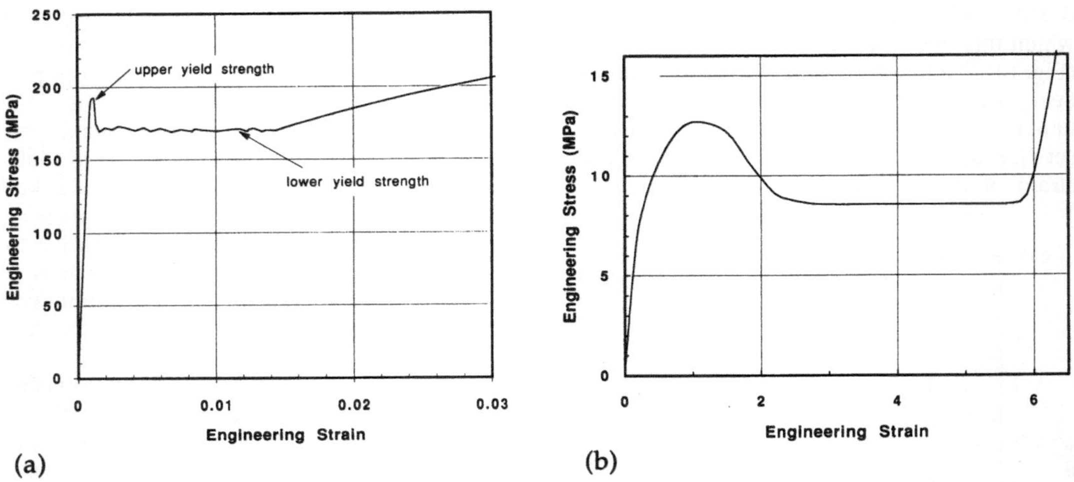

Fig. 7 Inhomogeneous yielding of a low-carbon steel (a) and a linear polymer (b). After the initial stress maxima, the deformation occurs within a narrow band, which propagates along the entire length of the gage section before the stress rises again.

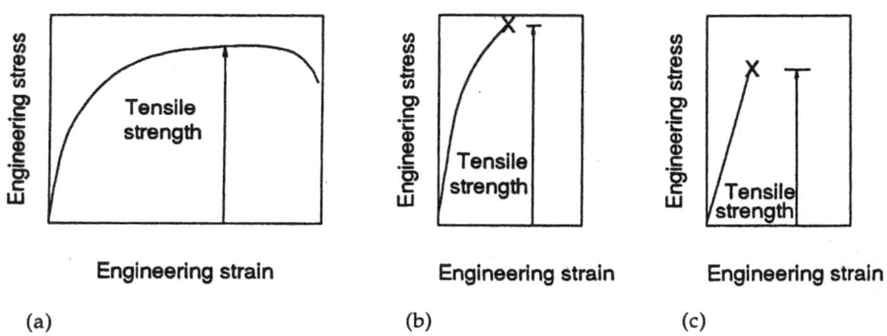

Fig. 8 Stress-strain curves showing that the tensile strength is the maximum engineering stress regardless of whether the specimen necks (a) or fractures before necking (b and c).

ameter), the necking elongation can account for most of the total elongation.

For round bars, this problem has been remedied by standardizing the ratio of gage length to diameter to 4:1. Within a series of bars, all with the same gage-length-to-diameter ratio, the necking elongation will be the same fraction of the total elongation. However, there is no simple way to make meaningful comparisons of percent elongation from such standardized bars with that measured on sheet tensile specimens or wire. With sheet tensile specimens, a portion of the elongation occurs during diffuse necking, and this could be standardized by maintaining the same ratio of width to gage length. However, a portion of the elongation also occurs during what is called localized necking, and this depends on the sheet thickness. For tensile testing of wire, it is impractical to have a reduced section, and so the ratio of gage length to diameter is necessarily very large. Necking elongation contributes very little to the total elongation.

Percent reduction of area, as a measure of ductility, has the disadvantage that with very ductile materials it is often difficult to measure the final cross-sectional area at fracture. This is particularly true of sheet specimens.

True Stress and Strain

If the results of tensile testing are to be used to predict how a metal will behave under other forms of loading, it is desirable to plot the data in terms of true stress and true strain. True stress, σ, is defined as

$$\sigma = F/A \quad \text{(Eq 8)}$$

where A is the cross-sectional area at the time that the applied force is F. Up to the point at which necking starts, true strain, ε, is defined as

$$\varepsilon = \ln(L/L_0) \quad \text{(Eq 9)}$$

This definition arises from taking an increment of true strain, $d\varepsilon$, as the incremental change in length, dL, divided by the length, L, at the time, $d\varepsilon = dL/L$, and integrating. As long as the deformation is uniform along the gage section, the true stress and strain can be calculated from the engineering quantities. With constant volume and uniform deformation, $LA = L_0A_0$:

$$A_0/A = L/L_0 \quad \text{(Eq 10)}$$

Thus, according to Eq 2, $A_0/A = 1 + e$. Equation 8 can be rewritten as

$$\sigma = (F/A_0)(A_0/A)$$

and, with substitution for A_0/A and F/A_0, as

$$\sigma = s(1 + e) \quad \text{(Eq 11)}$$

Substitution of $L/L_0 = 1 + e$ into the expression for true strain (Eq 9) gives

$$\varepsilon = \ln(1 + e) \quad \text{(Eq 12)}$$

At very low strains, the differences between true and engineering stress and strain are very small. It does not really matter whether Young's modulus is defined in terms of engineering or true stress strain.

It must be emphasized that these expressions are valid only as long as the deformation is uniform. Once necking starts, Eq 8 for true stress is still valid, but the cross-sectional area at the base of the neck must be measured directly rather than being inferred from the length measurements. Because the true stress, thus calculated, is the true stress at the base of the neck, the corresponding true strain should also be at the base of the neck. Equation 9 could still be used if the L and L_0 values were known for an extremely short gage section centered on the middle of the neck (one so short that variations of area along it would be negligible). Of course, there will be no such gage section, but if there were, Eq 10 would be valid. Thus the true strain can be calculated as

$$\varepsilon = \ln(A_0/A) \quad \text{(Eq 13)}$$

Figure 9 shows a comparison of engineering and true stress-strain curves for the same material.

Other Factors Influencing the Stress-Strain Curve

There are a number of factors not previously discussed in this chapter that have an effect on the shape of the stress-strain curve. These include strain rate, temperature, and anisotropy. For information on these subjects, the reader should refer to Chapters 2 and 3 listed in the introduction to this chapter as well as Chapter 12, "Hot Tensile Testing" and Chapter 15, "High Strain Rate Tensile Testing."

Test Methodology and Data Analysis

This section reviews some of the more important considerations involved in tensile testing. These include:

- Sample selection
- Sample preparation
- Test set-up
- Test procedure
- Data recording and analysis
- Reporting

Sample Selection. When a material is tested, the objective usually is to determine whether or not the material is suitable for its intended use.

The sample to be tested must fairly represent the body of material in question. In other words, it must be from the same source and have undergone the same processing steps.

It is often difficult to match exactly the test samples to the structure made from the material. A common practice for testing of large castings, forgings, and composite layups is to add extra material to the part for use as "built-in" test samples. This material is cut from the completed part after processing and is made into test specimens that have been subjected to the same processing steps as the bulk of the part.

In practice, these specimens may not exactly match the bulk of the part in certain important details, such as the grain patterns in critical areas of a forging. One or more complete parts may be sacrificed to obtain test samples from the most critical areas for comparison with the "built-in" samples. Thus, it may be determined how closely the "built-in" samples represent the material in question.

There is a special case in which the object of the test is to evaluate not the material, but the test itself. Here, the test specimens must be as nearly identical as possible so the differences in the test results represent, as far as possible, only the variability in the testing process.

Sample Preparation. It should be remembered that a "sample" is a quantity of material that represents a larger lot. The sample usually is made into multiple "specimens" for testing. Test samples must be prepared properly to achieve accurate results. The following rules are suggested for general guidance.

First, as each sample is obtained, it should be identified as to material description, source, location and orientation with respect to the body of material, processing status at the time of sampling, and the data and time of day that the sample was obtained.

Second, test specimens must be made carefully, with attention to several details. The specimen axis must be properly aligned with the material rolling direction, forging grain pattern, or composite layup. Cold working of the test section must be minimized. The dimensions of the specimen must be held within the allowable tolerances established by the test procedure. The attachment areas at each end of the specimen must be aligned with the axis of the bar (see Fig. 10). Each specimen must be identified as belonging to the original sample. If total elongation is to be measured after the specimen breaks, the gage length must be marked on the reduced section of the bar prior to testing.

The test set-up requires that equipment be properly matched to the test at hand. There are

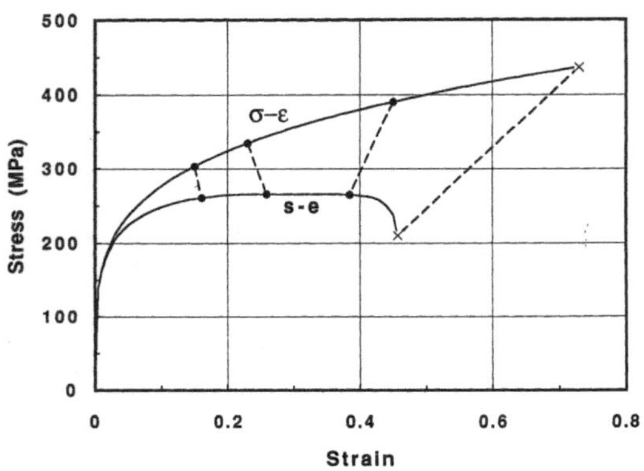

Fig. 9 Comparison of engineering and true stress-strain curves. Prior to necking, a point on the σ-ε curve can be constructed from a point on the *s-e* curve using Eq 11 and 12. Subsequently, the cross section must be measured to find true stress and strain.

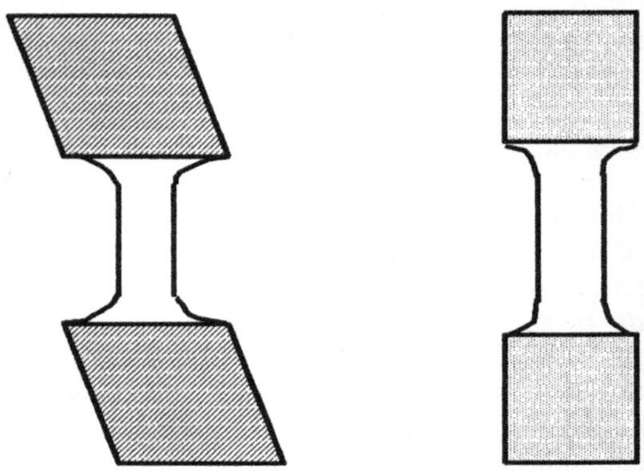

Fig. 10 Improper (left) and proper (right) alignment of specimen attachment areas with axis of specimen

three requirements of the testing machine: force capacity sufficient to break the specimens to be tested; control of test speed (or strain rate or load rate), as required by the test specification; and precision and accuracy sufficient to obtain and record properly the load and extension information generated by the test. This precision and accuracy should be ensured by current calibration certification.

For grips, of which many types are in common use in tensile testing, only two rules apply: the grips must properly fit the specimens (or vice versa), and they must have sufficient force capacity so that they are not damaged during testing.

As described earlier in the section "Tensile Specimens and Testing Machines," there are several techniques for installing the specimen in the grips. With wedge grips, placement of the specimen in the grips is critical to proper alignment (see Fig. 11). Ideally, the grip faces should be of the same width as the tab ends of the test bar; otherwise, lateral alignment is dependent only on the skill of the technician. The wedge grip inserts should be contained within the grip body or crosshead, and the specimen tabs should be fully engaged by the grips (see Fig. 12).

Other types of grips have perhaps fewer traps for the inexperienced technician, but an obvious one is that, with threaded grips, a length of threads on the specimen equal to at least one diameter should be engaged in the threaded grips.

There are several potential problems that must be watched for during the test set-up, including specimen misalignment and worn grips. The physical alignment of the two points of attachment of the specimen is important, because any off-center loading will exert bending loads on the specimen. This is critical in testing of brittle materials, and may cause problems even for ductile materials. Alignment will be affected by the testing-machine loadframe, any grips and fixtures used, and the specimen itself. Misalignment may also induce load-measurement errors due to the passage of bending forces through the load-measuring apparatus. Such errors may be reduced by the use of spherical seats or "U-joints" in the set-up.

Worn grips may contribute to off-center loading. Uneven tooth marks across the width of the specimen tab are an indication of trouble in wedge grips. Split-collar grips may also cause off-center loading. Uneven wear of grips and mismatching of split-shell insert pairs are potential problem areas.

Strain measurements are required for many tests. They are commonly made with extensometers, but strain gages are frequently used—especially on small specimens or where Pois-

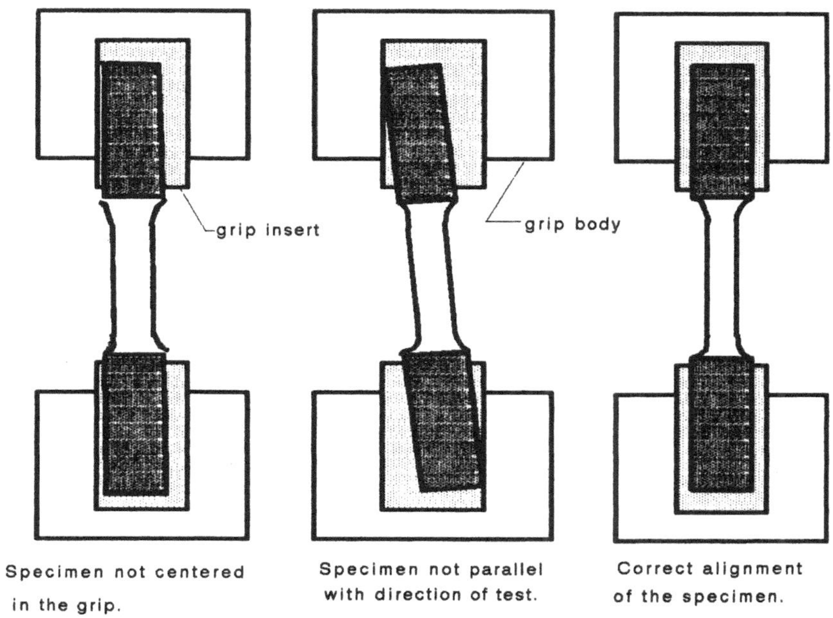

Fig. 11 Improper (left, center) and proper (right) alignment of specimen in wedge grips

son's ratio is to be measured. If strain measurements are required, appropriate strain-measuring instruments must be properly installed. The technician should pay particular attention to setting of the extensometer gage length (mechanical zero). The zero of the strain readout should repeat consistently if the mechanical zero is set properly. In other words, once the extensometer has been installed and zeroed, subsequent installations should require minimal readjustment of the zero.

Test Procedure. The following general rules for test procedure may be applied to almost every tensile test.

Load and strain ranges should be selected so that the test will fit the range. The maximum values to be recorded should be as close to the top of the selected scale as convenient without running the risk of going past full scale. Ranges may be selected using past experience for a particular test, or specification data for the material (if available). Note that many computer-based testing systems have automatic range selection and will capture data even if the range initially selected is too small.

The identity of each specimen should be verified, and pertinent identification should be accurately recorded for the test records and report.

The dimensions needed to calculate the cross-sectional area of the reduced section should be measured and recorded. These measurements should be repeated for every specimen; it should not be assumed that sample preparation is perfectly consistent.

The load-indicator zero and the plot-load-axis zero, if applicable, should be set before the specimen is placed in the grips. Zeroes should never be reset after the specimen is in place.

The specimen is placed in the grips and is secured by closing the grips. If preload is to be removed before the test is started, it should be physically unloaded by moving the loading mechanism. The zero adjustment should never be used for this purpose. Note that, in some cases, preload may be desirable and may be deliberately introduced. For materials for which the initial portion of the curve is linear, the strain zero may be corrected for preload by extending the initial straight portion of the stress-strain curve to zero load and measuring strain from that point. The strain valve at the zero-load intercept is commonly called the "foot correction" and is subtracted from readings taken from strain scale (see Fig. 10 in Chapter 3, "Uniaxial Tensile Testing").

When the extensometer, if applicable, is installed, the technician should be sure to set the mechanical zero correctly. The strain-readout zero should be set after the extensometer is in place on the specimen.

The test procedure should be in conformance with the published test specification and should

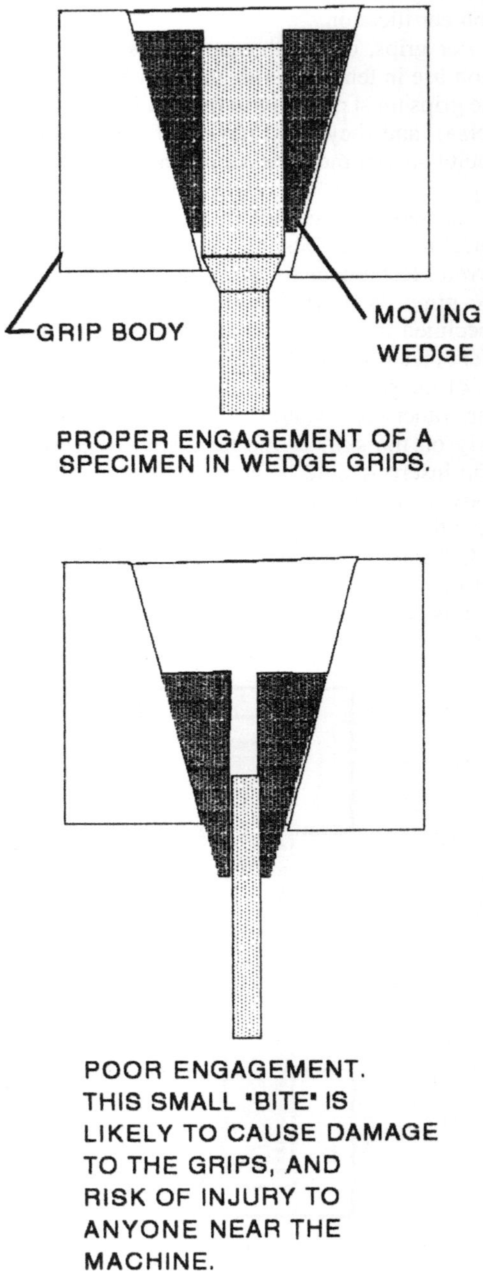

Fig. 12 Proper and improper engagement of a specimen in wedge grips

be repeated consistently for every test. It is important that the test specification be followed for speed of testing. Some materials are sensitive to test speed, and different speeds will give different results. Also, many testing machine load- and strain-measuring instruments are not capable of responding fast enough for accurate recording of test results if an excessive test speed is used.

The technician should monitor the test closely and be alert for problems. One common sign of trouble is a load-versus-strain plot in which the initial portion of the curve is not straight. This may indicate off-center loading of the specimen, improper installation of the extensometer, or the specimen was not straight to begin with.

Another potential trouble sign is a sharp drop in indicated load during the test. Such a drop may be characteristic of the material, but it also can indicate problems such as slippage between the specimen and the grips or stick-slip movement of the wedge grip inserts in the grip body. Slippage may be caused by worn inserts with dull teeth, particularly for hard, smooth specimens.

The stick-slip action in wedge grips is more common in testing of resilient materials, but it also can occur in testing of metals. Specimens cut from the wall of a pipe or tube may have curved tab ends that flatten with increasing force, allowing the inserts to move relative to the grip body. Short tab ends on round specimens also may be crushed by the wedge grips, with the same result. If the sliding faces are not lubricated, they may move in unpredictable steps accompanied by drops in the load reading. Dry-film molybdenum disulfide lubricants are effective in solving stick-slip problems in wedge grips, particularly when testing is done at elevated temperature.

When wedge grips are used, the specimen must be installed so that the clamping force is contained within the grip body. Placing the specimen too near the open end of the grip body results in excessive stress on the grip body and inserts and is a common cause of grip failure. WARNING: Grip failures are dangerous and may cause injury to personnel and damage to equipment.

Data generally may be grouped into "raw data," meaning the observed readings of the measuring instruments, and "calculated data," meaning the test results obtained after the first step of analysis.

In the most simple tensile test, the raw data comprise a single measurement of peak force and the dimensional measurements taken to determine the cross-sectional area of the test specimen. The first analysis step is to calculate the "tensile strength," defined as the force per unit area required to fracture the specimen. More complicated tests will require more information, which typically takes the form of a graph of force versus extension. Computer-based testing machines can display the graph without paper, and can save the measurements associated with the graph by electronic means.

A permanent record of the raw test data is important, because it allows additional analyses to be performed later, if desired, and because it allows errors in analysis to be found and corrected by reference to the original data.

Data Recording. Test records may be needed by many departments within an organization, including metallurgy, engineering, commercial, and legal departments.

Engineering and metallurgy departments typically are most interested in material properties, but may use raw data for error checking or additional analyses. The metallurgy department wants to know how variations in raw materials or processing change the properties of the product being produced and tested, and the engineering department wants to know the properties of the material for design purposes.

Shipping, receiving, and accounting departments need to know whether or not the material meets the specifications for shipping, acceptance, and payment. The sales department needs information for advertising and for advising prospective customers.

If a product incorporating the tested material later fails—particularly if persons are injured—the legal department may need test data as evidence in legal proceedings. In this case, a record of the raw data will be important for support of the original analysis and test report.

Analysis of test data is done at several levels. First, the technician observes the test in progress, and may see that a grip is slipping or that the specimen fractures outside the gage section. These observations may be sufficient to determine that a test is invalid.

Immediately after the test, a first-level analysis is performed according to the calculation requirements of the test procedure. ASTM test specifications typically show the necessary equations with an explanation and perhaps an example. This analysis may be as simple as dividing peak force by cross-sectional area, or it may require more complex calculations. The outputs of this first level of analysis are the mechanical properties of the material being tested.

Upon completion of the group of tests performed on the sample, a statistical analysis may be made. The statistical analysis produces average (mean or median) values for representation of the sample in the subsequent database and also provides information about the uniformity of the material and the repeatability of the test.

The results of tests on each sample of material may be stored in a database for future use. The database allows a wide range of analyses to be performed using statistical methods to correlate the mechanical-properties data with other information about the material. For example, it may allow determination of whether or not there is a significant difference between the material tested and similar material obtained from a different supplier or through a different production path.

Reporting. The test report usually contains the results of tests performed on one sample composed of several specimens.

When ASTM specifications are used for testing, the requirements for reporting are defined by the specification. The needs of a particular user probably will determine the form for identification of the material, but the reported results will most likely be as given in the ASTM test specification.

The information contained in the test report generally should include identification of the testing equipment, the material tested, and the test procedure; the raw and calculated data for each specimen; and a brief statistical summary for the sample.

Each piece of test equipment used for the test should be identified, including serial numbers, capacity or range used, and date of certification or date due for certification.

Identification of the material tested should include the type of material (alloy, part number, etc.); the specific batch, lot, order, heat, or coil from which the sample was taken; the point in the processing sequence (condition, temper, etc.) at which the sample was taken; and any test or pretest conditions (test temperature, aging, etc.).

Identification of the test procedure usually will be reported by reference to a standard test procedure such as those published by ASTM or perhaps to a proprietary specification originating within the testing organization.

The raw data for each specimen are recorded, or a reference to the raw data is included so that the data can be obtained from a file if and when they are needed. Frequently, only a portion of the raw data—dimensions, for example—is recorded, and information on the force-versus-extension graph is referenced.

A tabulation of the properties calculated for each specimen is recorded. The calculations at this stage are the first level of data analysis. The calculations required usually are defined in the test procedure or specification.

A brief statistical summary for the sample is a feature that is becoming more common with the proliferation of computerized testing systems, because the computations required can be done automatically without added operator workload. The statistical summary may include the average (mean) value, median value, standard deviation, highest value, lowest value, range, etc. The average or median value would be used to represent this sample at the next level of analysis, which is the material database.

Examination of this initial statistical information can tell a great deal about the test as well as the material. A low standard deviation or range indicates that the material in the sample has uniform properties (each of several specimens has nearly the same values for the measured properties) and that the test is producing consistent results. Conversely, a high standard deviation or range indicates that a problem of inconsistent material or testing exists and needs to be investigated.

A continuing record of the average properties and the associated standard deviation and range information is the basis for statistical process control, which systematically interprets this information so as to provide the maximum information about both the material and the test process.

ACKNOWLEDGMENTS

This chapter was adapted from:

- W.F. Hosford, Overview of Tensile Testing, *Tensile Testing,* P. Han, Ed., ASM International, 1992, p 1–24
- P.M. Mumford, Test Methodology and Data Analysis, *Tensile Testing,* P. Han, Ed., ASM International, 1992, p 49–60
- R. Gedney, Guide To Testing Metals Under Tension, *Advanced Materials & Processes,* February, 2002, p 29–31

CHAPTER 2

Mechanical Behavior of Materials under Tensile Loads

THE MECHANICAL BEHAVIOR OF MATERIALS is described by their deformation and fracture characteristics under applied stresses (for example, tensile, compressive, or multiaxial stresses). Determination of this mechanical behavior is influenced by several factors that include metallurgical/material variables, test methods, and the nature of the applied stresses.

This chapter focuses on mechanical behavior under conditions of uniaxial tension during tensile testing. As stated in other chapters, the engineering tensile test is widely used to provide basic design information on the strength of materials and as an acceptance test for the specification of materials. In this test procedure, a specimen is subjected to a continually increasing uniaxial load (force), while simultaneous observations are made of the elongation of the specimen. In this chapter, emphasis is placed on the interpretation of these observations rather than on the procedures for conducting the tests. Test procedures are described in Chapter 3, "Uniaxial Tensile Testing."

Emphasis has also been placed in this chapter on the response of *metallic materials* to tensile stresses. Additional information can be found in Chapter 7, "Tensile Testing of Metals and Alloys." The mechanical behaviors of *nonmetallic materials* under tension are discussed in Chapters 8 (plastics), 9 (elastomers), 10 (ceramics and ceramic-matrix composites), and 11 (fiber-reinforced composites).

Engineering Stress-Strain Curve

In the conventional engineering tensile test, an engineering stress-strain curve is constructed from the load-elongation measurements made on the test specimen (Fig. 1). The engineering stress (s) used in this stress-strain curve is the average longitudinal stress in the tensile specimen. It is obtained by dividing the load (P) by the original area of the cross section of the specimen (A_0):

$$s = \frac{P}{A_0} \qquad \text{(Eq 1)}$$

The strain, e, used for the engineering stress-strain curve is the average linear strain, which is obtained by dividing the elongation of the gage length of the specimen (δ) by its original length (L_0):

$$e = \frac{\delta}{L_0} = \frac{\Delta L}{L_0} = \frac{L - L_0}{L_0} \qquad \text{(Eq 2)}$$

Because both the stress and the strain are obtained by dividing the load and elongation by constant factors, the load-elongation curve has the same shape as the engineering stress-strain curve. The two curves frequently are used interchangeably.

The shape and magnitude of the stress-strain curve of a metal depend on its composition, heat treatment, prior history of plastic deformation, and the strain rate, temperature, and state of stress imposed during the testing. The parameters that are used to describe the stress-strain curve of a metal are the tensile strength, yield strength or yield point, percent elongation, and reduction in area. The first two are strength parameters; the last two indicate ductility.

The general shape of the engineering stress-strain curve (Fig. 1) requires further explanation. In the elastic region, stress is linearly proportional to strain. When the stress exceeds a value corresponding to the yield strength, the specimen undergoes gross plastic deformation. If the load is subsequently reduced to zero, the specimen will remain permanently deformed. The stress required to produce continued plastic deformation increases with increasing plastic strain; that is, the metal strain hardens. The volume of the specimen (area × length) remains constant during plastic deformation, $AL = A_0L_0$, and as the specimen elongates, its cross-sectional area decreases uniformly along the gage length.

Initially, the strain hardening more than compensates for this decrease in area, and the engineering stress (proportional to load P) continues to rise with increasing strain. Eventually, a point is reached where the decrease in specimen cross-sectional area is greater than the increase in deformation load arising from strain hardening. This condition will be reached first at some point in the specimen that is slightly weaker than the rest. All further plastic deformation is concentrated in this region, and the specimen begins to neck or thin down locally. Because the cross-sectional area now is decreasing far more rapidly than the deformation load is increased by strain hardening, the actual load required to deform the specimen falls off, and the engineering stress defined in Eq 1 continues to decrease until fracture occurs.

The tensile strength, or ultimate tensile strength (s_u) is the maximum load divided by the original cross-sectional area of the specimen:

$$s_u = \frac{P_{max}}{A_0} \quad \text{(Eq 3)}$$

The tensile strength is the value most frequently quoted from the results of a tension test. Actually, however, it is a value of little fundamental significance with regard to the strength of a metal. For ductile metals, the tensile strength should be regarded as a measure of the maximum load that a metal can withstand under the very restrictive conditions of uniaxial loading. This value bears little relation to the useful strength of the metal under the more complex conditions of stress that usually are encountered.

For many years, it was customary to base the strength of members on the tensile strength, suitably reduced by a factor of safety. The current trend is to use the more rational approach of basing the static design of ductile metals on the yield strength. However, due to the long practice of using the tensile strength to describe the strength of materials, it has become a familiar property, and as such, it is a useful identification of a material in the same sense that the chemical composition serves to identify a metal or alloy. Furthermore, because the tensile strength is easy to determine and is a reproducible property, it is useful for the purposes of specification and for quality control of a product. Extensive empirical correlations between tensile strength and properties such as hardness and fatigue strength are

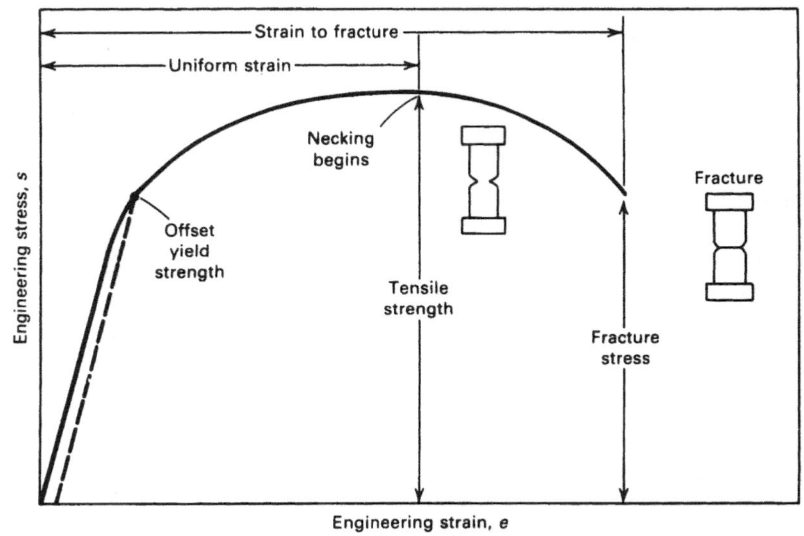

Fig. 1 Engineering stress-strain curve. Intersection of the dashed line with the curve determines the offset yield strength. See also Fig. 2 and corresponding text.

often useful. For brittle materials, the tensile strength is a valid design criterion.

Measures of Yielding. The stress at which plastic deformation or yielding is observed to begin depends on the sensitivity of the strain measurements. With most materials, there is a gradual transition from elastic to plastic behavior, and the point at which plastic deformation begins is difficult to define with precision. In tests of materials under uniaxial loading, three criteria for the initiation of yielding have been used: the elastic limit, the proportional limit, and the yield strength.

Elastic limit, shown at point A in Fig. 2, is the greatest stress the material can withstand without any measurable permanent strain remaining after the complete release of load. With increasing sensitivity of strain measurement, the value of the elastic limit is decreased until it equals the true elastic limit determined from microstrain measurements. With the sensitivity of strain typically used in engineering studies (10^{-4} in./in.), the elastic limit is greater than the proportional limit. Determination of the elastic limit requires a tedious incremental loading-unloading test procedure. For this reason, it is often replaced by the proportional limit.

Proportional limit, shown at point A′ in Fig. 2, is the highest stress at which stress is directly proportional to strain. It is obtained by observing the deviation from the straight-line portion of the stress-strain curve.

The yield strength, shown at point B in Fig. 2, is the stress required to produce a small specified amount of plastic deformation. The usual definition of this property is the offset yield strength determined by the stress corresponding to the intersection of the stress-strain curve offset by a specified strain (see Fig. 1 and 2). In the United States, the offset is usually specified as a strain of 0.2 or 0.1% (e = 0.002 or 0.001):

$$s_0 = \frac{P_{(\text{strain offset}=0.002)}}{A_0} \qquad \text{(Eq 4)}$$

Offset yield strength determination requires a specimen that has been loaded to its 0.2% offset yield strength and unloaded so that it is 0.2% longer than before the test. The offset yield strength is often referred to in Great Britain as the proof stress, where offset values are either 0.1 or 0.5%. The yield strength obtained by an offset method is commonly used for design and specification purposes, because it avoids the practical difficulties of measuring the elastic limit or proportional limit.

Some materials have essentially no linear portion to their stress-strain curve, for example, soft copper, gray cast iron, and many polymers. For these materials, the offset method cannot be used, and the usual practice is to define the yield strength as the stress to produce some total strain, for example, e = 0.005.

Some metals, particularly annealed low-carbon steel, show a localized, heterogeneous type of transition from elastic to plastic deformation that produces a yield point in the stress-strain curve. Rather than having a flow curve with a gradual transition from elastic to plastic behavior, such as Fig. 1 and 2, metals with a yield point produce a flow curve or a load-elongation diagram similar to Fig. 3. The load increases steadily with elastic strain, drops suddenly, fluctuates about some approximately constant value of load, and then rises with further strain.

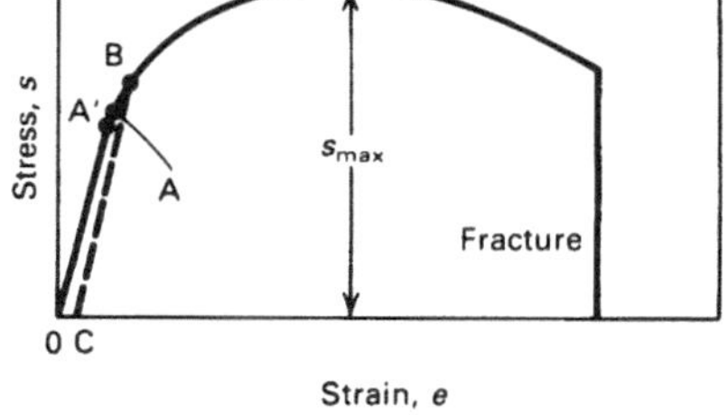

Fig. 2 Typical tensile stress-strain curve for ductile metal indicating yielding criteria. Point A, elastic limit; point A′, proportional limit; point B, yield strength or offset (0 to C) yield strength; 0, intersection of the stress-strain curve with the strain axis

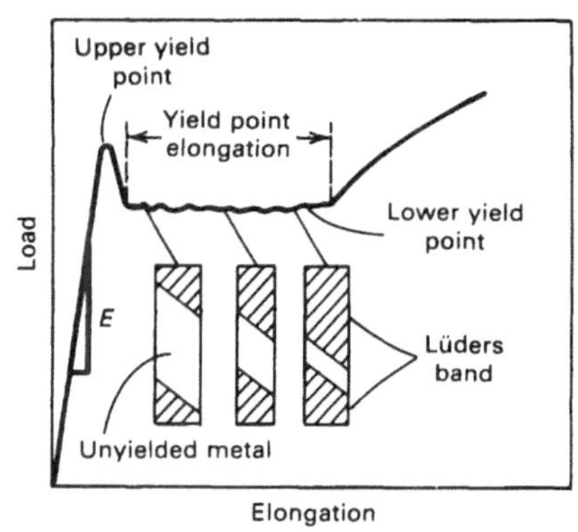

Fig. 3 Typical yield-point behavior of low-carbon steel. The slope of the initial linear portion of the stress-strain curve, designated by E, is the modulus of elasticity.

The load at which the sudden drop occurs is called the upper yield point. The constant load is called the lower yield point, and the elongation that occurs at constant load is called the yield-point elongation. The deformation occurring throughout the yield-point elongation is heterogeneous. At the upper yield point, a discrete band of deformed metal, often readily visible, appears at a stress concentration, such as a fillet. Coincident with the formation of the band, the load drops to the lower yield point. The band then propagates along the length of the specimen, causing the yield-point elongation. A similar behavior occurs with some polymers and superplastic metal alloys, where a neck forms but grows in a stable manner, with material being fed into the necked region from the thicker adjacent regions. This type of deformation in polymers is called "drawing."

In typical cases, several bands form at several points of stress concentration. These bands are generally at approximately 45° to the tensile axis. They are usually called Lüders bands or stretcher strains, and this type of deformation is sometimes referred to as the Piobert effect. When several Lüders bands are formed, the flow curve during the yield-point elongation is irregular, each jog corresponding to the formation of a new Lüders band. After the Lüders bands have propagated to cover the entire length of the specimen test section, the flow will increase with strain in the typical manner. This marks the end of the yield-point elongation. Lüders bands formed on a rimmed 1008 steel are shown in Fig. 4.

Measures of Ductility. Currently, ductility is considered a qualitative, subjective property of a material. In general, measurements of ductility are of interest in three respects (Ref 1):

- To indicate the extent to which a metal can be deformed without fracture in metalworking operations, such as rolling and extrusion
- To indicate to the designer the ability of the metal to flow plastically before fracture. A high ductility indicates that the material is "forgiving" and likely to deform locally without fracture should the designer err in the stress calculation or the prediction of severe loads.
- To serve as an indicator of changes in impurity level or processing conditions. Ductility measurements may be specified to assess material quality, even though no direct relationship exists between the ductility measurement and performance in service.

The conventional measures of ductility that are obtained from the tension test are the engineering strain at fracture (e_f) (usually called the elongation) and the reduction in area at fracture (q). Elongation and reduction in area usually are expressed as a percentage. Both of these properties are obtained after fracture by putting the specimen back together and taking measurements of the final length, L_f, and final specimen cross section, A_f:

$$e_f = \frac{L_f - L_0}{L_0} \quad \text{(Eq 5)}$$

$$q = \frac{A_0 - A_f}{A_0} \quad \text{(Eq 6)}$$

Because an appreciable fraction of the plastic deformation will be concentrated in the necked region of the tension specimen, the value of e_f will depend on the gage length (L_0) over which the measurement was taken (see the section of this article on ductility measurement in tension testing). The smaller the gage length, the greater the contribution to the overall elongation from the necked region and the higher the value of e_f. Therefore, when reporting values of percentage elongation, the gage length should always be given.

Reduction in area does not suffer from this difficulty. These values can be converted into an equivalent zero-gage-length elongation (e_0).

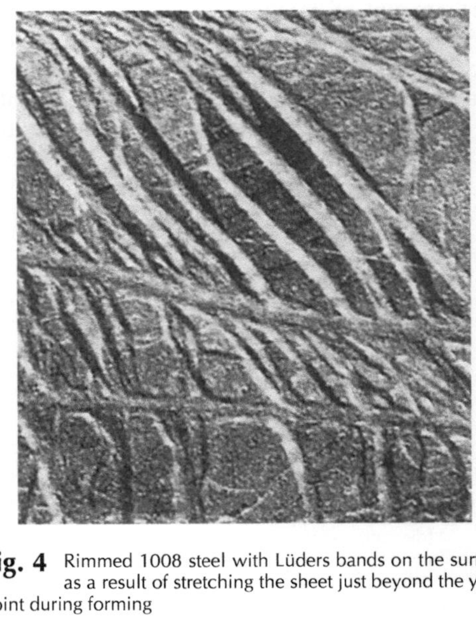

Fig. 4 Rimmed 1008 steel with Lüders bands on the surface as a result of stretching the sheet just beyond the yield point during forming

From the constancy of volume relationship for plastic deformation, $AL = A_0L_0$:

$$\frac{L}{L_0} = \frac{A_0}{A} = \frac{1}{1-q}$$

$$e_0 = \frac{L - L_0}{L_0} = \frac{A_0}{A} - 1 = \frac{1}{1-q} - 1$$

$$= \frac{q}{1-q} \quad \text{(Eq 7)}$$

This represents the elongation based on a very short gage length near the fracture.

Another way to avoid the complications resulting from necking is to base the percentage elongation on the uniform strain out to the point at which necking begins. The uniform elongation (e_u), correlates well with stretch-forming operations. Because the engineering stress-strain curve often is quite flat in the vicinity of necking, it may be difficult to establish the strain at maximum load without ambiguity. In this case, the method suggested in Ref 2 is useful.

Modulus of Elasticity. The slope of the initial linear portion of the stress-strain curve is the modulus of elasticity, or Young's modulus, as shown in Fig. 3. The modulus of elasticity (E) is a measure of the stiffness of the material. The greater the modulus, the smaller the elastic strain resulting from the application of a given stress.

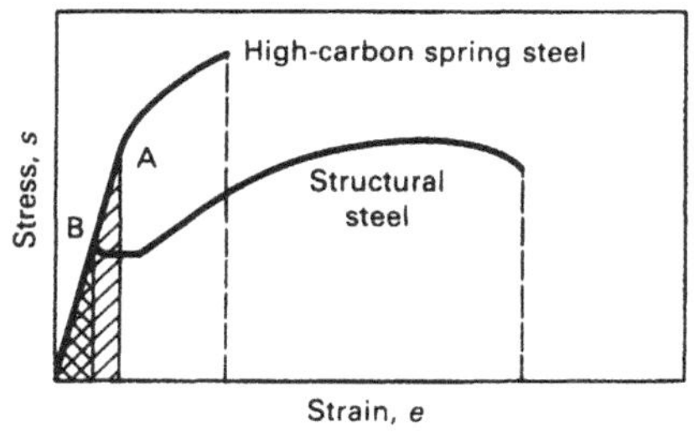

Fig. 5 Comparison of stress-strain curves for high- and low-toughness steels. Cross-hatched regions in this curve represent the modulus of resilience (U_R) of the two materials. The U_R is determined by measuring the area under the stress-strain curve up to the elastic limit of the material. Point A represents the elastic limit of the spring steel; point B represents that of the structural steel.

Because the modulus of elasticity is needed for computing deflections of beams and other members, it is an important design value.

The modulus of elasticity is determined by the binding forces between atoms. Because these forces cannot be changed without changing the basic nature of the material, the modulus of elasticity is one of the most structure-insensitive of the mechanical properties. Generally, it is only slightly affected by alloying additions, heat treatment, or cold work (Ref 3). However, increasing the temperature decreases the modulus of elasticity. At elevated temperatures, the modulus is often measured by a dynamic method (Ref 4). Typical values of the modulus of elasticity for common engineering metals at different temperatures are given in Table 1.

Resilience. The ability of a material to absorb energy when deformed elastically and to return it when unloaded is called resilience. This property usually is measured by the modulus of resilience, which is the strain energy per unit volume (U_0) required to stress the material from zero stress to the yield stress (σ_0). The strain energy per unit volume for uniaxial tension is:

$$U_0 = \frac{1}{2}\sigma_x e_x \quad \text{(Eq 8)}$$

From the above definition, the modulus of resilience (U_R) is:

$$U_R = \frac{1}{2} s_0 e_0 = \frac{1}{2} s_0 \frac{s_0}{E} = \frac{s_0^2}{2E} \quad \text{(Eq 9)}$$

This equation indicates that the ideal material for resisting energy loads in applications where the material must not undergo permanent distortion, such as in mechanical springs, is one having a high yield stress and a low modulus of elasticity.

For various grades of steel, the modulus of resilience ranges from 100 to 4500 kJ/m^3 (14.5–650 lbf · in./in.^{3}), with the higher values representing steels with higher carbon or alloy contents (Ref 5). The cross-hatched regions in Fig. 5 indicate the modulus of resilience for two

Table 1 Typical values of modulus of elasticity at different temperatures

	Modulus of elasticity GPa (10^6 psi), at:				
Material	Room temperature	250 °C (400 °F)	425 °C (800 °F)	540 °C (1000 °F)	650 °C (1200 °F)
Carbon steel	207 (30.0)	186 (27.0)	155 (22.5)	134 (19.5)	124 (18.0)
Austenitic stainless steel	193 (28.0)	176 (25.5)	159 (23.0)	155 (22.5)	145 (21.0)
Titanium alloys	114 (16.5)	96.5 (14.0)	74 (10.7)	70 (10.0)	...
Aluminum alloys	72 (10.5)	65.5 (9.5)	54 (7.8)	...	...

steels. Due to its higher yield strength, the high-carbon spring steel has the greater resilience.

The toughness of a material is its ability to absorb energy in the plastic range. The ability to withstand occasional stresses above the yield stress without fracturing is particularly desirable in parts such as freight-car couplings, gears, chains, and crane hooks. Toughness is a commonly used concept that is difficult to precisely define. Toughness may be considered to be the total area under the stress-strain curve. This area, which is referred to as the modulus of toughness (U_T) is an indication of the amount of work per unit volume that can be done on the material without causing it to rupture.

Figure 5 shows the stress-strain curves for high- and low-toughness materials. The high-carbon spring steel has a higher yield strength and tensile strength than the medium-carbon structural steel. However, the structural steel is more ductile and has a greater total elongation. The total area under the stress-strain curve is greater for the structural steel; therefore, it is a tougher material. This illustrates that toughness is a parameter that comprises both strength and ductility.

Several mathematical approximations for the area under the stress-strain curve have been suggested. For ductile metals that have a stress-strain curve like that of the structural steel, the area under the curve can be approximated by:

$$U_T \approx s_u e_f \qquad \text{(Eq 10)}$$

or

$$U_T \approx \frac{s_0 + s_u}{2} e_f \qquad \text{(Eq 11)}$$

For brittle materials, the stress-strain curve is sometimes assumed to be a parabola, and the area under the curve is given by:

$$U_T \approx \frac{2}{3} s_u e_f \qquad \text{(Eq 12)}$$

Typical Stress-Strain Curves. Figure 6 compares the engineering stress-strain curves in tension for three materials. The 0.8% carbon eutectoid steel is representative of a material with low ductility. The annealed 0.2% carbon mild steel shows a pronounced upper and lower yield point. The polycarbonate engineered polymer has no well-defined linear modulus, and a large strain to fracture. Note the pronounced difference in stress level at which yielding is defined, as well as the quite different shape of the stress-strain curves.

True Stress-True Strain Curve

The engineering stress-strain curve does not give a true indication of the deformation characteristics of a metal because it is based entirely on the original dimensions of the specimen, and these dimensions change continuously during the test. Also, ductile metal that is pulled in tension becomes unstable and necks down during the course of the test. Because the cross-sectional area of the specimen is decreasing rapidly at this stage in the test, the load required to continue deformation falls off.

The average stress based on the original area likewise decreases, and this produces the fall-off in the engineering stress-strain curve beyond the point of maximum load. Actually, the metal continues to strain harden to fracture, so that the stress required to produce further deformation should also increase. If the true stress, based on the actual cross-sectional area of the specimen, is used, the stress-strain curve increases continuously to fracture. If the strain measurement is also based on instantaneous measurement, the curve that is obtained is known as true stress-true strain curve. This is also known as a flow curve because it represents the basic plastic-flow characteristics of the material.

Any point on the flow curve can be considered the yield stress for a metal strained in tension by the amount shown on the curve. Thus, if the load is removed at this point and then reapplied, the material will behave elastically throughout the entire range of reloading.

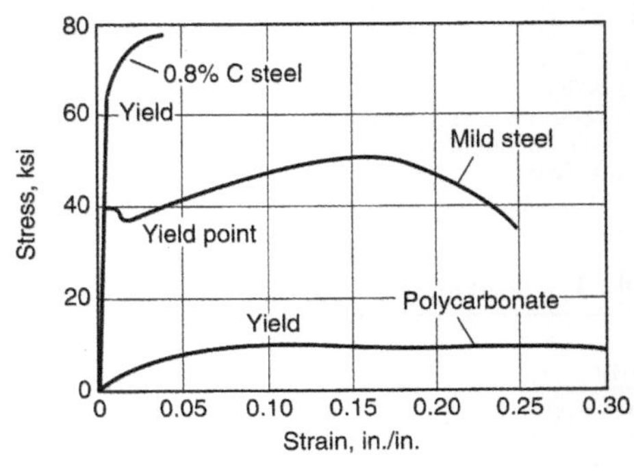

Fig. 6 Typical engineering stress-strain curves

The true stress (σ) is expressed in terms of engineering stress (s) by:

$$\sigma = \frac{P}{A_0}(e + 1) = s(e + 1) \quad \text{(Eq 13)}$$

The derivation of Eq 13 assumes both constancy of volume and a homogeneous distribution of strain along the gage length of the tension specimen. Thus, Eq 13 should be used only until the onset of necking. Beyond the maximum load, the true stress should be determined from actual measurements of load and cross-sectional area.

$$\sigma = \frac{P}{A} \quad \text{(Eq 14)}$$

The true strain, ε, may be determined from the engineering or conventional strain (e) by:

$$\varepsilon = \ln(e + 1) = \ln \frac{L}{L_0} \quad \text{(Eq 15)}$$

This equation is applicable only to the onset of necking for the reasons discussed above. Beyond maximum load, the true strain should be based on actual area or diameter (D) measurements:

$$\varepsilon = \ln \frac{A_0}{A} = \ln \frac{\left(\frac{\pi}{4}\right)D_0^2}{\left(\frac{\pi}{4}\right)D^2} = 2 \ln \frac{D_0}{D} \quad \text{(Eq 16)}$$

Figure 7 compares the true stress-true strain curve with its corresponding engineering stress-strain curve. Note that because of the relatively large plastic strains, the elastic region has been compressed into the y-axis. In agreement with Eq 13 and 15; the true stress-true strain curve is always to the left of the engineering curve until the maximum load is reached.

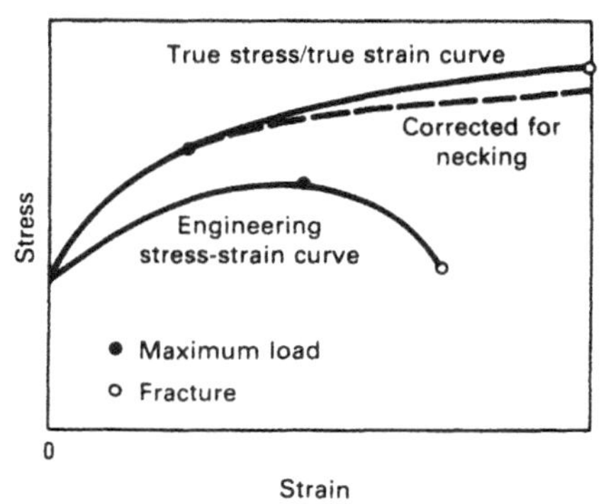

Fig. 7 Comparison of engineering and true stress-true strain curves

However, beyond maximum load, the high, localized strains in the necked region that are used in Eq 16 far exceed the engineering strain calculated from Eq 2. Frequently, the flow curve is linear from maximum load to fracture, while in other cases its slope continuously decreases to fracture. The formation of a necked region or mild notch introduces triaxial stresses that make it difficult to determine accurately the longitudinal tensile stress from the onset of necking until fracture occurs. This concept is discussed in greater detail in the section "Instability in Tension" later in this chapter. The following parameters usually are determined from the true stress-true strain curve.

The true stress at maximum load corresponds to the true tensile strength. For most materials, necking begins at maximum load at a value of strain where the true stress equals the slope of the flow curve. Let σ_u and ε_u denote the true stress and true strain at maximum load when the cross-sectional area of the specimen is A_u. The ultimate tensile strength can be defined as:

$$s_u = \frac{P_{max}}{A_0} \quad \text{(Eq 17)}$$

and

$$\sigma_u = \frac{P_{max}}{A_u} \quad \text{(Eq 18)}$$

Eliminating P_{max} yields:

$$\sigma_u = s_u \frac{A_0}{A_u} \quad \text{(Eq 19)}$$

and

$$\sigma_u = s_u e^{\varepsilon_u} \quad \text{(Eq 20)}$$

The true fracture stress is the load at fracture divided by the cross-sectional area at fracture. This stress should be corrected for the triaxial state of stress existing in the tensile specimen at fracture. Because the data required for this correction frequently are not available, true fracture stress values are frequently in error.

The true fracture strain, ε_f, is the true strain based on the original area (A_0) and the area after fracture (A_f):

$$\varepsilon_f = \ln \frac{A_0}{A_f} \quad \text{(Eq 21)}$$

This parameter represents the maximum true strain that the material can withstand before fracture and is analogous to the total strain to fracture of the engineering stress-strain curve. Because Eq 15 is not valid beyond the onset of necking, it is not possible to calculate ε_f from measured values of e_f. However, for cylindrical tensile specimens, the reduction in area (q) is related to the true fracture strain by:

$$\varepsilon_f = \ln \frac{1}{1-q} \quad \text{(Eq 22)}$$

The true uniform strain ε_u, is the true strain based only on the strain up to maximum load. It may be calculated from either the specimen cross-sectional area (A_u) or the gage length (L_u) at maximum load. Equation 15 may be used to convert conventional uniform strain to true uniform strain. The uniform strain frequently is useful in estimating the formability of metals from the results of a tension test:

$$\varepsilon_u = \ln \frac{A_0}{A_u} \quad \text{(Eq 23)}$$

The true local necking strain (ε_n) is the strain required to deform the specimen from maximum load to fracture:

$$\varepsilon_n = \ln \frac{A_u}{A_f} \quad \text{(Eq 24)}$$

Mathematical Expressions for the Flow Curve

The flow curve of many metals in the region of uniform plastic deformation can be expressed by the simple power curve relation:

$$\sigma = K\varepsilon^n \quad \text{(Eq 25)}$$

where n is the strain-hardening exponent, and K is the strength coefficient. A log-log plot of true stress and true strain up to maximum load will result in a straight line if Eq 25 is satisfied by the data (Fig. 8).

The linear slope of this line is n, and K is the true stress at $\varepsilon = 1.0$ (corresponds to $q = 0.63$). As shown in Fig. 9, the strain-hardening exponent may have values from $n = 0$ (perfectly plastic solid) to $n = 1$ (elastic solid). For most metals, n has values between 0.10 and 0.50 (see Table 2).

The rate of strain hardening $d\sigma/d\varepsilon$ is not identical to the strain-hardening exponent. From the definition of n:

$$n = \frac{d(\log \sigma)}{d(\log \varepsilon)} = \frac{d(\ln \sigma)}{d(\ln \varepsilon)} = \frac{\varepsilon}{\sigma}\frac{d\sigma}{d\varepsilon}$$

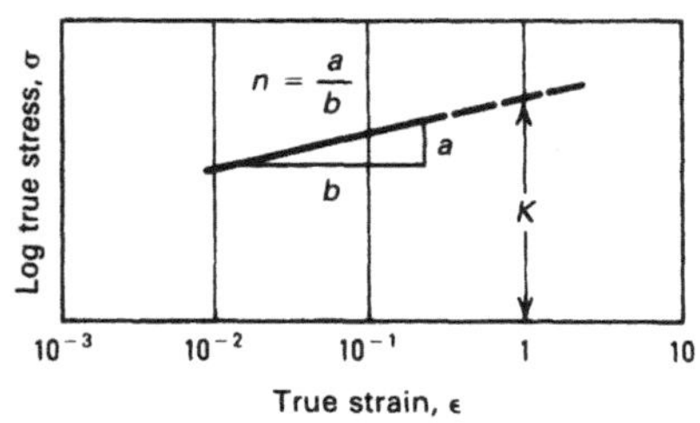

Fig. 8 Log-log plot of true stress-true strain curve n is the strain-hardening exponent; K is the strength coefficient.

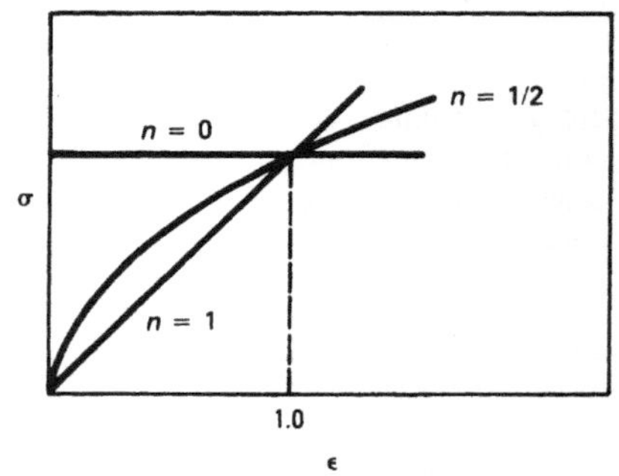

Fig. 9 Various forms of power curve $\sigma = K\varepsilon^n$

Table 2 Values for *n* and *K* for metals at room temperature

Metals	Condition	n	K MPa	K ksi	Ref
0.05% carbon steel	Annealed	0.26	530	77	6
SAE 4340 steel	Annealed	0.15	641	93	6
0.6% carbon steel	Quenched and tempered at 540 °C (1000 °F)	0.10	1572	228	7
0.6% carbon steel	Quenched and tempered at 705 °C (1300 °F)	0.19	1227	178	7
Copper	Annealed	0.54	320	46.4	6
70/30 brass	Annealed	0.49	896	130	7

or

$$\frac{d\sigma}{d\varepsilon} = n\frac{\sigma}{\varepsilon} \quad \text{(Eq 26)}$$

Deviations from Eq 25 frequently are observed, often at low strains (10^{-3}) or high strains ($\varepsilon \approx 1.0$). One common type of deviation is for a log-log plot of Eq 25 to result in two straight lines with different slopes. Sometimes data that do not plot according to Eq 25 will yield a straight line according to the relationship:

$$\sigma = K(\varepsilon_0 + \varepsilon)^n \quad \text{(Eq 27)}$$

ε_0 can be considered to be the amount of strain that the material received prior to the tensile test (Ref 8). Another common variation on Eq 25 is the Ludwik equation:

$$\sigma = \sigma_0 + K\varepsilon^n \quad \text{(Eq 28)}$$

where σ_0 is the yield stress, and K and n are the same constants as in Eq 25. This equation may be more satisfying than Eq 25, because the latter implies that at zero true strain the stress is zero. It has been shown that σ_0 can be obtained from the intercept of the strain-hardening portion of the stress-strain curve and the elastic modulus line by (Ref 9):

$$\sigma_0 = \left(\frac{K}{E^n}\right)^{1/(1-n)} \quad \text{(Eq 29)}$$

The true stress-true strain curve of metals such as austenitic stainless steel, which deviate markedly from Eq 25 at low strains (Ref 10), can be expressed by:

$$\sigma = K\varepsilon^n + e^{K_1} + e^{K_1}e^{n_1\varepsilon} \quad \text{(Eq 30)}$$

where e^{K_1} is approximately equal to the proportional limit, and n_1 is the slope of the deviation of stress from Eq 25 plotted against ε. Other expressions for the flow curve are available (Ref 11, 12). The true strain term in Eq 25 to 28 properly should be the plastic strain, $\varepsilon_p = \varepsilon_{total} - \varepsilon_E = \varepsilon_{total} - \sigma/E$, where ε_E represents elastic strain.

Effect of Strain Rate and Temperature

The rate at which strain is applied to the tensile specimen has an important influence on the stress-strain curve. Strain rate is defined as $\dot{\varepsilon} = d\varepsilon/dt$. It is expressed in units of s^{-1}. The range of strain rates encompassed by various tests is shown in Table 3.

Increasing strain rate increases the flow stress. Moreover, the strain-rate dependence of strength increases with increasing temperature. The yield stress and the flow stress at lower values of plastic strain are more affected by strain rate than the tensile strength.

If the crosshead velocity of the testing machine is $v = dL/dt$, then the strain rate expressed in terms of conventional engineering strain is:

$$\dot{e} = \frac{de}{dt} = \frac{d(L - L_0)/L_0}{dt} = \frac{1}{L_0}\frac{dL}{dt} = \frac{v}{L_0} \quad \text{(Eq 31)}$$

The engineering strain rate is proportional to the crosshead velocity. In a modern testing machine, in which the crosshead velocity can be set accurately and controlled, it is a simple matter to carry out tensile tests at a constant engineering strain rate.

The true strain rate is given by:

$$\dot{\varepsilon} = \frac{d\varepsilon}{dt} = \frac{d[\ln(L/L_0)]}{dt} = \frac{1}{L}\frac{dL}{dt} = \frac{v}{L} \quad \text{(Eq 32)}$$

Equation 32 shows that for a constant crosshead velocity the true strain rate will decrease as the specimen elongates or cross-sectional area shrinks. To run tensile tests at a constant true strain rate requires monitoring the instantaneous cross section of the deforming region, with closed-loop control feed back to increase the crosshead velocity as the area decreases. The true strain rate is related to the engineering strain rate by the following equation:

$$\dot{\varepsilon} = \frac{v}{L} = \frac{L_0}{L}\frac{de}{dt} = \frac{1}{1+e}\frac{de}{dt} = \frac{\dot{e}}{1+e} \quad \text{(Eq 33)}$$

Table 3 Range of strain rates in common mechanical property tests

Range of strain rate	Type of test
10^{-8} to 10^{-5} s^{-1}	Creep test at constant load or stress
10^{-5} to 10^{-1} s^{-1}	Tension test with hydraulic or screw driven machines
10^{-1} to 10^{2} s^{-1}	Dynamic tension or compression tests
10^{2} to 10^{4} s^{-1}	High-speed testing using impact bars
10^{4} to 10^{8} s^{-1}	Hypervelocity impact using gas guns or explosively driven projectiles

The strain-rate dependence of flow stress at constant strain and temperature is given by:

$$\sigma = C(\dot{\varepsilon})^m|_{\varepsilon,T} \quad \text{(Eq 34)}$$

The exponent in Eq 34, m, is known as the strain-rate sensitivity, and C is the strain hardening coefficient. It can be obtained from the slope of a plot of log σ versus log $\dot{\varepsilon}$. However, a more sensitive way to determine m is with a rate-change test (Fig. 10). A tensile test is carried out at strain rate $\dot{\varepsilon}_1$ and at a certain flow stress, σ_1, the strain rate is suddenly increased to $\dot{\varepsilon}_2$. The flow stress quickly increases to σ_2. The strain-rate sensitivity, at constant strain and temperature, can be determined from:

$$m = \left(\frac{\partial \ln \sigma}{\partial \ln \dot{\varepsilon}}\right)_{\varepsilon,T} = \frac{\dot{\varepsilon}}{\sigma}\left(\frac{\partial \sigma}{\partial \dot{\varepsilon}}\right) = \frac{\Delta \log \sigma}{\Delta \log \dot{\varepsilon}}$$
$$= \frac{\log \sigma_2 - \log \sigma_1}{\log \dot{\varepsilon}_2 - \log \dot{\varepsilon}} = \frac{\log(\sigma_2/\sigma_1)}{\log(\dot{\varepsilon}_2/\dot{\varepsilon}_1)} \quad \text{(Eq 35)}$$

The strain-rate sensitivity of metals is quite low (<0.1) at room temperature, but m increases with temperature. At hot-working temperatures, $T/T_m > 0.5$, m values of 0.1 to 0.2 are common in metals. Polymers have much higher values of m, and may approach $m = 1$ in room-temperature tests for some polymers.

The temperature dependence of flow stress can be represented by:

$$\sigma = C_2\, e^{Q/RT}|_{\varepsilon,\dot{\varepsilon}} \quad \text{(Eq 36)}$$

where Q is an activation energy for plastic flow, cal/g · mol; R is universal gas constant, 1.987 cal/K · mol; and T is testing temperature in kelvin. From Eq 36, a plot of ln σ versus $1/T$ will give a straight line with a slope Q/R.

Instability in Tension

Necking generally begins at maximum load during the tensile deformation of a ductile metal. An exception to this is the behavior of cold-rolled zirconium tested at 200 to 370 °C (390–700 °F), where necking occurs at a strain of twice the strain at maximum load (Ref 13). An ideal plastic material in which no strain hardening occurs would become unstable in tension and begin to neck as soon as yielding occurred. However, an actual metal undergoes strain hardening, which tends to increase the load-carrying capacity of the specimen as deformation increases.

This effect is opposed by the gradual decrease in the cross-sectional area of the specimen as it elongates. Necking or localized deformation begins at maximum load, where the increase in stress due to decrease in the cross-sectional area of the specimen becomes greater than the increase in the load-carrying ability of the metal due to strain hardening. This condition of instability leading to localized deformation is defined by the condition $dP = 0$:

$$P = \sigma A \quad \text{(Eq 37)}$$

$$dP = \sigma dA + A d\sigma = 0 \quad \text{(Eq 38)}$$

From the constancy-of-volume relationship:

$$\frac{dL}{L} = -\frac{dA}{A} = d\varepsilon \quad \text{(Eq 39)}$$

and from the instability condition, Eq 38:

$$-\frac{dA}{A} = \frac{d\sigma}{\sigma} \quad \text{(Eq 40)}$$

so that at a point of tensile instability:

$$\frac{d\sigma}{d\varepsilon} = \sigma \quad \text{(Eq 41)}$$

Therefore, the point of necking at maximum load can be obtained from the true stress-true strain curve by finding the point on the curve having a subtangent of unity (Fig. 11a), or the point where the rate of strain hardening equals the stress (Fig. 11b). The necking criterion can

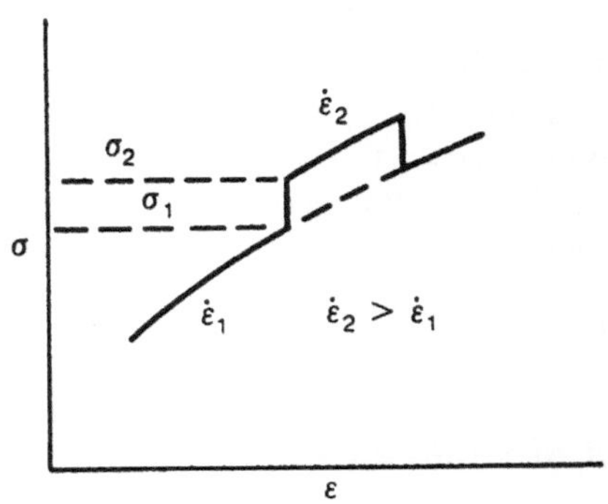

Fig. 10 Strain-rate change test, used to determine strain-rate sensitivity, m. See text for discussion.

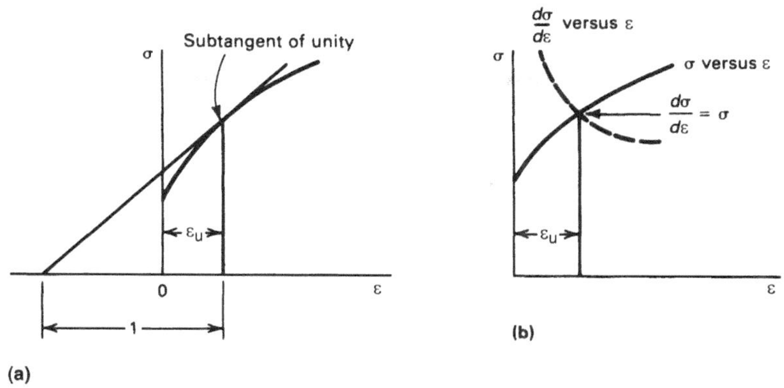

Fig. 11 Graphical interpretation of necking criterion. The point of necking at maximum load can be obtained from the true stress-true strain curve by finding (a) the point on the curve having a subtangent of unity or (b) the point where $d\sigma/d\varepsilon = \sigma$.

be expressed more explicitly if engineering strain is used. Starting with Eq 41:

$$\frac{d\sigma}{de} = \frac{d\sigma}{d\varepsilon}\frac{de}{d\varepsilon} = \frac{d\sigma}{de}\frac{\frac{dL}{L_0}}{\frac{dL}{L}} = \frac{d\sigma}{de}\frac{L}{L_0}$$

$$= \frac{d\sigma}{de}(1 + e) = \sigma$$

$$\frac{d\sigma}{de} = \frac{\sigma}{1 + e} \qquad \text{(Eq 42)}$$

Equation 42 permits an interesting geometrical construction for the determination of the point of maximum load (Ref 14). In Fig. 12, the stress-strain curve is plotted in terms of true stress against engineering strain. Let point A represent a negative strain of 1.0. A line drawn from point A, which is tangent to the stress-strain curve, will establish the point of maximum load because, according to Eq 42, the slope at this point is $\sigma/(1 + e)$.

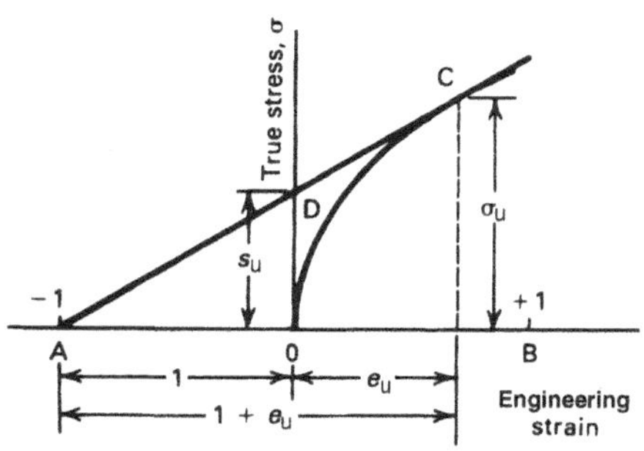

Fig. 12 Considére's construction for the determination of the point of maximum load. Source: Ref 14

By substituting the necking criterion given in Eq 41 into Eq 26, a simple relationship for the strain at which necking occurs is obtained:

$$\varepsilon_u = n \qquad \text{(Eq 43)}$$

Although Eq 26 is based on the assumption that the flow curve is given by Eq 25, it has been shown that $\varepsilon_u = n$ does not depend on this power law behavior (Ref 15).

Stress Distribution at the Neck

The formation of a neck in the tensile specimen introduces a complex triaxial state of stress in that region. The necked region is in effect a mild notch. A notch under tension produces radial stress (σ_r) and transverse stress (σ_t) which raise the value of longitudinal stress required to cause the plastic flow. Therefore, the average true stress at the neck, which is determined by dividing the axial tensile load by the minimum cross-sectional area of the specimen at the neck, is higher than the stress that would be required to cause flow if simple tension prevailed.

Figure 13 illustrates the geometry at the necked region and the stresses developed by this localized deformation. R is the radius of curvature of the neck, which can be measured either by projecting the contour of the necked region on a screen or by using a tapered, conical radius gage.

Bridgman made a mathematical analysis that provides a correction to the average axial stress to compensate for the introduction of transverse stresses (Ref 16). This analysis was based on the following assumptions:

- The contour of the neck is approximated by the arc of a circle.
- The cross section of the necked region remains circular throughout the test.
- The von Mises criterion for yielding applies.
- The strains are constant over the cross section of the neck.

According to this analysis, the uniaxial flow stress corresponding to that which would exist in the tensile test if necking had not introduced triaxial stresses is:

$$\sigma = \frac{(\sigma_x)_{avg}}{\left(\frac{1 + 2R}{a}\right)\left[\ln\left(1 + \frac{a}{2R}\right)\right]} \quad \text{(Eq 44)}$$

where $(\sigma_x)_{avg}$ is the measured stress in the axial direction (load divided by minimum cross section) and a is the minimum radius at the neck. Figure 7 shows how the application of the Bridgman correction changes the true stress-true strain curve. A correction for the triaxial stresses in the neck of a flat tensile specimen has been considered (Ref 17). The values of a/R needed for the analysis can be obtained either by straining a specimen a given amount beyond necking and unloading to measure a and R directly, or by measuring these parameters continuously past necking using photography or a tapered ring gage (Ref 18).

To avoid these measurements, Bridgman presented an empirical relation between a/R and the true strain in the neck. Figure 14 shows that this gives close agreement for steel specimens, but not for other metals with widely different necking strains. A much better correlation is obtained between the Bridgman correction and the true strain in the neck minus the true strain at necking, ε_u (Ref 20).

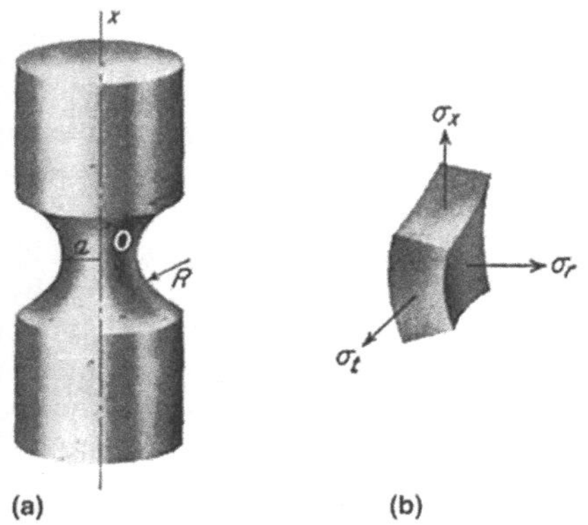

Fig. 13 Stress distribution at the neck of a tensile specimen. (a) Geometry of necked region. R is the radius of curvature of the neck; a is the minimum radius at the neck. (b) Stresses acting on element at point O. σ_x is the stress in the axial direction; σ_r is the radial stress; σ_t is the transverse stress.

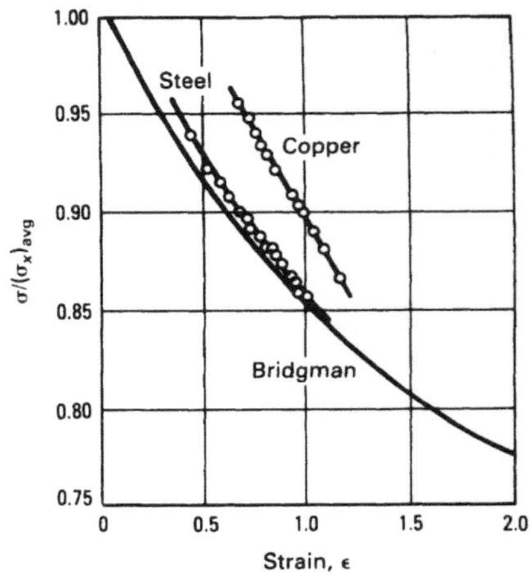

Fig. 14 Relationship between Bridgman correction factor $\sigma/(\sigma_x)_{avg}$ and true tensile strain. Source: Ref 19

Dowling (Ref 21) has shown that the Bridgman correction factor B can be estimated from:

$$B = 0.83 - 0.186 \log \varepsilon \; (0.15 \leq \varepsilon \geq 3) \quad \text{(Eq 45)}$$

where $B = \sigma/(\sigma_x)_{avg}$.

Ductility Measurement in Tensile Testing

The measured elongation from a tensile specimen depends on the gage length of the specimen, or the dimensions of its cross section. This is because the total extension consists of two components: the uniform extension up to necking and the localized extension once necking begins. The extent of uniform extension depends on the metallurgical condition of the material (through n) and the effect of specimen size and shape on the development of the neck.

Figure 15 illustrates the variation of the local elongation, as defined in Eq 7, along the gage length of a prominently necked tensile specimen. The shorter the gage length, the greater the influence of localized deformation at the neck on the total elongation of the gage length. The extension of a specimen at fracture can be expressed by:

$$L_f - L_0 = \alpha + e_u L_0 \quad \text{(Eq 46)}$$

where α is the local necking extension, and $e_u L_0$ is the uniform extension. The tensile elongation is then:

$$e_f = \frac{L_f - L_0}{L_0} = \frac{\alpha}{L_0} + e_u \quad \text{(Eq 47)}$$

This clearly indicates that the total elongation is a function of the specimen gage length. The shorter the gage length, the greater the percent elongation.

Numerous attempts have been made to rationalize the strain distribution in the tensile test. Perhaps the most general conclusion that can be drawn is that geometrically similar specimens develop geometrically similar necked regions. According to Barba's law (Ref 22), $\alpha = \beta\sqrt{A_0}$, and the elongation equation becomes:

$$e_f = \beta \frac{\sqrt{A_0}}{L_0} + e_u \quad \text{(Eq 48)}$$

where β is a coefficient of proportionality.

To compare elongation measurements of different sized specimens, the specimens must be geometrically similar. Equation 48 shows that the critical geometrical factor for which similitude must be maintained is $L_0/\sqrt{A_0}$ for sheet specimens, or L_0/D_0 for round bars. In the United States, the standard round tensile specimen has a 12.8 mm (0.505 in.) diameter and a 50 mm (2 in.) gage length. Subsize specimens have the following respective diameter and gage length: 9.06 and 35.6 mm (0.357 and 1.4 in.), 6.4 and 25 mm (0.252 and 1.0 in.), and 4.06 and 16.1 mm (0.160 and 0.634 in.). Different values of $L_0/\sqrt{A_0}$ are specified for sheet specimens by the standardizing agencies in different countries. In the United States, ASTM recommends a $L_0/\sqrt{A_0}$ value of 4.5 for sheet specimens and a L_0/D_0 value of 4.0 for round specimens.

Generally, a given elongation will be produced in a material if $\sqrt{A_0}/L_0$ is maintained constant as predicted by Eq 48. Thus, at a constant value of elongation $\sqrt{A_1}/L_1 = \sqrt{A_2}/A_2$, where A and L are the areas and gage lengths of two different specimens, 1 and 2, of the same metal. To predict elongation using gage length L_2 on a specimen with area A_2 by means of measurements on a specimen with area A_1, it only is necessary to adjust the gage length of specimen 1 to conform with $L_1 = L_2\sqrt{A_1/A_2}$. For example, suppose that a 3.2 mm (0.125 in.) thick sheet is available, and one wishes to predict the elongation with a 50 mm (2 in.) gage length for the identical material but in 2.0 mm (0.080 in.) thickness. Using 12.7 mm (0.5 in.) wide sheet specimens, a test specimen with a gage length L = 50 mm (3.2 mm/2.0 mm)$^{1/2}$ = 63 mm, or 2 in. (0.125 in./0.080 in.)$^{1/2}$ = 2.5 in., made from the 3.2 mm (0.125 in.) sheet would be predicted to give the same elongation as a 50 mm (2 in.) gage length in 2.0 mm (0.080 in.) thick sheet. Experimental verification for this procedure has been shown in Ref 23.

The occurrence of necking in the tensile test, however, makes any quantitative conversion between elongation and reduction in area impossible. Although elongation and reduction in area usually vary in the same way—for example, as a function of test temperature, tempering temperature, or alloy content—this is not always the case. Generally, elongation and reduction in area measure different types of material behavior. Provided the gage length is not too short, percent elongation is primarily influenced by uniform elongation, and thus it is dependent on the strain-hardening capacity of the material.

Reduction in area is more a measure of the deformation required to produce fracture, and its chief contribution results from the necking process. Because of the complicated stress state in the neck, values of reduction in area are dependent on specimen geometry and deformation behavior, and they should not be taken as true material properties. However, reduction in area is the most structure-sensitive ductility parameter, and as such, it is useful in detecting quality changes in the material.

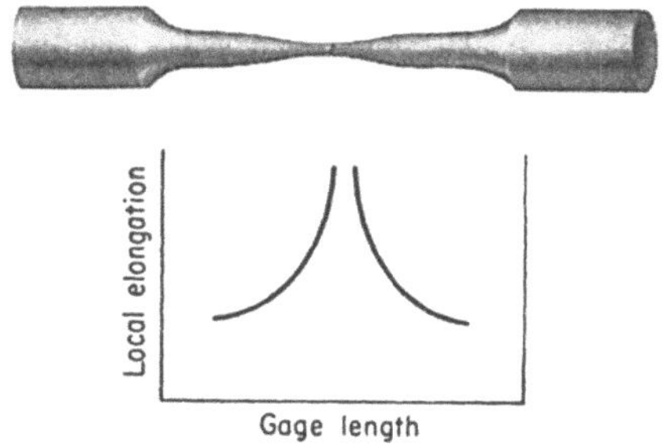

Fig. 15 Variation of local elongation with position along gage length of tensile specimen

Sheet Anisotropy (Ref 24)

If the tensile tests are performed on specimens cut from sheet material at different orientations to the prior rolling direction (Fig. 16), there may not be much difference between the stress-strain curves. However, the lack of variation of the stress-strain curves with direction does not indicate that the material is isotropic. The parameter that is commonly used to characterize the anisotropy of sheet metal is the *strain ratio* or *r-value* defined as the ratio of the contractile strains measured in a tensile test before necking occurs:

$$r = \frac{\varepsilon_w}{\varepsilon_t} \qquad \text{(Eq 49)}$$

where ε_w is the width strain, ln (w/w_0), and ε_t is the thickness strain, ln (t/t_0). The value of r would be equal to 1 for an isotropic material. Often, however, r is either greater or less than 1. For thin sheets, accurate direct measurement of the thickness strain is difficult. Therefore, the thickness strain is often deduced from the constant-volume relationship, $\varepsilon_t = -\varepsilon_w - \varepsilon_l$, where ε_l is the length strain, ln (l/l_0). By substitution,

$$r = \frac{-\varepsilon_w}{\varepsilon_w + \varepsilon_l} \qquad \text{(Eq 50)}$$

To avoid constraints from the test specimen grips, the strains should be measured on a gage section that is removed from the enlarged ends by a distance at least equal to the width of the specimen. Some workers suggest that the strains be measured when the elongation is about 15% as long as this is less than the strain at which necking starts.

Although the r-value usually does not change much during the tensile test, the strains at 15% are large enough to be measured with reasonable accuracy. The measurement of r is subject to greater error than may at first be apparent. If the accuracy of measuring strains were ± 0.01, the error in r would be $\pm 25\%$. Consider, for example, a material for which $r = 1$ ($\varepsilon_t = \varepsilon_w$). At an elongation strain of 15% ($\varepsilon_l = 0.14$), the values of ε_t and ε_w should be 0.07. Measurement errors of ± 0.01 could lead to $r = 0.08/0.06 = 1.33$ or $r = 0.06/0.08 = 0.75$. Even if the accuracy were ± 0.002, the limits on r would be $0.072/0.068 = 1.06$ and $0.068/0.072 = 0.94$. Measurements of r at lower elongations are even less accurate.

The value of r often depends on the angle at which the specimen is cut from the sheet (Fig. 17). In this case an average r-value, $\bar{R}$, is often quoted, where $\bar{r}$ is given by:

$$\bar{r} = \frac{r_0 + r_{90} + 2r_{45}}{4} \qquad \text{(Eq 51)}$$

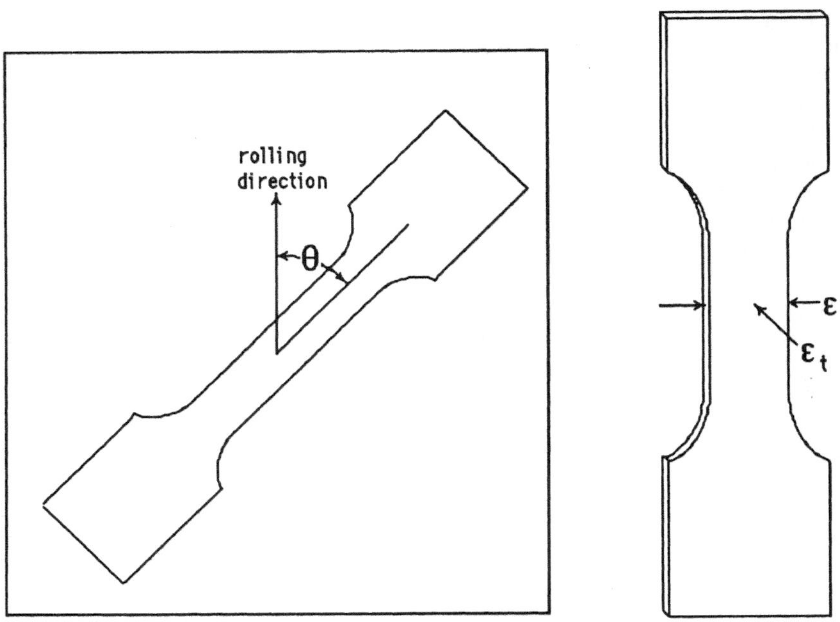

Fig. 16 Tensile specimen cut from a rolled sheet (left). The r-value is the ratio of $\varepsilon_w/\varepsilon_t$ during extension (right). Source: Ref 24

The subscripts refer to the angles between the tensile axis and the rolling direction (Fig. 17). The $\bar{r}$ value describes the degree of *normal anisotropy,* reflecting the difference between plastic properties in and normal to the plane of the sheet. Typical $\bar{r}$ values for metals and alloys are given in Table 4.

Other properties are averaged in an analogous way. For example, for n and K in Eq 25:

$$\bar{n} = \frac{n_0 + n_{90} + 2n_{45}}{4} \quad \text{(Eq 52)}$$

and

$$\bar{K} = \frac{K_0 + K_{90} + 2K_{45}}{4} \quad \text{(Eq 53)}$$

The degree of anisotropy in the plane of the sheet (*planar anisotropy*) can be described by the parameter:

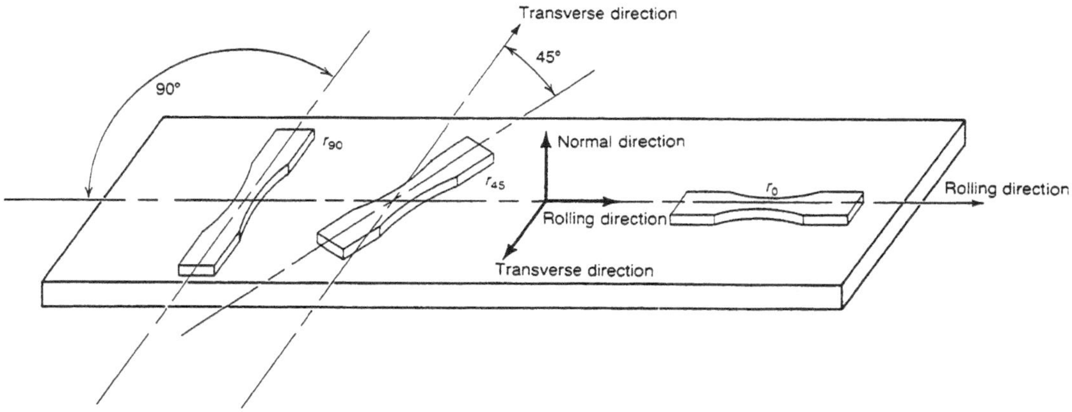

Fig. 17 Tensile specimen orientation to determine r_0, r_{45}, and r_{90} in rolled sheet.

Table 4 Plastic anisotropy factor *r* for selected alloys

Material	Condition	$\bar{r}$	r_0	r_{90}
Aluminum				
1100	H14	0.56	0.43	0.89
5082	H19	1.16	0.39	1.81
2024	O	0.81	0.81	0.63
3003	O	0.75	...	...
5052	O	0.58	...	...
Copper and copper alloys				
Copper	Annealed	0.95	...	...
65/35 brass	Annealed	0.74	...	...
70/30 brass	Annealed	0.70	...	...
Steel				
Rimmed	Annealed in hydrogen	1.01–1.11	...	...
Killed	Annealed	1.4–1.62	...	...
Killed, draw quality	Annealed	1.79	...	...
Killed, draw quality	Temper rolled	1.23	...	...
Sheet, draw quality	...	0.7–2.8	...	...
Titanium				
Ti-6Al-4V	Annealed	...	2.57	3.0
Ti-5Al-2.5Sn	Annealed	...	8.1	9.0
Magnesium				
AZ31B-H24	0.004 plastic strain	...	1.0	2.9
AZ31B-H24	0.10 plastic strain	...	2.0	0.67
Zirconium				
Zircaloy-2	Heat treated	...	0.13	0.13
Zircaloy-2	Heat treated and cross rolled	...	2.29	0.67

Source: Ref 25

$$\Delta r = \frac{r_0 + r_{90} - 2r_{45}}{2} \qquad \text{(Eq 54)}$$

The degree of earing in deep drawing correlates well with Δr.

Notch Tensile Test

Ductility measurements on standard smooth tensile specimens do not always reveal metallurgical or environmental changes that lead to reduced local ductility. The tendency for reduced ductility in the presence of a triaxial stress field and steep stress gradients (such as occur at a notch) is called notch sensitivity. A common way of evaluating notch sensitivity is a tensile test using a notched specimen.

The notch tensile test has been used extensively for investigating the properties of high-strength steels, for studying hydrogen embrittlement in steels and titanium, and for investigating the notch sensitivity of high-temperature alloys. More recently, notched tensile specimens have been used for fracture mechanics measurements. Notch sensitivity can also be investigated with the notched-impact test.

The most common notch tensile specimen uses a 60° notch with a root radius 0.025 mm (0.001 in.) or less introduced into a round (circumferential notch) or flat (double-edge notch) tensile specimen. Usually, the depth of the notch is such that the cross-sectional area at the root of the notch is one half of the area in the unnotched section. The specimen is aligned carefully and loaded in tension until fracture occurs. The notch strength is defined as the maximum load divided by the original cross-sectional area at the notch. Because of the plastic constraint at the notch, this value will be higher than the tensile strength of an unnotched specimen if the material possesses some ductility. Therefore, the common way of detecting notch brittleness (or high notch sensitivity) is by determining the notch-strength ratio, NSR:

$$\text{NSR} = \frac{s_{\text{net}}(\text{for notched specimen at maximum load})}{s_{\text{u}}(\text{tensile strength for unnotched specimen})} \qquad \text{(Eq 55)}$$

If the NSR is less than unity, the material is notch brittle. The other property that is measured in the notch tensile test is the reduction in area at the notch.

As strength, hardness, or some metallurgical variable restricting plastic flow increases, the metal at the root of the notch is less able to flow, and fracture becomes more likely. Notch brittleness may be considered to begin at the strength level where the notch strength begins to fall or, more conventionally, at the strength level where the NSR becomes less than unity.

The sensitivity of notch strength for detecting metallurgical embrittlement is illustrated in Fig. 18. Note that the conventional elongation measured on a smooth specimen was unable to detect the fall in notch strength produced by tempering in the 330 to 480 °C (600–900 °F) range. For a more detailed review of notch tensile testing, see Ref 27.

Tensile Test Fractures (Ref 28)

Tensile test specimens can exhibit either ductile or brittle types of fracture. Ductile and brittle are terms that describe the amount of macroscopic plastic deformation that precedes fracture. Ductile fractures are characterized by tearing of metal accompanied by appreciable gross plastic deformation and expenditure of considerable energy. Ductile tensile fractures in most materials have a gray, fibrous appearance and are

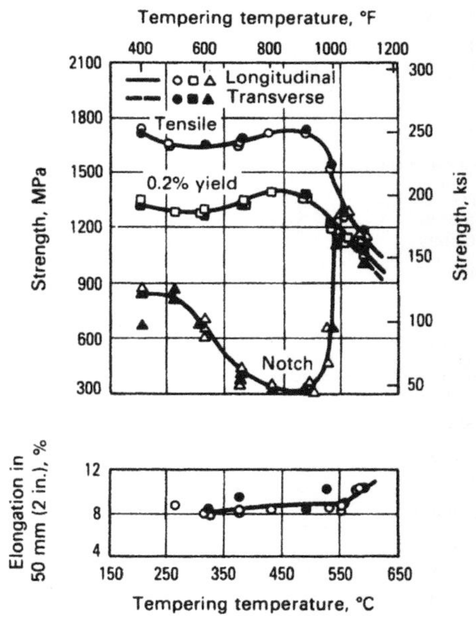

Fig. 18 Notched and unnotched tensile properties of an alloy steel as a function of tempering temperature. Source: Ref 26

classified on a macroscopic scale as either flat (perpendicular to the maximum tensile stress) or shear (at a 45° slant to the maximum tensile stress) fractures.

Brittle fractures are characterized by rapid crack propagation with less expenditure of energy than with ductile fractures and without appreciable gross plastic deformation. Brittle tensile fractures have a bright, granular appearance and exhibit little or no necking. They are generally of the flat type, that is, normal (perpendicular) to the direction of the maximum tensile stress. A chevron pattern may be present on the fracture surface, pointing toward the origin of the crack, especially in brittle fractures in flat platelike components.

It must be pointed out, however, that these terms can also be applied, and are applied, to fracture on a microscopic level. Ductile fractures are those that occur by microvoid formation and coalescence, whereas brittle fractures may occur by either transgranular (cleavage or quasi-cleavage) or intergranular cracking.

Clearly, the classic cup-and-cone fracture shown in Fig. 19(a) has occurred as a result of appreciable plastic deformation and thus is a ductile fracture, whereas the fracture shown in Fig. 19(b) is a brittle fracture. The cup-and-cone

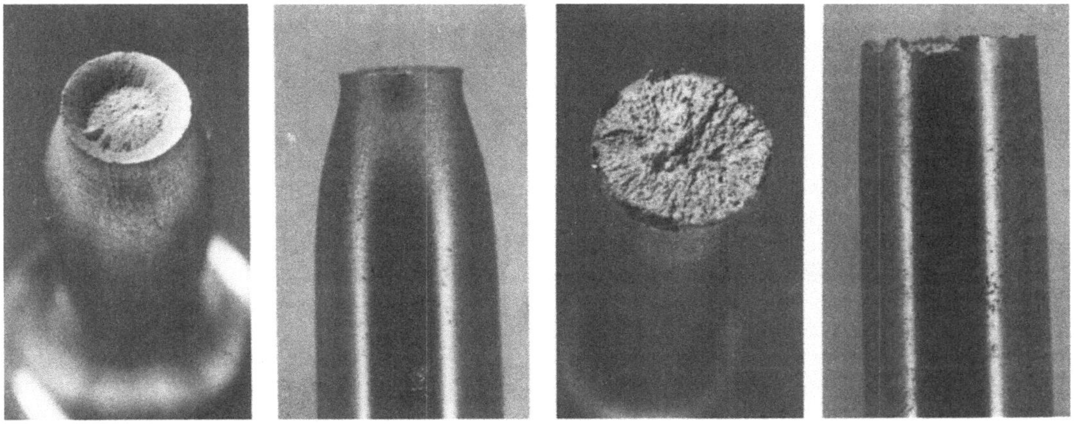

(a) (b)

Fig. 19 Appearance of ductile (a) and brittle (b) tensile fractures. Source: Ref 28

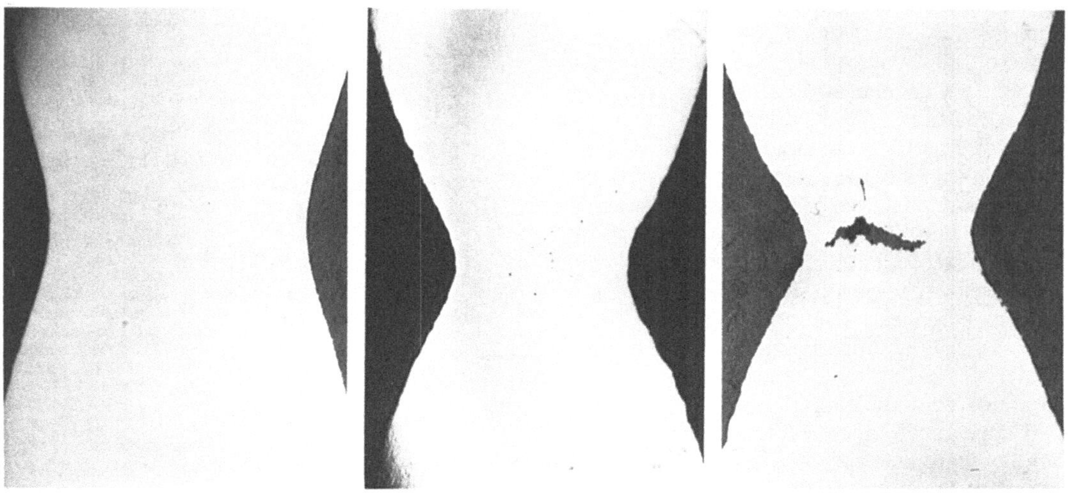

Fig. 20 Sections of a tensile specimen at various stages of formation during development of a cup-and-cone fracture. Note that the fracture is initiating internally. 7×. Source: Ref 28

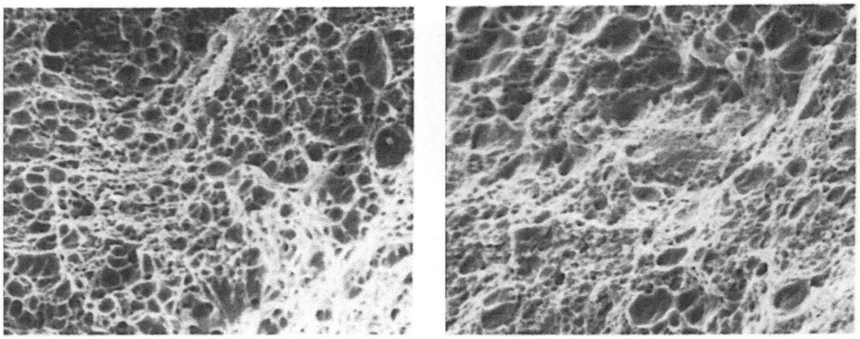

(a) (b)

Fig. 21 Fractographs of a ductile cup-and-cone fracture surface. (a) Bottom of the cup. (b) Sidewall of the cup. SEM. 650×. Source: Ref 28

tensile fracture exhibits three zones: the inner flat fibrous zone where the fracture begins, an intermediate radial zone, and the outer shear-lip zone where the fracture terminates. Figure 19(a) shows each of these zones; the flat brittle fracture shown in Fig. 19(b) exhibits little or no shear-lip zone.

Ductile Fracture. The sequence of events that culminates in a cup-and-cone fracture is illustrated in Fig. 20, which shows the development of voids within the necked region (triaxial tensile stresses) of a tensile specimen and the coalescence of the voids to produce an internal crack by normal rupture. Final separation of the cross section occurs by shear rupture, which produces the wall of the cup. Figure 21 shows scanning electron microscopy (SEM) fractographs of the bottom and the sidewall of the cup. On the microscopic level, a crack is formed by coalescence of microvoids that form as a result of particle-matrix decohesion or cracking of second-phase particles; the microvoids and the associated particles are shown at high magnification in Fig. 22. The process of microvoid formation and coalescence involves considerable localized plastic deformation and requires the expenditure of a large amount of energy, which is the basis of selection of a material with good fracture toughness. The reduction of area of ultrahigh-purity aluminum and copper approaches 100% because of the absence within these materials of void-nucleating particles. In their visual appearance, ductile fractures have a matte or silky texture.

Brittle Fracture. Regarding the brittle fracture shown in Fig. 19(b), it will be noted that the fracture surface is characterized by radial ridges that emanate from the center of the fracture surface. The ridges run parallel to the direction of crack propagation, and a ridge is produced when two cracks that are not coplanar become connected by tearing of the intermediate material. The cracks, which propagate predominantly by quasi-cleavage, move rapidly toward the periphery of the specimen cross section and, as shown in Fig. 19(b), penetrate the external surface of the specimen by shear rupture along a relatively small shear lip. The shear lip develops as a result of the change in the state of stress from one of triaxial tension to one of plane stress. The extent or width of the shear lip de-

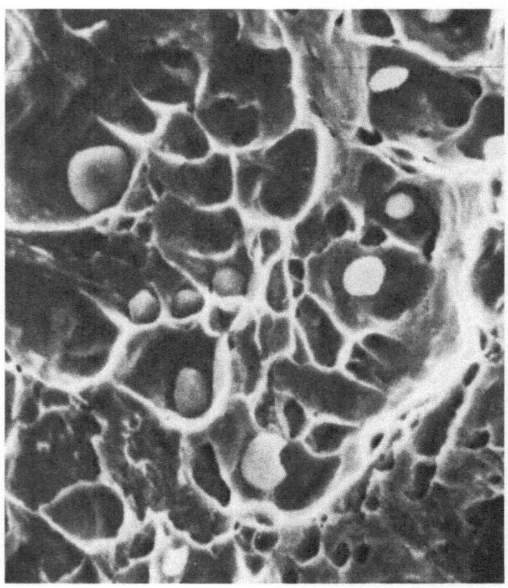

Fig. 22 Large and small sulfide inclusions in a ductile dimple fracture. SEM. 5000×.

pends on the temperature at which fracture occurs, formation of a shear lip being favored by higher temperatures.

ACKNOWLEDGMENTS

This chapter was adapted from:

- G.E. Dieter, Mechanical Behavior Under Tensile and Compressive Loads, *Mechanical Testing and Evaluation,* Vol 8, *ASM Handbook,* ASM International, 2000, p 99–108
- W.F. Hosford, Overview of Tensile Testing, *Tensile Testing,* P. Han, Ed., ASM International, 1992, p 2–24
- W.T. Becker, Special Applications of Tension and Compression Testing, Course 12, Lesson 5, *Mechanical Testing of Metals,* American Society for Metals, 1983, p 19

REFERENCES

1. G.E. Dieter, Introduction to Ductility, in *Ductility,* American Society for Metals, 1968
2. P.G. Nelson and J. Winlock, *ASTM Bull.,* Vol 156, Jan 1949, p 53
3. D.J. Mack, *Trans. AIME,* Vol 166, 1946, p 68–85
4. P.E. Armstrong, Measurement of Elastic Constants, in *Techniques of Metals Research,* Vol V, R.F. Brunshaw, Ed., Interscience, New York, 1971
5. H.E. Davis, G.E. Troxell, and G.F.W. Hauck, *The Testing of Engineering Materials,* McGraw-Hill, New York, 1964, p 33
6. J.R. Low and F. Garofalo, *Proc. Soc. Exp. Stress Anal.,* Vol 4 (No. 2), 1947, p 16–25
7. J.R. Low, *Properties of Metals in Materials Engineering,* American Society for Metals, 1949
8. J. Datsko, *Material Properties and Manufacturing Processes,* John Wiley & Sons, New York, 1966, p 18–20
9. W.B. Morrison, *Trans. ASM,* Vol 59, 1966, p 824
10. D.C. Ludwigson, *Metall. Trans.,* Vol 2, 1971, p 2825–2828
11. H.J. Kleemola and M.A. Nieminen, *Metall. Trans.,* Vol 5, 1974, p 1863–1866
12. C. Adams and J.G. Beese, *Trans. ASME,* Series H, Vol 96, 1974, p 123–126
13. J.H. Keeler, *Trans. ASM,* Vol 47, 1955, p 157–192
14. A. Considére, *Ann. Ponts Chaussées,* Vol 9, 1885, p 574–775
15. G.W. Geil and N.L. Carwile, *J. Res. Natl. Bur. Stand.,* Vol 45, 1950, p 129
16. P.W. Bridgman, *Trans. ASM,* Vol 32, 1944, p 553
17. J. Aronofsky, *J. Appl. Mech.,* Vol 18, 1951, p 75–84
18. T.A. Trozera, *Trans. ASM,* Vol 56, 1963, p 280–282
19. E.R. Marshall and M.C. Shaw, *Trans. ASM,* Vol 44, 1952, p 716
20. W.J. McG. Tegart, *Elements of Mechanical Metallurgy,* Macmillan, New York, 1966, p 22
21. N.E. Dowling, *Mechanical Behavior of Materials,* Prentice-Hall, Englewood Cliffs, NJ, 1993, p 165
22. M.J. Barba, *Mem. Soc. Ing. Civils,* Part I, 1880, p 682
23. E.G. Kula and N.N. Fahey, *Mater. Res. Stand.,* Vol 1, 1961, p 631
24. W.F. Hosford, Overview of Tensile Testing, *Tensile Testing,* P. Han, Ed., ASM International, 1992, p 2–24
25. W.T. Becker, Special Applications of Tension and Compression Testing, Course 12, Lesson 5, *Mechanical Testing of Metals,* American Society for Metals, 1983, p 19
26. G.B. Espey, M.H. Jones, and W.F. Brown, Jr., *ASTM Proc.,* Vol 59, 1959, p 837
27. J.D. Lubahn, *Trans. ASME,* Vol 79, 1957, p 111–115
28. Ductile and Brittle Fractures, *Failure Analysis and Prevention,* Vol 11, *Metals Handbook,* 9th ed., American Society for Metals, 1986, p 82–101

CHAPTER 3

Uniaxial Tensile Testing

THE TENSILE TEST is one of the most commonly used tests for evaluating materials. In its simplest form, the tensile test is accomplished by gripping opposite ends of a test piece (specimen) within the load frame of a test machine. A tensile force is applied by the machine, resulting in the gradual elongation and eventual fracture of the test piece. During the process, force-extension data, a quantitative measure of how the test piece deforms under the applied tensile force, usually are monitored and recorded. When properly conducted, the tensile test provides force-extension data that can quantify several important mechanical properties of a material. These mechanical properties determined from tensile tests include, but are not limited to, the following:

- Elastic deformation properties, such as the modulus of elasticity (Young's modulus) and Poisson's ratio
- Yield strength and ultimate tensile strength
- Ductility properties, such as elongation and reduction in area
- Strain-hardening characteristics

These material characteristics from tensile tests are used for quality control in production, for ranking performance of structural materials, for evaluation of newly developed alloys, and for dealing with the static-strength requirements of design.

The basic principle of the tensile test is quite simple, but numerous variables affect results. General sources of variation in mechanical-test results include several factors involving materials, namely, methodology, human factors, equipment, and ambient conditions, as shown in the "fish-bone" diagram in Fig. 1. This chapter discusses the methodology of the tensile test and the effect of some of the variables on the tensile

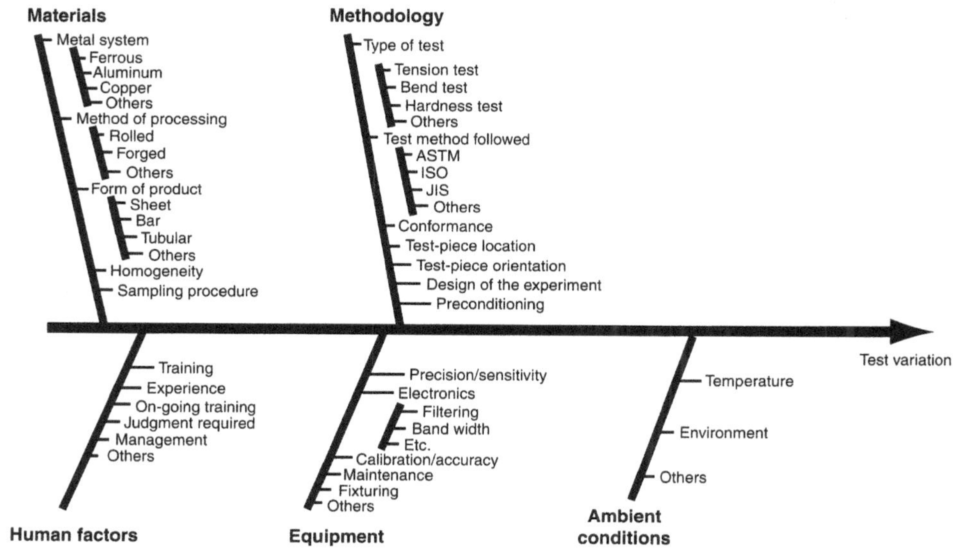

Fig. 1 "Fish-bone" diagram of sources of variability in mechanical-test results

properties determined. The following methodology and variables are discussed:

- Shape of the item being tested
- Method of gripping the item
- Method of applying the force
- Determination of strength properties other than the maximum force required to fracture the test item
- Ductility properties to be determined
- Speed of force application or speed of elongation (e.g., control of stress rate or strain rate)
- Test temperature

The main focus of this chapter is on the methodology of tensile tests as it applies to metallic materials. Factors associated with test machines and their method of force application are described in more detail in Chapter 4, "Tensile Testing Equipment and Strain Sensors."

This chapter does not address the tensile testing of nonmetallic materials, such as plastics, elastomers, or ceramics. Although uniaxial tensile testing is used in the mechanical evaluation of these materials, other test methods often are used for mechanical-property evaluation. The general concept of tensile properties is very similar for these nonmetallic materials, but there are also some very important differences in their behavior and the required test procedures for these materials:

- *Tensile-test results for plastics* depend more strongly on the strain rate because plastics are viscoelastic materials that exhibit time-dependent deformation (i.e., creep) during force application. Plastics are also more sensitive to temperature than metals. Thus, control of strain rates and temperature are more critical with plastics, and sometimes tensile tests are run at more than one strain and/or temperature. The ASTM standard for tension testing of plastics is D 638. See Chapter 8, "Tensile Testing of Plastics," for further details.
- *Tensile testing of ceramics* requires more attention to alignment and gripping of the test piece in the test machine because ceramics are brittle materials that are extremely sensitive to bending strains and because the hard surface of ceramics reduces the effectiveness of frictional gripping devices. The need for large gripping areas thus requires the use of larger test pieces (Ref 1). The ASTM standard for tensile testing of monolithic ceramics at room temperature is C 1273. The standard for continuous fiber-reinforced advanced ceramics at ambient temperatures is C 1275. See Chapter 10, "Tensile Testing of Ceramics and Ceramic-Matrix Composites," for further details.
- *Tensile testing of elastomers* is described in ASTM D 412 with specific instructions about test-piece preparation, equipment, and test conditions. Tensile properties of elastomers vary widely, depending on the particular formulation, and scatter both within and between laboratories is appreciable compared with the scatter of tensile-test results of metals (Ref 2). The use of tensile-test results of elastomers is limited principally to comparison of compound formulations. See Chapter 9, "Tensile Testing of Elastomers," for further details.

Definitions and Terminology

The basic results of a tensile test and other mechanical tests are quantities of stress and strain that are measured. These basic terms and their units are briefly defined here, along with discussions of basic stress-strain behavior and the differences between related terms, such as stress and force and strain and elongation.

Load (or force) typically refers to the force acting on a body. However, there is currently an effort within the technical community to replace the word load with the more precise term *force,* which has a distinct meaning for any type of force applied to a body. Load applies, in a strict sense, only to the gravitational force that acts on a mass. Nonetheless, the two terms are often used interchangeably.

Force is usually expressed in units of pounds-force, lbf, in the English system. In the metric system, force is expressed in units of newtons (N), where one newton is the force required to give a 1 kg mass an acceleration of 1 m/s^2 (1 N = 1 kgm/s^2). Although newtons are the preferred metric unit, force is also expressed as kilogram force, kgf, which is the gravitational force on a 1 kg mass on the surface of the earth. The numerical conversions between the various units of force are as follows:

- 1 lbf = 4.448222 N or 1 N = 0.2248089 lbf
- 1 kgf = 9.80665 N

In some engineering disciplines, such as civil engineering, the quantity of 1000 lbf is also expressed in units of kip, such that 1 kip = 1000 lbf.

Stress is simply the amount of force that acts over a given cross-sectional area. Thus, stress is expressed in units of force per area units and is obtained by dividing the applied force by the cross-sectional area over which it acts. Stress is an important quantity because it allows strength comparison between tests conducted using test pieces of different sizes and/or shapes. When discussing strength values in terms of force, the load (force) carrying capacity of a test piece is a function of the size of the test piece. However, when material strength is defined in terms of stress, the size or shape of the test piece has little or no influence on stress measurements of strength (provided the cross section contains at least 10 to 15 metallurgical grains).

Stress is typically denoted by either the Greek symbol sigma, σ, or by s, unless a distinction is being made between true stress and nominal (engineering) stress as discussed in this article. The units of stress are typically lbf/in.^{2} (psi) or thousands of psi (ksi) in the English system and a pascal (Pa) in the metric system. Engineering stresses in metric units are also expressed in terms of newtons per area (i.e., N/m^2 or N/mm^2) or as kilopascals (kPA) and megapascals (MPa). Conversions between these various units of stress are as follows:

- 1 Pa = 1.45×10^{-4} psi
- 1 Pa = 1 N/m^2
- 1 kPa = 10^3 Pa or 1 kPa = 0.145 psi
- 1 MPa = 10^6 Pa or 1 MPa = 0.145 ksi
- 1 N/mm^2 = 1 MPa

Strain and elongation are similar terms that define the amount of deformation from a given amount of applied stress. In general terms, strain is defined (by ASTM E 28) as "the change per unit length due to force in an original linear dimension." The phrase *change per unit length* means that a change in length, ΔL, is expressed as a ratio of the original length, L_0. This change in length can be expressed in general terms as a strain or as elongation of gage length, as described subsequently in the context of a tension test.

Strain is a general term that can be expressed mathematically, either as engineering strain or as true strain. Nominal (or engineering) strain is often represented by the letter e, and logarithmic (or true) strain is often represented by the Greek letter ε. The equation for engineering strain, e, is based on the nominal change in length (ΔL) where:

$$e = \Delta L/L_0 = (L - L_0)/L_0$$

The equation for true strain, ε, is based on the instantaneous change in length (dl) where:

$$\varepsilon = \int_{L_0}^{L} \frac{dl}{l} = \ln\left(\frac{L}{L_0}\right)$$

These two basic expressions for strain are interrelated, such that:

$$\varepsilon = \ln(1 + e)$$

In a tensile test, the typical measure of strain is engineering strain, e, and the units are inches per inch (or millimeter per millimeter and so on). Often, however, no units are shown because strain is the ratio of length in a given measuring system.

This chapter refers to only engineering strain unless otherwise specified. In a tensile test, true strain is based on the change in the cross-sectional area of the test piece as it is loaded. It is not further discussed herein, but a detailed discussion is found in Chapter 2, "Mechanical Behavior of Materials under Tensile Loads."

Elongation is a term that describes the amount that the test piece stretches during a tensile test. This stretching or elongation can be defined either as the total amount of stretch, ΔL, that a part undergoes or the increase in gage length per the initial gage length, L_0.

The latter definition is synonymous with the meaning of engineering strain, $\Delta L/L_0$, while the first definition is the total amount of extension. Because two definitions are possible, it is imperative that the exact meaning of elongation be understood each time it is used.

This chapter uses the term elongation, e, to mean nominal or engineering strain (i.e., $e = \Delta L/L_0$). The amount of stretch is expressed as extension, or the symbol ΔL. In many cases, elongation, e, is also reported as a percentage change in gage length as a measure of ductility (i.e., percent elongation), $(\Delta L/L_0) \times 100$. This convention is used in Chapter 1, "Introduction to Tensile Testing."

Engineering Stress and True Stress. Along with the previous descriptions of engineering strain and true strain, it is also possible to define stress in two different ways as engineering stress and true stress. As is intuitive, when a tensile force stretches a test piece, the cross-sectional area must decrease (because the overall volume of the test piece remains essentially constant). Hence, because the cross section of the test piece becomes smaller during a test, the value of stress depends on whether it is calculated based on the area of the unloaded test piece (the initial area) or on the area resulting from that applied force (the instantaneous area). Thus, in this context, there are two ways to define stress:

- *Engineering stress, s:* The force at any time during the test divided by the initial area of the test piece; $s = F/A_0$ where F is the force, and A_0 is the initial cross section of a test piece.
- *True stress,* σ: The force at any time divided by the instantaneous area of the test piece; $\sigma = F/A_i$ where F is the force, and A_i is the instantaneous cross section of a test piece.

Because an increasing force stretches a test piece, thus decreasing its cross-sectional area, the value of true stress will always be greater than the nominal, or engineering, stress.

These two definitions of stress are further related to one another in terms of the strain that occurs when the deformation is assumed to occur at a constant volume (as it frequently is). As previously noted, strain can be expressed as either engineering strain (e) or true strain, where the two expressions of strain are related as $\varepsilon = \ln(1 + e)$. When the test-piece volume is constant during deformation (i.e., $A_iL_i = A_0L_0$), then the instantaneous cross section, A_i, is related to the initial cross section, A_0, where

$$\begin{aligned} A &= A_0 \exp\{-\varepsilon\} \\ &= A_0/(1 + e) \end{aligned}$$

If these expressions for instantaneous and initial cross sections are divided into the applied force to obtain values of true stress (at the instantaneous cross section, A_i) and engineering stress (at the initial cross section, A_0), then:

$$\sigma = s \exp\{\varepsilon\} = s(1 + e)$$

Typically, engineering stress is more commonly considered during uniaxial tension tests. All discussions in this article are based on nominal engineering stress and strain unless otherwise noted. More detailed discussions on true stress and true strain are in Chapter 2, "Mechanical Behavior of Materials under Tensile Loads."

Stress-Strain Behavior

During a tensile test, the force applied to the test piece and the amount of elongation of the test piece are measured simultaneously. The applied force is measured by the test machine or by accessory force-measuring devices. The amount of stretching (or extension) can be measured with an extensometer. An extensometer is a device used to measure the amount of stretch that occurs in a test piece. Because the amount of elastic stretch is quite small at or around the onset of yielding (in the order of 0.5% or less for steels), some manner of magnifying the stretch is required. An extensometer may be a mechanical device, in which case the magnification occurs by mechanical means. An extensometer may also be an electrical device, in which case the magnification may occur by mechanical means, electrical means, or by a combination of both. Extensometers generally have fixed gage lengths. If an extensometer is used only to obtain a portion of the stress-strain curve sufficient to determine the yield properties, the gage length of the extensometer may be shorter than the gage length required for the elongation-at-fracture measurement. It may also be longer, but in general, the extensometer gage length should not exceed approximately 85 to 90% of the length of the reduced section or of the distance between the grips for test pieces without reduced sections. This ratio for some of the most common test configurations with a 2 in. gage length and 2¼ in. reduced section is 0.875%.

The applied force, F, and the extension, ΔL, are measured and recorded simultaneously at regular intervals, and the data pairs can be converted into a stress-strain diagram as shown in Fig. 2. The conversion from force-extension data to stress-strain properties is shown schematically in Fig. 2(a). Engineering stress, s, is obtained by dividing the applied force by the original cross-sectional area, A_0, of the test piece, and strain, e, is obtained by dividing the amount of extension, ΔL, by the original gage length, L. The basic result is a stress-strain curve (Fig. 2b) with regions of elastic deformation and permanent (plastic) deformation at stresses greater than those of the elastic limit (EL in Fig. 2b).

Typical stress-strain curves for three types of steels, aluminum alloys, and plastics are shown in Fig. 3 (Ref 3). Stress-strain curves for some structural steels are shown in Fig. 4(a) (Ref 4) for elastic conditions and for small amounts of plastic deformation. The general shape of the stress-strain curves can be described for deformation in this region. However, as plastic deformation occurs, it is more difficult to generalize about the shape of the stress-strain curve. Figure 4(b) shows the curves of Fig. 4(a) continued to fracture.

Elastic deformation occurs in the initial portion of a stress-strain curve, where the stress-strain relationship is initially linear. In this region, the stress is proportional to strain. Mechanical behavior in this region of stress-strain curve is defined by a basic physical property called the modulus of elasticity (often abbreviated as E). The modulus of elasticity is the slope of the stress-strain line in this linear re-

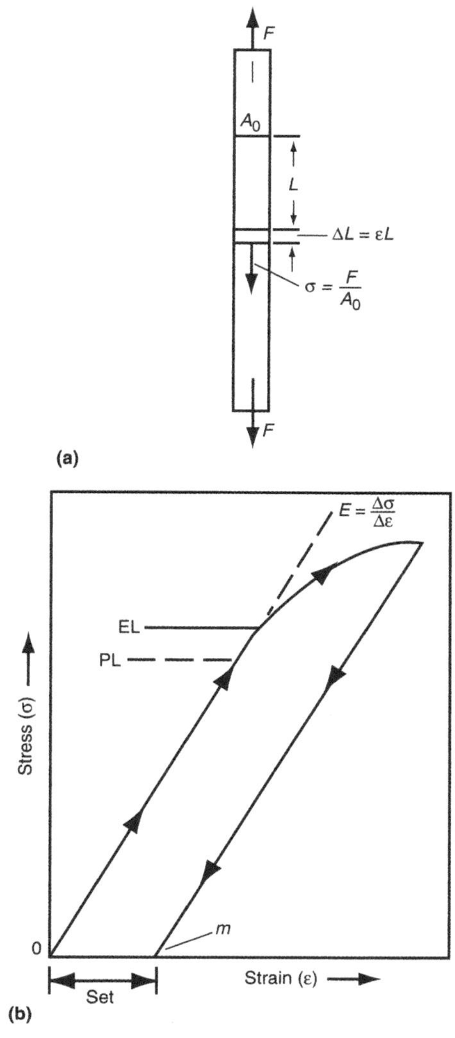

Fig. 2 Stress-strain behavior in the region of the elastic limit. (a) Definition of σ and ε in terms of initial test piece length, L, and cross-sectional area, A_0, before application of a tensile force, F. (b) Stress-strain curve for small strains near the elastic limit (EL)

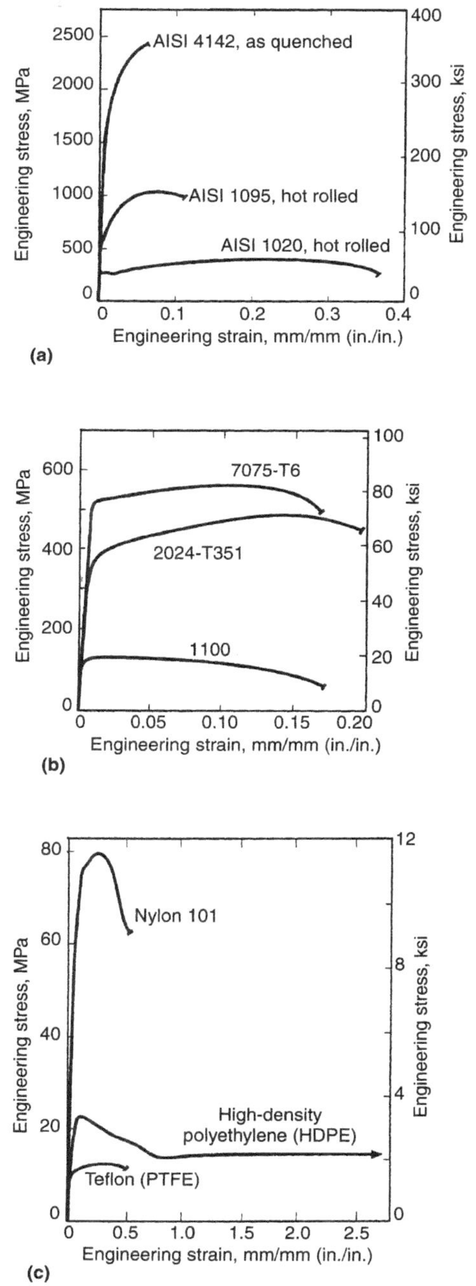

Fig. 3 Typical engineering stress-strain curves from tensile tests on (a) three steels, (b) three aluminum alloys, and (c) three plastics. PTFE, polytetrafluoroethylene. Source: Ref 3

gion, and it is a basic physical property of all materials. It essentially represents the spring constant of a material.

The modulus of elasticity is also called Hooke's modulus or Young's modulus after the scientists who discovered and extensively studied the elastic behavior of materials. The behavior was first discovered in the late 1600s by the English scientist Robert Hooke. He observed that a given force would always cause a repeatable, elastic deformation in all materials. He further discovered that there was a force above which the deformation was no longer elastic; that is, the material would not return to its original length after release of the force. This limiting force is called the elastic limit (EL in Fig. 2b). Later, in the early 1800s, Thomas Young, an English physicist, further investigated and de-

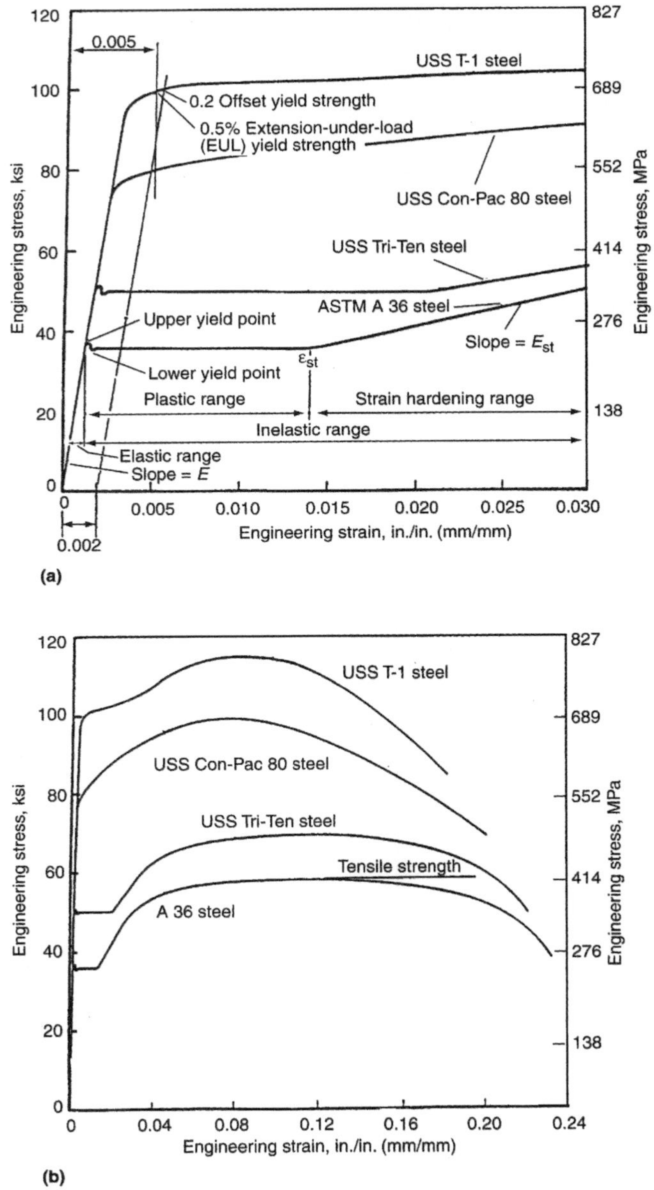

Fig. 4 Typical stress-strain curves for structural steels having specified minimum tensile properties. (a) Portions of the stress-strain curves in the yield-strength region. (b) Stress-strain curves extended through failure. Source: Ref 4

scribed this elastic phenomenon, and so his name is associated with it.

The proportional limit (PL) is a point in the elastic region where the linear relationship between stress and strain begins to break down. At some point in the stress-strain curve (PL in Fig. 2b), linearity ceases, and small increase in stress causes a proportionally larger increase in strain. This point is referred to as the proportional limit (PL) because up to this point, the stress and strain are proportional. If an applied force below the PL point is removed, the trace of the stress and strain points returns along the original line. If the force is reapplied, the trace of the stress and strain points increases along the original line. (When an exception to this linearity is observed, it usually is due to mechanical hysteresis in the extensometer, the force indicating system, the recording system, or a combination of all three.)

The elastic limit (EL) is a very important property when performing a tensile test. If the applied stresses are below the elastic limit, then the test can be stopped, the test piece unloaded, and the test restarted without damaging the test piece or adversely affecting the test results. For example, if it is observed that the extensometer is not recording, the force-elongation curve shows an increasing force, but no elongation. If the force has not exceeded the elastic limit, the test piece can be unloaded, adjustments made, and the test restarted without affecting the results of the test. However, if the test piece has been stressed above the EL, plastic deformation (set) will have occurred (Fig. 2b), and there will be a permanent change in the stress-strain behavior of the test piece in subsequent tension (or compression) tests.

The PL and the EL are considered identical in most practical instances. In theory, however, the EL is considered to be slightly higher than the PL, as illustrated in Fig. 2b. The measured values of EL or PL are highly dependent on the magnification and sensitivity of the extensometer used to measure the extension of the test piece. In addition, the measurement of PL and EL also highly depends on the care with which a test is performed.

Plastic Deformation (Set) from Stresses above the Elastic Limit. If a test piece is stressed (or loaded) and then unloaded, any retest proceeds along the unloading path whether or not the elastic limit was exceeded. For example, if the initial stress is less than the elastic limit, the load-unload-reload paths are identical. However, if a test piece is stressed in tension beyond the elastic limit, then the unload path is offset and parallel to the original loading path (Fig. 2b). Moreover, any subsequent tension measurements will follow the previous unload path parallel to the original stress-strain line. Thus, the application and removal of stresses above the elastic limit affect all subsequent stress-strain measurements.

The term *set* refers to the permanent deformation that occurs when stresses exceed the elastic limit (Fig. 2b). ASTM E 6 defines set as the strain remaining after the complete release of a load-producing deformation. Because set is permanent deformation, it affects subsequent stress-strain measurements whether the reloading occurs in tension or compression. Likewise, permanent set also affects all subsequent tests if the initial loading exceeds the elastic limit in compression. Discussions of these two situations follow.

Reloading after Exceeding the Elastic Limit in Tension. If a test piece is initially loaded in tension beyond the elastic limit and then unloaded, the unload path is parallel to the initial load path but offset by the set; on reloading in tension, the unloading path will be followed. Figure 5 illustrates a series of stress-strain curves obtained using a machined round test piece of steel. (The strain axis is not to scale.) In this figure, the test piece was loaded first to Point *A* and unloaded. The area of the test piece was again determined (A_2) and reloaded to Point

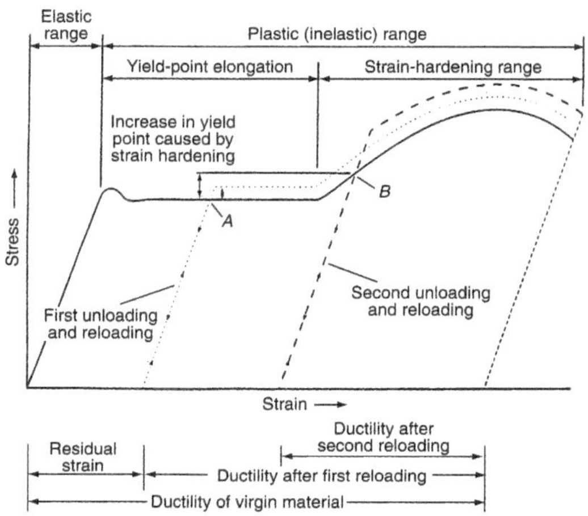

Fig. 5 Effects of prior tensile loading on tensile stress-strain behavior. Solid line, stress-strain curve based on dimensions of unstrained test piece (unloaded and reloaded twice); dotted line, stress-strain curve based on dimensions of test piece after first unloading; dashed line, stress-strain curve based on dimensions of test piece after second unloading. Note: Graph is not to scale.

B and unloaded. The area of the test piece was determined for a third time (A_3) and reloaded until fracture occurred. Because during each loading the stresses at Points *A* and *B* were in excess of the elastic limit, plastic deformation occurred. As the test piece is elongated in this series of tests, the cross-sectional area must decrease because the volume of the test piece must remain constant. Therefore, $A_1 > A_2 > A_3$.

The curve with a solid line in Fig. 5 is obtained for engineering stresses calculated using the applied forces divided by the original cross-sectional area. The curve with a dotted line is obtained from stresses calculated using the applied forces divided by the cross-sectional area, A_2, with the origin of this stress-strain curve located on the abscissa at the end point of the first unloading line. The curve represented by the dashed line is obtained from the stresses calculated using the applied forces divided by the cross-sectional area, A_3, with the origin of this stress-strain curve located on the abscissa at the end point of the second unloading line. This figure illustrates what happens if a test is stopped, unloaded, and restarted. It also illustrates one of the problems that can occur when testing pieces from material that has been formed into a part (or otherwise plastically strained before testing). An example is a test piece that was machined from a failed structure to determine the tensile properties. If the test piece is from a location that was subjected to tensile deformation during the failure, the properties obtained are probably not representative of the original properties of the material.

Bauschinger Effect. The other loading condition occurs when the test piece is initially loaded in compression beyond the elastic limit and then unloaded. The unload path is parallel to the initial load path but offset by the set; on reloading in tension, the elastic limit is much lower, and the shape of the stress-strain curve is significantly different. The same phenomenon occurs if the initial loading is in tension and the subsequent loading is in compression. This condition is called the Bauschinger effect, named for the German scientist who first described it around 1860. Again, the significance of this phenomenon is that if a test piece is machined from a location that has been subjected to plastic deformation, the stress-strain properties will be significantly different than if the material had not been so strained. This occurrence is illustrated in Fig. 6, where a machined round steel test piece was first loaded in tension to about 1% strain, unloaded, loaded in compression to about 1% strain, unloaded, and reloaded in tension. For this steel, the initial portion of tension and compression stress-strain curves are essentially identical.

Properties from Test Results

A number of tensile properties can be determined from the stress-strain diagram. Two of these properties, the tensile strength and the yield strength, are described in the next section of this article, "Strength Properties." In addition, total elongation (ASTM E 6), yield-point elongation (ASTM E 6), Young's modulus (ASTM E 111), and the strain-hardening exponent (ASTM E 646) are sometimes determined from the stress-strain diagram. Other tensile properties include the following:

- Poisson's ratio (ASTM E 132)
- Plastic-strain ratio (ASTM E 517)
- Elongation by manual methods (ASTM E 8)
- Reduction of area

These properties require more information than just the data pairs generating a stress-strain curve. None of these four properties can be determined from a stress-strain diagram.

Strength Properties

Tensile strength and yield strength are the most common strength properties determined in

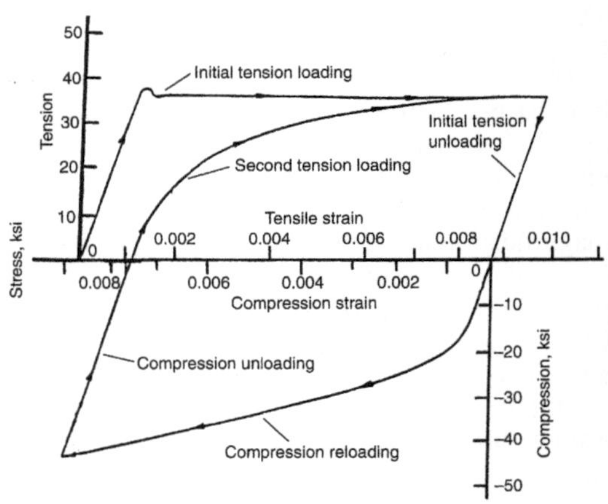

Fig. 6 Example of the Bauschinger effect and hysteresis loop in tension-compression-tension loading. This example shows initial tension loading to 1% strain, followed by compression loading to 1% strain, and then a second tension loading to 1% strain.

a tensile test. According to ASTM E 6, tensile strength is calculated from the maximum force during a tension test that is carried to rupture divided by the original cross-sectional area of the test piece. By this definition, it is a stress value, although some product specifications define the tensile strength as the force (load) sustaining ability of the product without consideration of the cross-sectional area. Fastener specifications, for example, often refer to tensile strength as the applied force (load-carrying) capacity of a part with specific dimensions.

The yield strength refers to the stress at which a small, but measurable, amount of inelastic or plastic deformation occurs. There are three common definitions of yield strength:

- Offset yield strength
- Extension-under-load (EUL) yield strength
- Upper yield strength (or upper yield point)

An upper yield strength (upper yield point) (Fig. 7a) usually occurs with low-carbon steels and some other metal systems to a limited degree. Often, the pronounced peak of the upper yield is suppressed due to slow testing speed or nonaxial loading (i.e., bending of the test piece), metallurgical factors, or a combination of these; in this case, a curve of the type shown in Fig. 7(b) is obtained. The other two definitions of yield strength, EUL and offset, were developed for materials that do not exhibit the yield-point behavior shown in Fig. 7. Stress-strain curves without a yield point are illustrated in Fig. 4(a) for USS Con-Pac 80 and USS T-1 steels. To determine either the EUL or the offset yield strength, the stress-strain curve must be determined during the test. In computer-controlled testing systems, this curve is often stored in memory and may not be charted or displayed.

Upper yield strength (or upper yield point) can be defined as the stress at which measurable strain occurs without an increase in the stress; that is, there is a horizontal region of the stress-strain curve (Fig. 7) where discontinuous yielding occurs. Before the onset of discontinuous yielding, a peak of maximum stress for yielding is typically observed (Fig. 7a). This pronounced yielding, of the type shown, is usually called yield-point elongation (YPE). This elongation is a diffusion-related phenomenon, where under certain combinations of strain rate and temperature as the material deforms, interstitial atoms are dragged along with dislocations, or dislocations can alternately break away and be repinned, with little or no increase in stress. Either or both of these actions cause serrations or discontinuous changes in a stress-strain curve, which are usually limited to the onset of yielding. This type of yield point is sometimes referred to as the upper yield strength or upper yield point. This type of yield point is usually associated with low-carbon steels, although other metal systems may exhibit yield points to some degree. For example, the stress-strain curves for A36 and USS Tri-Ten steels shown in Fig. 4(a) exhibit this behavior.

The yield point is easy to measure because the increase in strain that occurs without an increase in stress is visually apparent during the conduct of the test by observing the force-indicating system. As shown in Fig. 7, the yield point is usually quite obvious and thus can easily be determined by observation during a tensile test. It can be determined from a stress-strain curve or

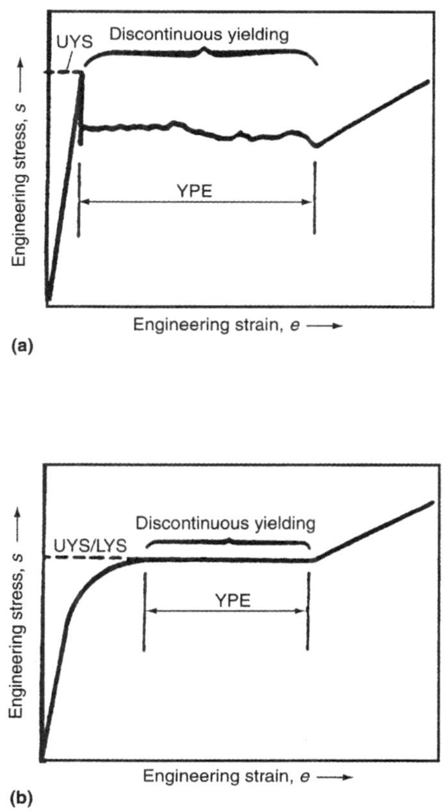

Fig. 7 Examples of stress-strain curves exhibiting pronounced yield-point behavior. Pronounced yielding, of the type shown, is usually called yield-point elongation (YPE). (a) Classic example of upper-yield-strength (UYS) behavior typically observed in low-carbon steels with a very pronounced upper yield strength. (b) General example of pronounced yielding without an upper yield strength. LYS, lower yield strength

by the halt of the dial when the test is performed on machines that use a dial to indicate the applied force. However, when watching the movement of the dial, sometimes a minimum value, recorded during discontinuous yielding, is noted. This value is sometimes referred to as the lower yield point. When the value is ascertained without instrumentation readouts, it is often referred to as the halt-of-dial or the drop-of-beam yield point (as an average usually results from eye readings). It is almost always the upper yield point that is determined from instrument readouts.

Extension-under-load (EUL) yield strength is the stress at which a specified amount of stretch has taken place in the test piece. The EUL is determined by the use of one of the following types of apparatus:

- Autographic devices that secure stress-strain data, followed by an analysis of this data (graphically or using automated methods) to determine the stress at the specified value of extension
- Devices that indicate when the specified extension occurs so that the stress at that point may be ascertained

Graphical determination is illustrated in Fig. 8. On the stress-strain curve, the specified amount of extension, 0-*m*, is measured along the strain axis from the origin of the curve and a vertical line, *m*-*n*, is raised to intersect the stress-strain curve. The point of intersection, *r*, is the EUL yield strength, and the value *R* is read from the stress axis. Typically, for many materials, the extension specified is 0.5%; however, other values may be specified. Therefore, when reporting the EUL, the extension also must be reported. For example, yield strength (EUL = 0.5%) = 52,500 psi is a correct way to report an EUL yield strength. The value determined by the EUL method may also be termed a yield point.

Offset yield strength is the stress that causes a specified amount of set to occur; that is, at this stress, the test piece exhibits plastic deformation (set) equal to a specific amount. To determine the offset yield strength, it is necessary to secure data (autographic or numerical) from which a stress-strain diagram may be constructed graphically or in computer memory. Figure 9 shows how to use these data; the amount of the specified offset 0-*m* is laid out on the strain axis. A line, *m*-*n*, parallel to the modulus of elasticity line, 0-*A*, is drawn to intersect the stress-strain curve. The point of intersection, *r*, is the offset yield strength, and the value, *R*, is read from the stress axis. Typically, for many materials, the offset specified is 0.2%; however, other values may be specified. Therefore, when reporting the offset yield strength, the amount of the offset also must be reported; for example, "0.2% offset yield strength = 52.8 ksi" or "yield strength (0.2% offset) = 52.8 ksi" are common formats used in reporting this information.

In Fig. 8 and 9, the initial portion of the stress-strain curve is shown in ideal terms as a straight line. Unfortunately, the initial portion of the stress-strain curve sometimes does not begin as

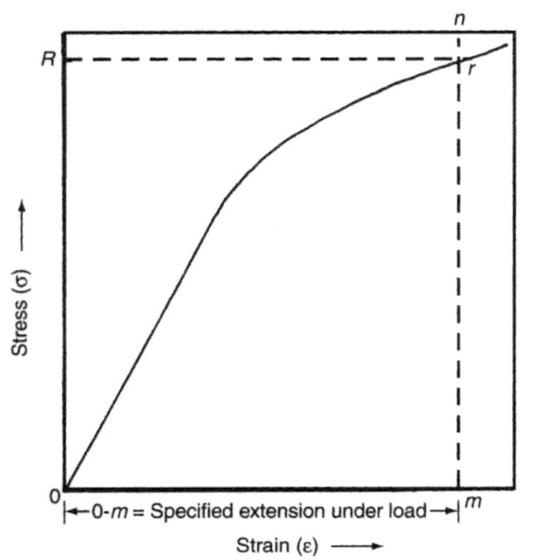

Fig. 8 Method of determining yield strength by the extension-under-load method (EUL). Source: adapted from ASTM E 8

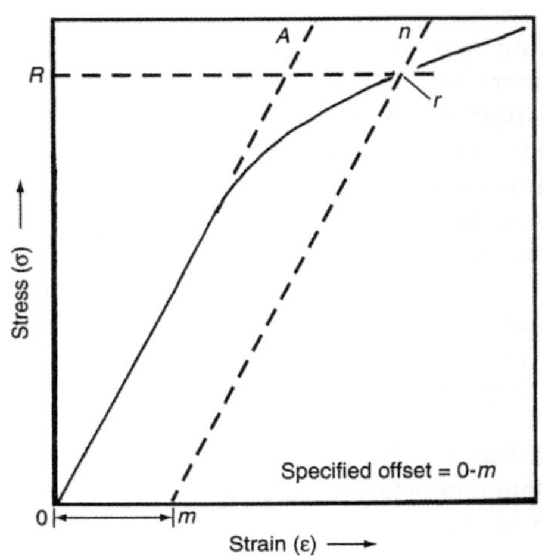

Fig. 9 Method of determining yield strength by the offset method. Source: adapted from ASTM E 8

a straight line but rather has either a concave or a convex foot (Fig. 10) (Ref 5). The shape of the initial portion of a stress-strain curve may be influenced by numerous factors such as, but not limited to, the following:

- Seating of the test piece in the grips
- Straightening of a test piece that is initially bent by residual stresses or bent by coil set
- Initial speed of testing

Generally, the aberrations in this portion of the curve should be ignored when fitting a modulus line, such as that used to determine the origin of the curve. As shown in Fig. 10, a "foot correction" may be determined by fitting a line, whether by eye or by using a computer program, to the linear portion and then extending this line back to the abscissa, which becomes point 0 in Fig. 8 and 9. As a rule of thumb, point *D* in Fig. 10 should be less than one-half the specified yield point or yield strength.

Tangent or Chord Moduli. For materials that do not have a linear relationship between stress and strain, even at very low stresses, the offset yield is meaningless without defining how to determine the modulus of elasticity. Often, a chord modulus or a tangent modulus is specified. A chord modulus is the slope of a chord between any two specified points on the stress-strain curve, usually below the elastic limit. A tangent modulus is the slope of the stress-strain curve at a specified value of stress or of strain. Chord and tangent moduli are illustrated in Fig. 11. Another technique that has been used is sketched in Fig. 12. The test piece is stressed to approximately the yield strength, unloaded to about 10% of this value, and reloaded. As previously discussed, the unloading line will be parallel to what would have been the initial modulus line, and the reloading line will coincide with the unloading line (assuming no hysteresis in any of the system components). The slope of this line is transferred to the initial loading line, and the offset is determined as before. The stress or strain at which the test piece is unloaded usually is not important. This technique is specified in the ISO standard for the tensile test of metallic materials, ISO 6892.

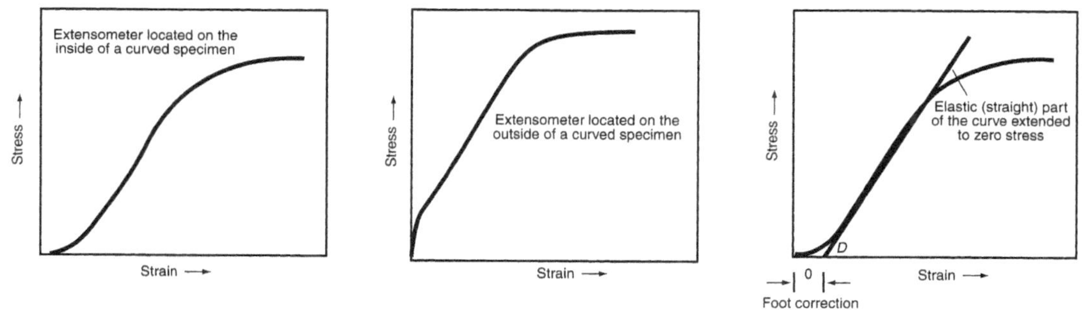

Fig. 10 Examples of stress-strain curves requiring foot correction. Point *D* is the point where the extension of the straight (elastic) part diverges from the stress-strain curve. Source: Ref 5

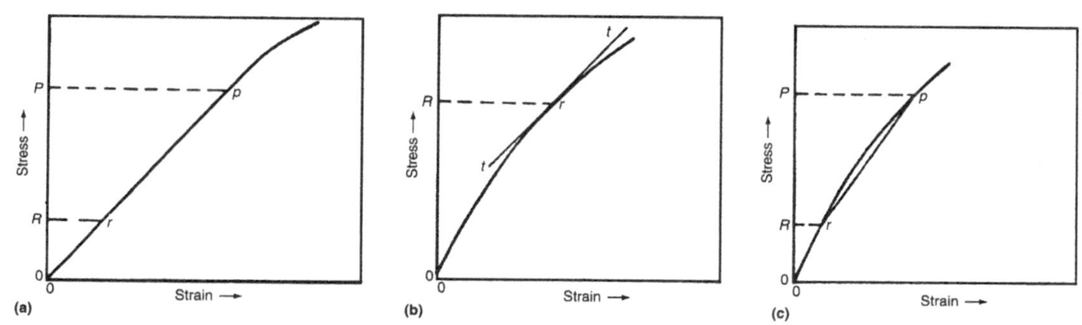

Fig. 11 Stress-strain curves showing straight lines corresponding to (a) Young's modulus between stress, *P*, below proportional limit and *R*, or preload; (b) tangent modulus at any stress, *R*; and (c) chord modulus between any two stresses, *P* and *R*. Source: Ref 6

Yield-strength-property values generally depend on the definition being used. As shown in Fig. 4(a) for the USS Con-Pac steel, the EUL yield is greater than the offset yield, but for the USS T-1 steel (Fig. 4a), the opposite is true. The amount of the difference between the two values is dependent upon the slope of the stress-strain curve between the two intersections. When the stress-strain data pairs are sampled by a computer, and a yield spike or peak of the type shown in Fig. 7(a) occurs, the EUL and the offset yield strength will probably be less than the upper yield point and will probably differ because the *m-n* lines of Fig. 8 and 9 will intersect at different points in the region of discontinuous yielding.

Ductility

Ductility is the ability of a material to deform plastically without fracturing. Figure 13 is a sketch of a test piece with a circular cross section that has been pulled to fracture. As indicated in this sketch, the test piece elongates during the tensile test and correspondingly reduces in cross-sectional area. The two measures of the ductility of a material are the amount of elongation and reduction of area that occurs during a tensile test.

Elongation, as previously noted, is defined in ASTM E 6 as the increase in the gage length of a test piece subjected to a tension force, divided by the original gage length on the test piece. Elongation usually is expressed as a percentage of the original gage length. ASTM E 6 further indicates the following:

- The increase in gage length may be determined either at or after fracture, as specified for the material under test.
- The gage length shall be stated when reporting values of elongation.
- Elongation is affected by test-piece geometry (gage length, width, and thickness of the gage section and of adjacent regions) and test procedure variables, such as alignment and speed of pulling.

The manual measurement of elongation on a tensile test piece can be done with the aid of gage marks applied to the unstrained reduced section. After the test, the amount of stretch between gage marks is measured with an appropriate device. The use of the term *elongation* in this instance refers to the total amount of stretch or extension. Elongation, in the sense of nominal engineering strain, *e*, is the value of gage extension divided by the original distance between the gage marks. Strain elongation is usually expressed as a percentage, where the nominal engineering strain is multiplied by 100 to obtain a percent value; that is:

$$e,\ \% = \left[\frac{(\text{final gage length} - \text{original gage length})}{\text{original gage length}}\right] \times 100$$

The final gage length at the completion of the test may be determined in two ways. Historically, it was determined manually by carefully fitting the two ends of the fractured test piece together (Fig. 13) and measuring the distance between the gage marks. However, some mod-

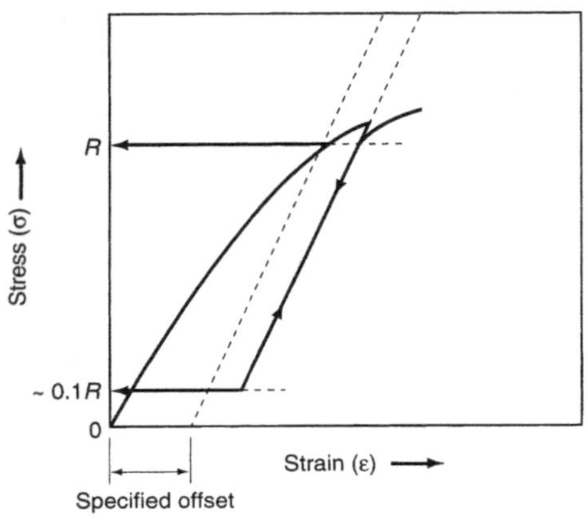

Fig. 12 Alternate technique for establishing Young's modulus for a material without an initial linear portion

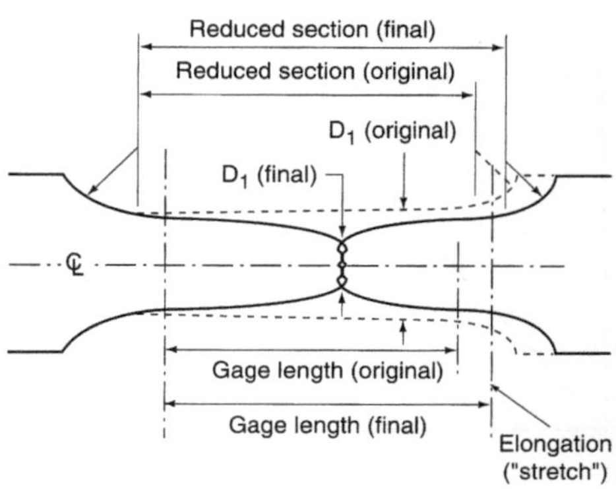

Fig. 13 Sketch of fractured, round tensile test piece. Dashed lines show original shape. Strain = elongation/gage length

ern computer-controlled testing systems obtain data from an extensometer that is left on the test piece through fracture. In this case, the computer may be programmed to report the elongation as the last strain value obtained prior to some event, perhaps the point at which the applied force drops to 90% of the maximum value recorded. There has been no general agreement about what event should be the trigger, and users and machine manufacturers find that different events may be appropriate for different materials (although some consensus has been reached, see ASTM E 8). The elongation values determined by these two methods are not the same; in general, the result obtained by the manual method is a couple of percent larger and is more variable because the test-piece ends do not fit together perfectly. It is strongly recommended that when disagreements arise about elongation results, agreement should be reached on which method will be used prior to any further testing.

Test methods often specify special conditions that must be followed when a product specification specifies elongation values that are small, or when the expected elongation values are small. For example, ASTM E 8 defines small as 3% or less.

Effect of Gage Length and Necking. Figure 14 (Ref 7) shows the effect of gage length on elongation values. Gage length is very important; however, as the gage length becomes quite large, the elongation tends to be independent of the gage length. The gage length must be specified prior to the test, and it must be shown in the data record for the test.

Figures 13 and 14 also illustrate considerable localized deformation in the vicinity of the fracture. This region of local deformation is often called a neck, and the occurrence of this deformation is termed necking. Necking occurs as the force begins to drop after the maximum force has been reached on the stress-strain curve. Up to the point at which the maximum force occurs, the strain is uniform along the gage length; that is, the strain is independent of the gage length. However, once necking begins, the gage length becomes very important. When the gage length is short, this localized deformation becomes the principal portion of measured elongation. For long gage lengths, the localized deformation is a much smaller portion of the total. For this reason, when elongation values are reported, the gage length must also be reported, for example, elongation = 25% (50 mm, or 2.00 in., gage length).

Effect of Test-Piece Dimensions. Test-piece dimensions also have a significant effect on elongation measurements. Experimental work has verified the general applicability of the following equation:

$$e = e_0(L/A^{1/2})^{-a}$$

where e_0 is the specific elongation constant; $L/A^{1/2}$ the slimness ratio, K, of gage length, L, and cross-sectional areas, A; and a is another material constant. This equation is known as the Bertella-Oliver equation, and it may be transformed into logarithmic form and plotted as shown in Fig. 15. In one study, quadruplet sets of machined circular test pieces (four different diameters ranging from 0.125 to 0.750 in.) and rectangular test pieces (½ in. wide with three thicknesses and 1½ in. wide with three thicknesses) were machined from a single plate. Multiple gage lengths were scribed on each test piece to produce a total of 40 slimness ratios. The results of this study, for one of the grades of steel tested, are shown in Fig. 16.

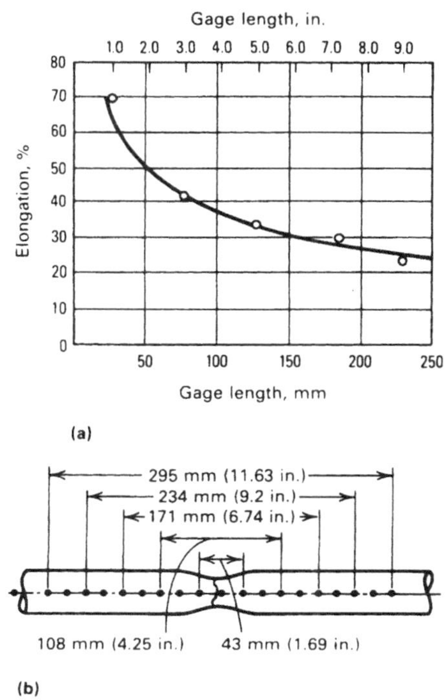

Fig. 14 Effect of gage length on the percent elongation. (a) Elongation, %, as a function of gage length for a fractured tensile test piece. (b) Distribution of elongation along a fractured tension test piece. Original spacing between gage marks, 12.5 mm (0.5 in.). Source: Ref 7

In order to compare elongation values of test pieces with different slimness ratios, it is necessary only to determine the value of the material constant, a. This calculation can be made by testing the same material with two different geometries (or the same geometry with different gage lengths) with different slimness ratios, K_1 and K_2, where

$$e_0 = e_1/K_1^{-a} = e_2/K_2^{-a}$$

solving for a, then:

$$(K_2/K_1)^{-a} = e_2/e_1$$

or:

$$-a = \frac{\ln(e_2/e_1)}{\ln(K_2/K_1)}$$

$$-a = \frac{\ln(e_2) - \ln(e_1)}{\ln(K_2) - \ln(K_1)}$$

The values of the e_0 and a parameters depend on the material composition, the strength, and the material condition and are determined empirically with a best-fit line plot around data points. Reference 8 specifies value $a = 0.4$ for carbon, carbon-manganese, molybdenum, and chromium-molybdenum steels within the tensile strength range of 275 to 585 MPa (40 to 85 ksi) and in the hot-rolled, in the hot-rolled and normalized, or in the annealed condition, with or without tempering. Materials that have been cold reduced require the use of a different value for a, and an appropriate value is not suggested. Reference 8 uses a value of $a = 0.127$ for annealed, austenitic stainless steels. However, Ref 8 states that "these conversions shall not be used where the width-to-thickness ratio, w/t, of the test piece exceeds 20." ISO 2566/1 (Ref 9) contains similar statements. In addition to the limit of $(w/t) < 20$, Ref 9 also specifies that the slimness ratio shall be less than 25.

Some tensile-test specifications do not contain standard test-piece geometries but require that the slimness ratio be either 5.65 or 11.3. For a round test piece, a slimness ratio of 5.65 produces a 5-to-1 relation between the diameter and the gage length, and a slimness ratio of 4.51 produces a 4-to-1 relation between the diameter and gage length (which is that of the test piece in ASTM E 8).

Reduction of area is another measure of the ductility of metal. As a test piece is stretched, the cross-sectional area decreases, and as long as the stretch is uniform, the reduction of area is proportional to the amount of stretch or extension. However, once necking begins to occur, proportionality is no longer valid.

According to ASTM E 6, reduction of area is defined as "the difference between the original cross-sectional area of a tension test piece and the area of its smallest cross section." Reduction of area is usually expressed as a percentage of the original cross-sectional area of the test piece. The smallest final cross section may be measured at or after fracture as specified for the material under test. The reduction of area (RA) is almost always expressed as a percentage:

$$\text{RA, \%} = \left[\frac{(\text{original area} - \text{final area})}{\text{original area}}\right] \times 100$$

Reduction of area is customarily measured only on test pieces with an initial circular cross

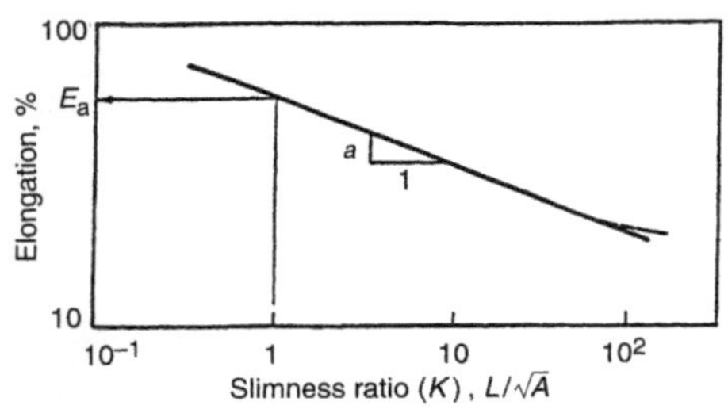

Fig. 15 Graphical form of the Bertella-Oliver equation

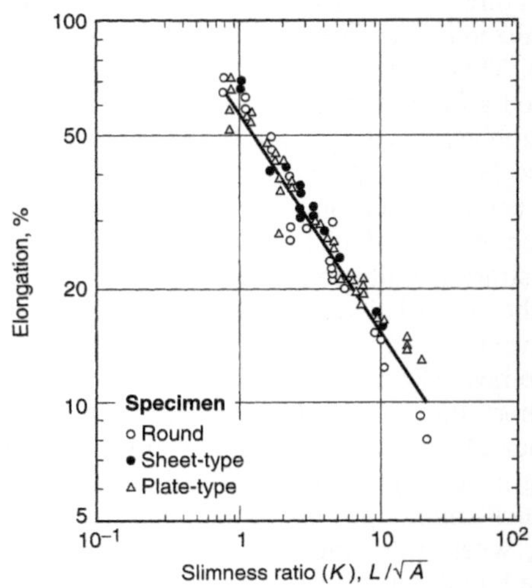

Fig. 16 Graphical form of the Bertella-Oliver equation showing actual data

section because the shape of the reduced area remains circular or nearly circular throughout the test for such test pieces. With rectangular test pieces, in contrast, the corners prevent uniform flow from occurring, and consequently, after fracture, the shape of the reduced area is not rectangular (Fig. 17). Although a number of expressions have been used in an attempt to describe the way to determine the reduced area, none has received general agreement. Thus, if a test specification requires the measurement of the reduction of area of a test piece that is not circular, the method of determining the reduced area should be agreed to prior to performing the test.

General Procedures

Numerous groups have developed standard methods for conducting the tensile test. In the United States, standards published by ASTM are commonly used to define tensile-test procedures and parameters. Of the various ASTM standards related to tensile tests (for example, those listed in "Selected References" at the end of this chapter), the most common method for tension testing of metallic materials is ASTM E 8 "Standard Test Methods for Tension Testing of Metallic Materials" (or the version using metric units, ASTM E 8M). Standard methods for conducting the tensile test are also available from other standards organizations, such as the Japanese Industrial Standards (JIS), the Deutsche Institut für Normung (DIN), and the International Organization for Standardization (ISO). Other domestic technical groups in the United States have developed standards, but in general, these are based on ASTM E 8.

With the increasing internationalization of trade, methods developed by other national standards organizations (such as JIS, DIN, or ISO standards) are increasingly being used in the United States. Although most tensile-test standards address the same concerns, they differ in the values assigned to variables. Thus, a tensile test performed in accordance with ASTM E 8 will not necessarily have been conducted in accordance with ISO 6892 or JIS Z2241, and so on, and vice versa. Therefore, it is necessary to specify the applicable testing standard for any test results or mechanical property data.

Unless specifically indicated otherwise, the values of all variables discussed hereafter are those related to ASTM E 8 "Standard Test Methods for Tension Testing of Metallic Materials."

A flow diagram of the steps involved when a tensile test is conducted in accordance with ASTM E 8 is shown in Fig. 18. The test consists of three distinct parts:

- Test-piece preparation, geometry, and material condition
- Test setup and equipment
- Test

The Test Piece

The test piece, also commonly referred to as the test specimen (see discussion below), is one of two basic types. Either it is a full cross section of the product form, or it is a small portion that has been machined to specific dimensions. Full-section test pieces consist of a part of the test unit as it is fabricated. Examples of full-section test pieces include bars, wires, and hot-rolled or extruded angles cut to a suitable length and then gripped at the ends and tested. In contrast, a machined test piece is a representative sample, such as one of the following:

- Test piece machined from a rough specimen taken from a coil or plate
- Test piece machined from a bar with dimensions that preclude testing a full-section test piece because a full-section test piece exceeds the capacity of the grips or the force capacity of the available testing machine or both
- Test piece machined from material of great monetary or technical value

In these cases, representative samples of the material must be obtained for testing. The descriptions of the tensile test in this chapter proceed from the point that a rough specimen (Fig. 19) has been obtained. That is, the rough specimen has been selected based on some criteria, usually a material specification or a test order issued for a specific reason.

In this chapter, the term *test piece* is used for what is often called a specimen. This ter-

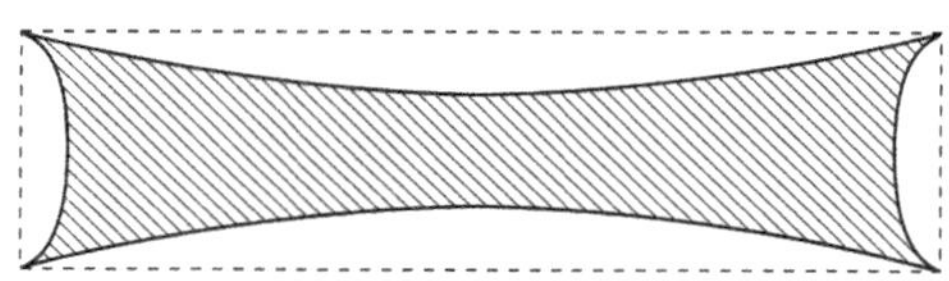

Fig. 17 Sketch of end view of rectangular test piece after fracture showing constraint at corners indicating the difficulty of determining reduced area

minology is based on the convention established by ISO Technical Committee 17, Steel in ISO 377-1, "Selection and Preparation of Samples and Test Pieces of Wrought Steel," where terms for a test unit, a sample product, sample, rough specimen, and test piece are defined as follows:

- *Test unit:* The quantity specified in an order that requires testing (for example, 10 tons of ¾ in. bars in random lengths)
- *Sample product:* Item (in the previous example, a single bar) selected from a test unit for the purpose of obtaining the test pieces
- *Sample:* A sufficient quantity of material taken from the sample product for the purpose of producing one or more test pieces. In some cases, the sample may be the sample product itself (i.e., a 2 ft length of the sample product).
- *Rough specimen:* Part of the sample having undergone mechanical treatment, followed

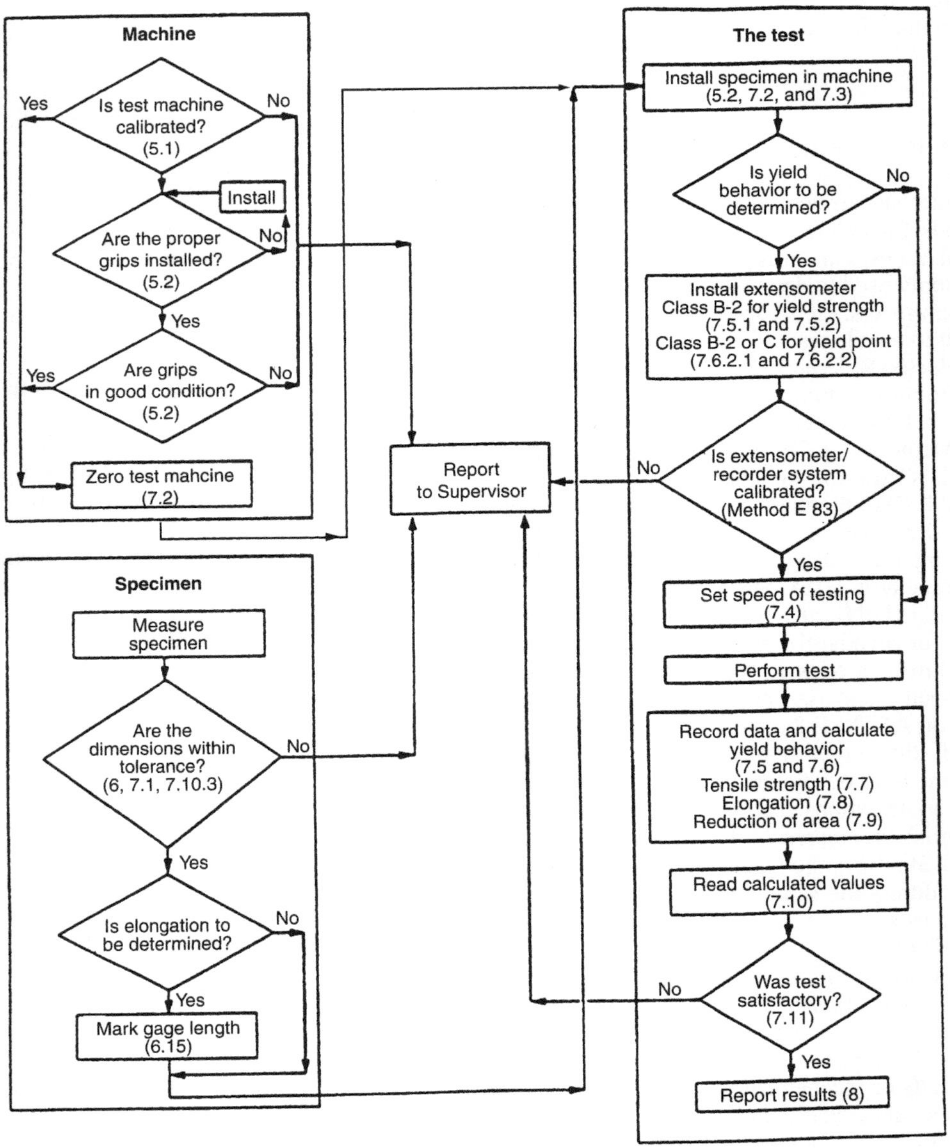

Fig. 18 General flow chart of the tensile test per procedures in ASTM E 8. Relevant paragraph numbers from ASTM E 8 are shown in parentheses.

by heat treatment where appropriate, for the purpose of producing test pieces; in the example, the sample is the rough specimen.

- *Test piece:* Part of the sample or rough specimen, with specified dimensions, machined or unmachined, brought to the required condition for submission to a given test. If a testing machine with sufficient force capacity is available, the test piece may be the rough specimen; if sufficient capacity is not available, or for another reason, the test piece may be machined from the rough specimen to dimensions specified by a standard.

These terms are shown graphically in Fig. 19. As can be seen, the test piece, or what is commonly called a specimen, is a very small part of the entire test unit.

Description of Test Material

Test-Piece Orientation. Orientation and location of a test material from a product can influence measured tensile properties. Although modern metal-working practices, such as cross rolling, have tended to reduce the magnitude of the variations in the tensile properties, it must not be neglected when locating the test piece within the specimen or the sample.

Because most materials are not isotropic, test-piece orientation is defined with respect to a set of axes as shown in Fig. 20. These terms for the orientation of the test-piece axes in Fig. 20 are based on the convention used by ASTM. This scheme is identical to that used by the ISO Technical Committee 164 "Mechanical Testing," although the L, T, and S axes are referred to as the X, Y, and Z axes, respectively, in the ISO documents.

When a test is being performed to determine conformance to a product standard, the product standard must state the proper orientation of the test piece with regard to the axis of prior working, (e.g., the rolling direction of a flat product). Because alloy systems behave differently, no general rule of thumb can be stated on how prior working may affect the directionality of properties. As can be seen in Table 1, the longitudinal strengths of steel are generally somewhat less than the transverse strength. However, for aluminum alloys, the opposite is generally true.

Many standards, such as ASTM A 370, E 8, and B 557, provide guidance in the selection of test-piece orientation relative to the rolling direction of the plate or the major forming axes of other types of products and in the selection of specimen and test-piece location relative to the surface of the product. Orientation is also important when characterizing the directionality of properties that often develops in the microstructure of materials during processing. For example, some causes of directionality include the fibering of inclusions in steels, the formation of crystallographic textures in most metals and alloys, and the alignment of molecular chains in polymers.

The location from which a test material is taken from the initial product form is important

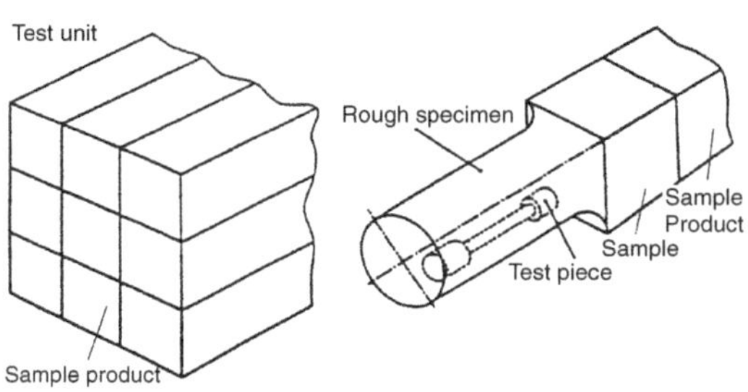

Fig. 19 Illustration of ISO terminology used to differentiate between sample, specimen, and test piece (see text for definitions of test unit, sample product, sample, rough specimen, and test piece). As an example, a test unit may be a 250-ton heat of steel that has been rolled into a single thickness of plate. The sample product is thus one plate from which a single test piece is obtained.

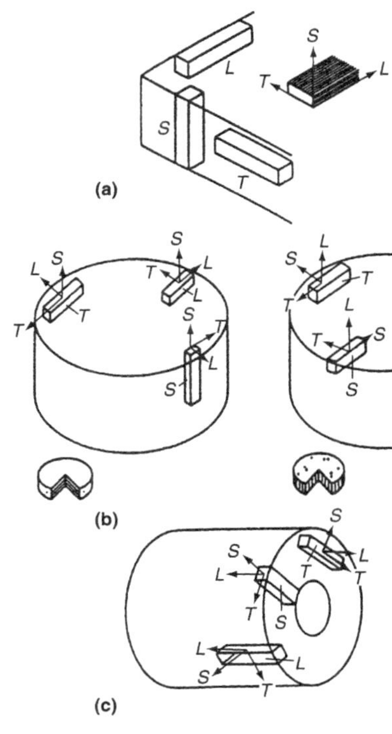

Fig. 20 System for identifying the axes of test-piece orientation in various product forms. (a) Flat-rolled products. (b) Cylindrical sections. (c) Tubular products

because the manner in which a material is processed influences the uniformity of microstructure along the length of the product as well as through its thickness properties. For example, the properties of metal cut from castings are influenced by the rate of cooling and by shrinkage stresses at changes in section. Generally, test pieces taken from near the surface of iron castings are stronger. To standardize test results relative to location, ASTM A 370 recommends that tensile test pieces be taken from midway between the surface and the center of round, square, hexagon, or octagonal bars. ASTM E 8 recommends that test pieces be taken from the thickest part of a forging from which a test coupon can be obtained, from a prolongation of the forging, or in some cases, from separately forged coupons representative of the forging.

Test-Piece Geometry

As previously noted, the item being tested may be either the full cross section of the item, or a portion of the item that has been machined to specific dimensions. This chapter focuses on tensile testing with test pieces that are machined from rough samples. Component testing is discussed in Chapter 12, "Tensile Testing of Components."

Test-piece geometry is often influenced by product form. For example, only test pieces with rectangular cross sections can be obtained from sheet products. Test pieces taken from thick plate may have either flat (plate-type) or round cross sections. Most tensile-test specifications show machined test pieces with either circular cross sections or rectangular cross sections. Nomenclature for the various sections of a machined test piece are shown in Fig. 21. Most tensile-test specifications present a set of dimensions, for each cross-section type, that are standard, as well as additional sets of dimensions for alternative test pieces. In general, the standard dimensions published by ASTM, ISO, JIS, and DIN are similar, but they are not identical.

Table 1 Effect of test-piece orientation on tensile properties

Orientation	Yield strength, ksi	Tensile strength, ksi	Elongation in 50 mm (2 in.), %	Reduction of area, %
ASTM A 572, Grade 50 (¾ in. thick plate, low sulfur level)				
Longitudinal	58.8	84.0	27.0	70.2
Transverse	59.8	85.2	28.0	69.0
ASTM A 656, Grade 80 (¾ in. thick plate, low sulfur level + controlled rolled)				
Longitudinal	81.0	102.3	25.8	71.2
Transverse	86.9	107.9	24.5	67.1
ASTM A 5414 (¾ in. thick plate, low sulfur level)				
Longitudinal	114.6	121.1	19.8	70.6
Transverse	116.3	122.2	19.5	69.9

Source: Courtesy of Francis J. Marsh

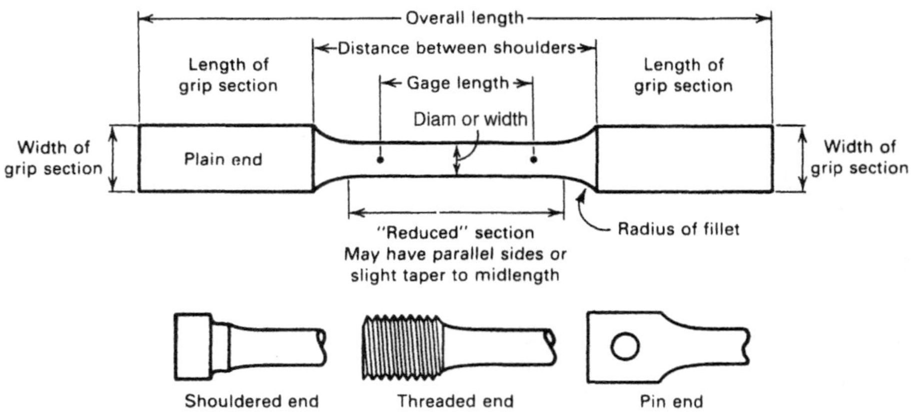

Fig. 21 Nomenclature for a typical tensile test piece

Gage lengths and standard dimensions for machined test pieces specified in ASTM E 8 are shown in Fig. 22(a) and (b) for rectangular and round test pieces. From this figure, it can be seen that the gage length is proportionally four times (4 to 1) the diameter (or width) of the test piece for the standard machined round test pieces and the sheet-type, rectangular test pieces. The length of the reduced section is also a *minimum* of 4½ times the diameter (or width) of these test-piece types. These relationships do not apply to plate-type rectangular test pieces.

Many specifications outside the United States require that the gage length of a test piece be a fixed ratio of the square root of the cross-sectional area, that is:

$$\text{Gage length} = \text{constant } x \text{ (cross-sectional area)}^{1/2}$$

The value of this constant is often specified as 5.65 or 11.3 and applies to both round and rectangular test pieces. For machined round test pieces, a value of 5.65 results in a 5-to-1 relationship between the gage length and the diameter.

Many tensile-test specifications permit a slight taper toward the center of the reduced section of machined test pieces so that the minimum cross section occurs at the center of the gage length and thereby tends to cause fracture to occur at the middle of the gage length. ASTM E 8 specifies that this taper cannot exceed 1% and requires that the taper is the same on both sides of the midlength.

When test pieces are machined, it is important that the longitudinal centerline of the reduced section be coincident with the longitudinal centerlines of the grip ends. In addition, for the rectangular test pieces, it is essential that the centers of the transition radii at each end of the reduced section are on common lines that are perpendicular to the longitudinal centerline. If any of these requirements is violated, bending will occur, which may affect test results.

The transition radii between the reduced section and the grip ends can be critical for test pieces from materials with very high strength or with very little ductility or both. This is discussed more fully in the section "Effect of Strain Concentrations" in this chapter.

Measurement of Initial Test-Piece Dimensions. Machined test pieces are expected to meet size specifications, but to ensure dimensional accuracy, each test piece should be measured prior to testing. Gage length, fillet radius, and cross-sectional dimensions are measured easily. Cylindrical test pieces should be measured for concentricity. Maintaining acceptable concentricity is extremely important in minimizing unintended bending stresses on materials in a brittle state.

Measurement of Cross-Sectional Dimensions. The test pieces must be measured to determine whether they meet the requirements of the test method. Test-piece measurements must also determine the initial cross-sectional area when it is compared against the final cross section after testing as a measure of ductility.

The precision with which these measurements are made is based on the requirements of the test method, or if none are given, on good engineering judgment. Specified requirements of ASTM E 8 are summarized as follows:

- For referee testing of test pieces under 3⁄16 in. in their least dimension, the dimensions should be measured where the least cross-sectional area is found.
- For cross sectional dimensions of 0.200 in. or more, cross-sectional dimensions should be measured and recorded to the nearest 0.001 in.
- For cross sectional dimensions from 0.100 in. but less than 0.200 in., cross-sectional dimensions should be measured and recorded to the nearest 0.0005 in.
- For cross sectional dimensions from 0.020 in. but less than 0.100 in., cross-sectional dimensions should be measured and recorded to the nearest 0.0001 in.
- When practical, for cross-sectional dimensions less than 0.020 in., cross-sectional dimensions should be measured to the nearest 1%, but in all cases, to at least the nearest 0.0001 in.

ASTM E 8 goes on to state how to determine the cross-sectional area of a test piece that has a nonsymmetrical cross section using the weight and density. When measuring dimensions of the test piece, ASTM E 8 makes no distinction between the shape of the cross section for standard test pieces.

Measurement of the Initial Gage Length. ASTM E 8 assumes that the initial gage length is within specified tolerance; therefore, it is necessary only to verify that the gage length of the test piece is within the tolerance.

Marking Gage Length. As shown in the flow diagram in Fig. 18, measurement of elongation requires marking the gage length of the test piece. The gage marks should be placed on the test piece in a manner so that when fracture occurs, the fracture will be located within the cen-

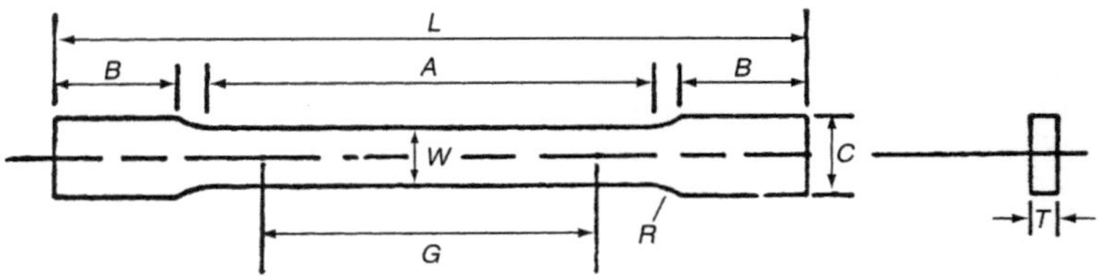

	Standard specimens, in.		Subsize specimen,
	Plate type, 1 1/2 in. wide	Sheet type, 1/2 in. wide	1/4 in. wide, in.
G, gage length(a)(b)	8.00 ± 0.01	2.00 ± 0.005	1.000 ± 0.003
W, width(c)(d)	1 1/2 + 1/8 − 1/4	0.500 ± 0.010	0.250 ± 0.005
T, thickness(e)	< 3/16	0.005 ≤ T ≤ 3/4	0.005 ≤ T ≤ 1/4
R, radius of fillet, min(f)	1	1/2	1/4
L, overall length, min(b)(g)	18	8	4
A, length of reduced section, min	9	2 1/4	1 1/4
B, length of grip section, min(h)	3	2	1 1/4
C, width of grip section, approximate(d)(i)	2	3/4	3/8

Note: (a) For the 1 1/2 in. wide specimen, punch marks for measuring elongation after fracture shall be made on the flat or on the edge of the specimen and within the reduced section. Either a set of nine or more punch marks 1 in. apart or one or more pairs of punch marks 8 in. apart may be used. (b) When elongation measurements of 1 1/2 in. wide specimens are not required, a minimum length of reduced section (*A*) of 2 1/4 in. may be used with all other dimensions similar to those of the plate-type specimen. (c) For the three sizes of specimens, the ends of the reduced section shall not differ in width by more than 0.004, 0.002, or 0.001 in., respectively. Also, there may be a gradual decrease in width from the ends to the center, but the width at each end shall not be more than 0.015, 0.005, or 0.003 in., respectively, larger than the width at the center. (d) For each of the three sizes of specimens, narrower widths (*W* and *C*) may be used when necessary. In such cases the width of the reduced section should be as large as the width of the material being tested permits; however, unless stated specifically, the requirements for elongation in a product specification shall not apply when these narrower specimens are used. (e) The dimension *T* is the thickness of the test specimen as provided for in the applicable material specifications. Minimum thickness of 1 1/2 in. wide specimens shall be 3/16 in. Maximum thickness of 1/2 in. and 1/4 in. wide specimens shall be 3/4 in. and 1/4 in., respectively. (f) For the 1 1/2 in. wide specimen, a 1/2 in. minimum radius at the ends of the reduced section is permitted for steel specimens under 100,000 psi in tensile strength when a profile cutter is used to machine the reduced section. (g) To aid in obtaining axial force application during testing of 1/4 in. wide specimens, the overall length should be as large as the material will permit, up to 8.00 in. (h) It is desirable, if possible, to make the length of the grip section large enough to allow the specimen to extend into the grips a distance equal to two-thirds or more of the length of the grips. If the thickness of 1/2 in. wide specimens is over 3/8 in., longer grips and correspondingly longer grip sections of the specimen may be necessary to prevent failure in the grip section. (i) For the three sizes of specimens, the ends of the specimen shall be symmetrical in width with the enter line of the reduced section within 0.10, 0.05, and 0.005 in., respectively. However, for referee testing and when required by product specifications, the ends of the 1/2 in. wide specimen shall be symmetrical within 0.01 in.

(a)

Fig. 22(a) Example of rectangular (flat) tensile test pieces per ASTM E 8.

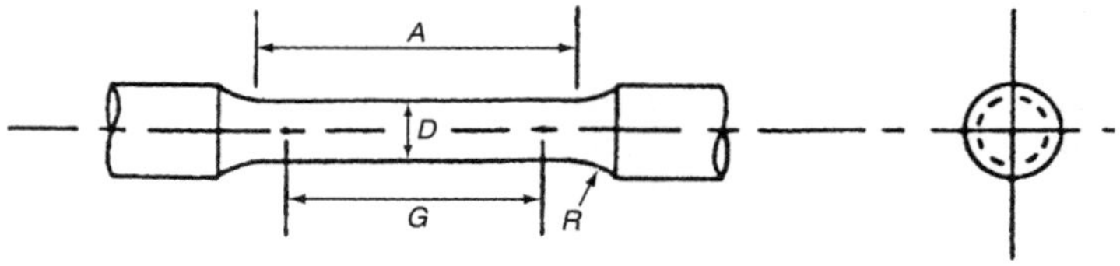

	Standard specimen, in., at nominal diameter:		Small-size specimens proportional to standard, in., at nominal diameter:		
	0.500	0.350	0.250	0.160	0.113
G, gage length	2.000 ± 0.005	1.400 ± 0.005	1.000 ± 0.005	0.640 ± 0.005	0.450 ± 0.005
D, diameter(a)	0.500 ± 0.010	0.350 ± 0.007	0.250 ± 0.005	0.160 ± 0.003	0.113 ± 0.002
R, radius of fillet, min	3/8	1/4	3/16	5/32	3/32
A, length of reduced section, min(b)	2 1/4	1 3/4	1 1/4	3/4	5/8

Note: (a) The reduced section may have a gradual taper from the ends toward the center, with the ends not more than 1% larger in diameter than the center (controlling dimension). (b) If desired, the length of the reduced section may be increased to accommodate an extensometer of any convenient gage length. Reference marks for the measurement of elongation should, nevertheless, be spaced at the indicated gage length.

(b)

Fig. 22(b) Example of a round tensile test piece per ASTM E 8.

ter one-third of the gage length (or within the center one-third of one of several sets of gage-length marks). For a test piece machined with a reduced-section length that is the minimum specified by ASTM E 8 and with a gage length equal to the maximum allowed for that geometry, a single set of marks is usually sufficient. However, multiple sets of gage lengths must be applied to the test piece to ensure that one set spans the fracture under any of the following conditions:

- Testing full-section test pieces
- Testing pieces with reduced sections significantly longer than the minimum
- Test requirements specify a gage length that is significantly shorter than the reduced section

For example, some product specifications require that the elongation be measured over a 2 in. gage length using the machined plate-type test piece with a 9 in. reduced section (Fig. 22a). In this case, it is recommended that a staggered series of marks (either in increments of 1 in. when testing to ASTM E 8 or in increments of 25.0 mm when testing to ASTM E 8M) be placed on the test piece such that, after fracture, the elongation can be measured using the set that best meets the center-third criteria. Many tensile-test methods permit a retest when the elongation is less than the minimum specified by a product specification if the fracture occurred outside the center third of the gage length. When testing full-section test pieces and determining elongation, it is important that the distance between the grips be greater than the specified gage length unless otherwise specified. As a rule of thumb, the distance between grips should be equal to at least the gage length plus twice the minimum dimension of the cross section.

The gage marks may be marks made with a center punch, or may be lines scribed using a sharp, pointed tool, such as a machinist's scribe (or any other means that will establish the gage length within the tolerance permitted by the test method). If scribed lines are used, a broad line or band may first be drawn along the length of the test piece using machinist's layout ink (or a similar substance), and the gage marks are made on this line. This practice is especially helpful to improve visibility of scribed gage marks after fracture. If punched marks are used, a circle around each mark or other indication made by ink may help improve visibility after fracture. Care must be taken to ensure that the gage marks, especially those made using a punch, are not deep enough to become stress raisers, which could cause the fracture to occur through them. This precaution is especially important when testing materials with high strength and low ductility.

Notched Test Pieces. Tensile test pieces are sometimes intentionally notched in the center of the gage length (Fig. 23). ASTM E 338 and E 602 describe procedures for testing notched test pieces. Results obtained using notched test pieces are useful for evaluating the response of a material to a localized stress concentration.

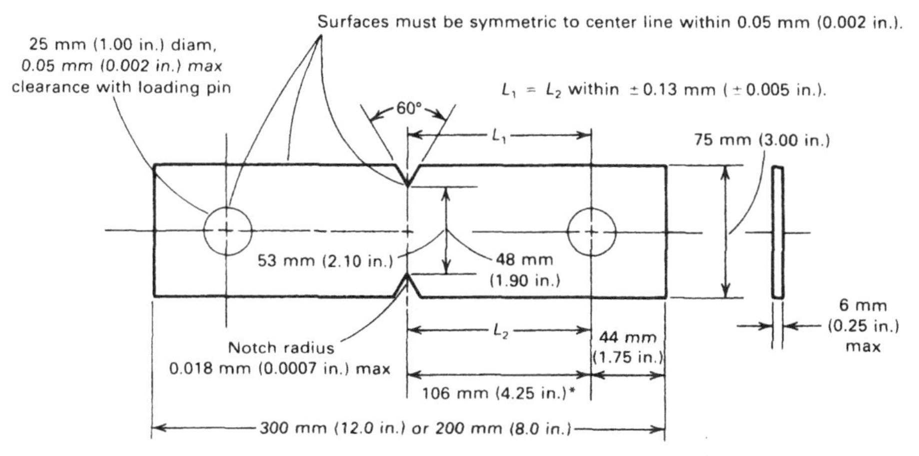

Fig. 23 Example of notched tensile-test test piece per ASTM E 338, "Standard Test Method of Sharp-Notch Tension Testing of High-Strength Sheet Materials"

Additional information on the notch tensile test and a discussion of the related material characteristics (notch sensitivity and notch strength) can be found in Chapter 2, "Mechanical Behavior of Materials under Tensile Loads." The effect of stress (or strain) concentrations is also discussed in the section "Effect of Strain Concentrations" in this chapter.

Surface Finish and Condition. The finish of machined surfaces usually is not specified in generic test methods (that is, a method that is not written for a specific item or material) because the effect of finish differs for different materials. For example, test pieces from materials that are not high strength or that are ductile are usually insensitive to surface finish effects. However, if surface finish in the gage length of a tensile test piece is extremely poor (with machine tool marks deep enough to act as stress-concentrating notches, for example), test results may exhibit a tendency toward decreased and variable strength and ductility.

It is good practice to examine the test piece surface for deep scratches, gouges, edge tears, or shear burrs. These discontinuities may sometimes be minimized or removed by polishing or, if necessary, by further machining; however, dimensional requirements often may no longer be met after additional machining or polishing. In all cases, the reduced sections of machined test pieces must be free of detrimental characteristics, such as cold work, chatter marks, grooves, gouges, burrs, and so on. Unless one or more of these characteristics is typical of the product being tested, an unmachined test piece must also be free of these characteristics in the portion of the test piece that is between the gripping devices. When rectangular test pieces are prepared from thin-gage sheet material by shearing (punching) using a die the shape of the test piece, ASTM E 8 states that the sides of the reduced section may need to be further machined to remove the cold work and shear burrs that occur when the test piece is sheared from the rough specimen. This method is impractical for material less than 0.38 mm (0.015 in.) thick. Burrs on test pieces can be virtually eliminated if punch-to-die clearances are minimized.

Test Setup

The setup of a tensile test involves the installation of a test piece in the load frame of a suitable test machine. Force capacity is the most important factor of a test machine. Other test machine factors, such as calibration and load-frame rigidity, are discussed in more detail in Chapter 4, "Tensile Testing Equipment and Strain Sensors." The other aspects of the test setup include proper gripping and alignment of the test piece, and the installation of extensometers or strain sensors when plastic deformation (yield behavior) of the piece is being measured, as described below.

Gripping Devices. The grips must furnish an axial connection between the test piece and the testing machine; that is, the grips must not cause bending in the test piece during loading. The choice of grip is primarily dependent on the geometry of the test piece and, to a lesser degree, on the preference of the test laboratory. That is, rarely do tension-test methods or requirements specify the method of gripping the test pieces.

Figure 24 shows several of the many grips that are in common use, but many other designs are also used. As can be seen, the gripping devices can be classified into several distinct types, wedges, threaded, button, and snubbing. Wedge grips can be used for almost any test-piece geometry; however, the wedge blocks must be designed and installed in the machine to ensure axial loading. Threaded grips and button grips are used only for machined round test pieces. Snubbing grips are used for wire (as shown) or for thin, rectangular test pieces, such as those made from foil.

As shown in Fig. 22, the dimensions of the grip ends for machined round test pieces are usually not specified, and only approximate dimensions are given for the rectangular test pieces. Thus, each test lab must prepare/machine grip ends appropriate for its testing machine. For machined-round test pieces, the grip end is often threaded, but many laboratories prefer either a plain end, which is gripped with the wedges in the same manner as a rectangular test piece, or with a button end that is gripped in a mating female grip. Because the principal disadvantage of a threaded grip is that the pitch of the threads tend to cause a bending moment, a fine-series thread is often used.

Bending stresses are normally not critical with test pieces from ductile materials. However, for test pieces from materials with limited ductility, bending stresses can be important, better alignment may be required. Button grips are often used, but adequate alignment is usually achieved with threaded test pieces. ASTM E 8 also recommends threaded gripping for brittle materials. The principal disadvantage of the button-end grip is that the diameter of the button or the base of the cone is usually at least twice the diameter of the reduced section, which necessitates a larger, rough specimen and more metal removal during machining.

Alignment of the Test Piece. The force-application axis of the gripping device must coincide with the longitudinal axis of symmetry of the test piece. If these axes do not coincide, the test piece will be subjected to a combination of axial loading and bending. The stress acting on the different locations in the cross section of the test piece then varies, from the sum of the axial and bending stresses on one side of the test piece, to the difference between the two stresses on the other side. Obviously, yielding will begin on the side where the stresses are additive and at a lower apparent stress than would be the case if only the axial stress were present. For this reason, the yield stress may be lowered, and the upper yield stress would appear suppressed in test pieces that normally exhibit an upper yield point. For ductile materials, the effect of bending is minimal, other than the suppression of the upper yield stress. However, if the material has little ductility, the increased strain due to bending may cause fracture to occur at a lower stress than if there were no bending.

Similarly, if the test piece is initially bent, for example, coil set in a machined-rectangular cross section or a piece of rod being tested in a

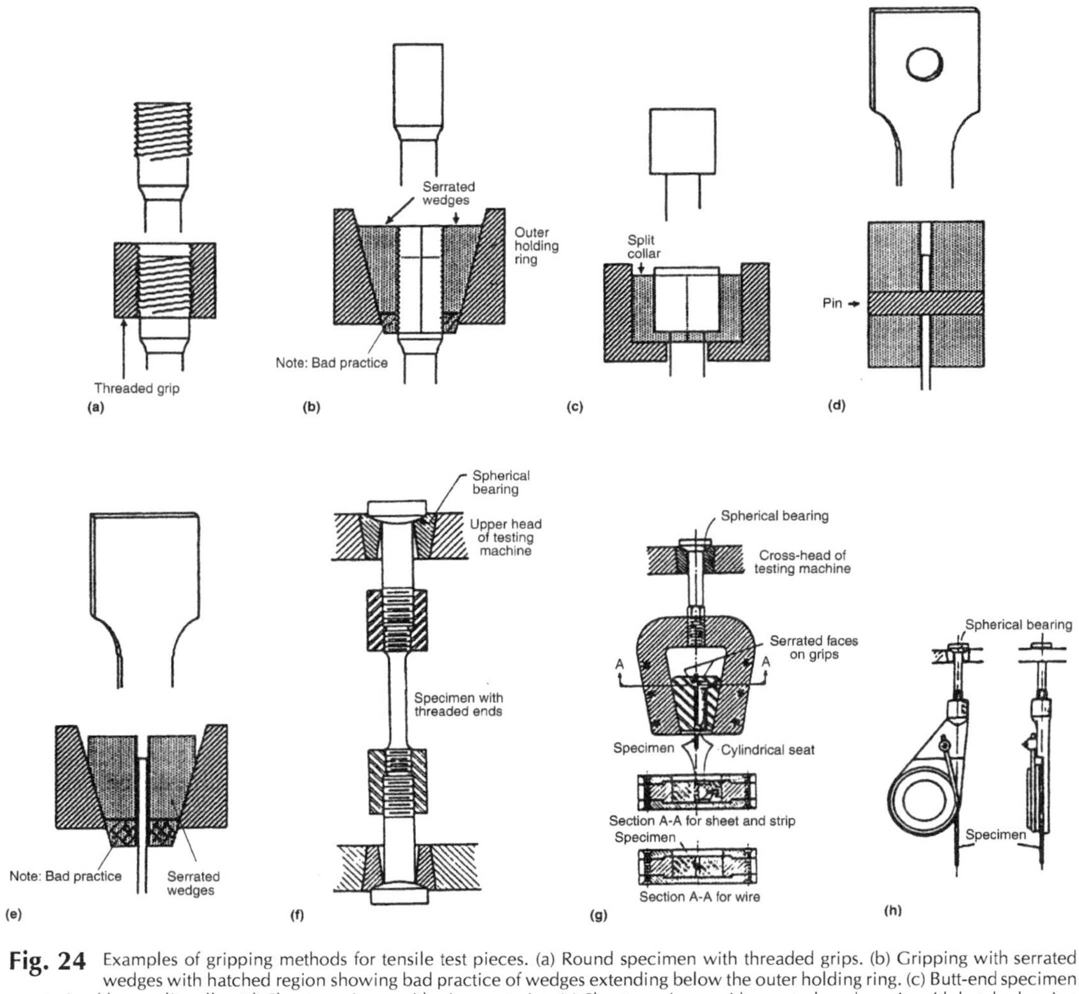

Fig. 24 Examples of gripping methods for tensile test pieces. (a) Round specimen with threaded grips. (b) Gripping with serrated wedges with hatched region showing bad practice of wedges extending below the outer holding ring. (c) Butt-end specimen constrained by a split collar. (d) Sheet specimen with pin constraint. (e) Sheet specimen with serrated-wedge grip with hatched region showing the bad practice of wedges extended below the outer holding ring. (f) Gripping device for threaded-end specimen. (g) Gripping device for sheet and wire. (h) Snubbing device for testing wire

full section, bending will occur as the test piece straightens, and the problems exist.

Methods for verification of alignment are described in ASTM E 1012.

Extensometers. When the tensile test requires the measurement of strain behavior (i.e., the amount of elastic and/or plastic deformation occurring during loading), extensometers must be attached to the test piece. The amount of strain can be quite small (e.g., approximately 0.5% or less for elastic strain in steels), and extensometers and other strain-sensing systems are designed to magnify strain measurement into a meaningful signal for data processing.

Several types of extensometers are available, as described in more detail in Chapter 4, "Tensile Testing Equipment and Strain Sensors." Extensometers generally have fixed gage lengths. If an extensometer is used only to obtain a portion of the stress-strain curve sufficient to determine the yield properties, the gage length of the extensometer may be shorter than the gage length required for the elongation-at-fracture measurement. It may also be longer, but in general, the extensometer gage length should not exceed approximately 85% of the length of the reduced section or the distance between the grips for test pieces without reduced sections. National and international standardization groups have prepared practices for the classification of extensometers, as described in Chapter 4. Extensometer classifications usually are based on error limits of a device, as in ASTM E 83 "Standard Practice for Verification and Classification of Extensometers."

Temperature Control. Tensile testing is sometimes performed at temperatures other than room temperature. ASTM E 21 describes standard procedures for elevated-temperature tensile testing of metallic materials, which is described further in Chapter 13, "Hot Tensile Testing." ASTM E 1450 describes standard procedures for tensile testing of structural alloys in liquid helium (cryogenic testing), which is described further in Chapter 14, "Tensile Testing at Low Temperatures."

Temperature gradients may occur in temperature-controlled systems, and gradients must be kept within tolerable limits. It is not uncommon to use more than one temperature-sensing device (e.g., thermocouples) when testing at other than room temperature. Besides the temperature-sensing device used in the control loop, auxiliary sensing devices may be used to determine whether temperature gradients are present along the gage length of the test piece.

Temperature control is also a factor during room-temperature tests because deformation of the test piece causes generation of heat within it. Test results have shown that the heating that occurs during the straining of a test piece can be sufficient to significantly change the properties that are determined because material strength typically decreases with an increase in the test temperature. When performing a test to duplicate the results of others, it is important to know the test speed and whether any special procedures were taken to remove the heat generated by straining the test piece.

Test Procedures

After the test piece has been properly prepared and measured and the test setup established, conducting the test is fairly routine. The test piece is installed properly in the grips, and if required, extensometers or other strain-measuring devices are fastened to the test piece for measurement and recording of extension data. Data acquisition systems also should be checked. In addition, it is sometimes useful to repetitively apply small initial loads and vibrate the load train (a metallographic engraving tool is a suitable vibrator) to overcome friction in various couplings, as shown in Fig. 25(a) and (b). A check can also be run to ensure that the test will run at the proper testing speed and temperature. The test is then begun by initiating force application.

Speed of Testing

The speed of testing is extremely important because mechanical properties are a function of strain rate, as discussed in the section "Effect of Strain Rate" in this chapter. It is, therefore, imperative that the speed of testing be specified in either the tension-test method or the product specification.

In general, a slow speed results in lower strength values and larger ductility values than a fast speed; this tendency is more pronounced for lower-strength materials than for higher-strength materials and is the reason that a tension test must be conducted within a narrow test-speed range.

In order to quantify the effect of deformation rate on strength and other properties, a specific definition of testing speed is required. A conventional (quasi-static) tensile test, for example, ASTM E 8, prescribes upper and lower limits on the deformation rate, as determined by one of the following methods during the test:

- Strain rate
- Stress rate (when loading is below the proportional limit)
- Cross-head separation rate (or free-running cross-head speed) during the test
- Elapsed time

These methods are listed in order of decreasing precision, except during the occurrence of upper-yield-strength behavior and yield point elongation (YPE) (where the strain rate may not necessarily be the most precise method). For some materials, elapsed time may be adequate, while for other materials, one of the remaining methods with higher precision may be necessary in order to obtain test values within acceptable limits. ASTM E 8 specifies that the test speed must be slow enough to permit accurate determination of forces and strains. Although the speeds specified by various test methods may differ somewhat, the test speeds for these methods are roughly equivalent in commercial testing.

Strain rate is expressed as the change in strain per unit time, typically expressed in units of min^{-1} or s^{-1} because strain is a dimensionless value expressed as a ratio of change in length per unit length. The strain rate can usually be dialed, or programmed, into the control settings of a computer-controlled system or paced or timed for other systems.

Stress rate is expressed as the change in stress per unit of time. When the stress rate is stipulated, ASTM E 8 requires that it not exceed 100 ksi/min. This number corresponds to an elastic strain rate of about 5×10^{-5} s^{-1} for steel or 15×10^{-5} s^{-1} for aluminum. As with strain rate, stress rate usually can be dialed or programmed into the control settings of computer-controlled test systems. However, because most older systems indicate force being applied, and not stress, the operator must convert stress to force and control this quantity. Many machines are equipped with pacing or indicating devices for the measurement and control of the stress rate, but in the absence of such a device, the average stress rate can be determined with a timing device by observing the time required to apply a known increment of stress. For example, for a test piece with a cross section of 0.500 in. by 0.250 in. and a specified stress rate of 100,000 psi/min, the maximum force application rate would be 12,500 lbf/min (force = stress rate × area = 100,000 psi/min × (0.500 in. × 0.250 in.)). A minimum rate of 1/10 of the maximum rate is usually specified.

Comparison between Strain-Rate and Stress-Rate Methods. Figure 26 compares strain-rate control with stress-rate control for de-

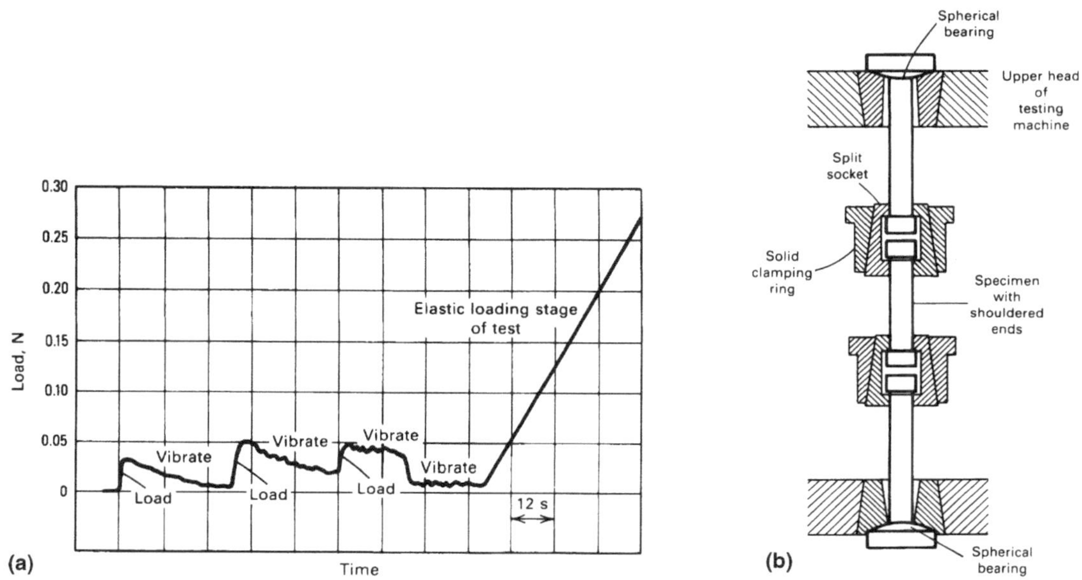

Fig. 25 (a) Effectiveness of vibrating the load train to overcome friction in the spherical ball and seat couplings shown in Fig. 25(b). (b) Spherically seated gripping device for shouldered tensile test piece

scribing the speed of testing. Below the elastic limit, the two methods are identical. However, as shown in Fig. 26, once the elastic limit is exceeded, the strain rate increases when a constant stress rate is applied. Alternatively, the stress rate decreases when a constant strain rate is specified. For a material with discontinuous yielding and a pronounced upper yield spike (Fig. 7a), it is a physical impossibility for the stress rate to be maintained in that region because, by definition, there is not a sustained increase in stress in this region. For these reasons, the test methods usually specify that the rate (whether stress rate or strain rate) is set prior to the elastic limit (EL), and the crosshead speed is not adjusted thereafter. Stress rate is not applicable beyond the elastic limit of the material. Test methods that specify rate of straining expect the rate to be controlled during yield; this minimizes effects on the test due to testing machine stiffness.

The rate of separation of the grips (or rate of separation of the cross heads or the cross-head speed) is a commonly used method of specifying the speed of testing. In ASTM A 370, for example, the specification of test speed is that "through the yield, the maximum speed shall not exceed 1⁄16 in. per inch of reduced section per minute; beyond yield or when determining tensile strength alone, the maximum speed shall not exceed ½ in. per inch of reduced section per minute. For both cases, the minimum speed shall be greater than 1⁄10 of this amount." This means that for a machined round test piece with a 2¼ in. reduced section, the rate prior to yielding can range from a maximum of 9⁄64 in./min (i.e., 2¼ in. reduced-section length × 1⁄16 in./min) down to 9⁄640 in./min (i.e., 2¼ in. reduced-section length × 1⁄160 in./min).

The elapsed time to reach some event, such as the onset of yielding or the tensile strength, or the elapsed time to complete the test, is sometimes specified. In this case, multiple test pieces are usually required so that the correct test speed can be determined by trial and error.

Many test methods permit any speed of testing below some percentage of the specified yield or tensile strength to allow time to adjust the force application mechanism, ensure that the extensometer is working, and so on. Values of 50 and 25%, respectively, are often used.

Post-Test Measurements

After the test has been completed, it is often required that the cross-sectional dimensions again be measured to obtain measures of ductility. ASTM E 8 states that measurements made after the test shall be to the same accuracy as the initial measurements.

Method E 8 also states that upon completion of the test, gage lengths 2 in. and under are to be measured to the nearest 0.01 in., and gage lengths over 2 in. are to be measured to the nearest 0.5%. The document goes on to state that a percentage scale reading to 0.5% of the gage length may be used. However, if the tensile test

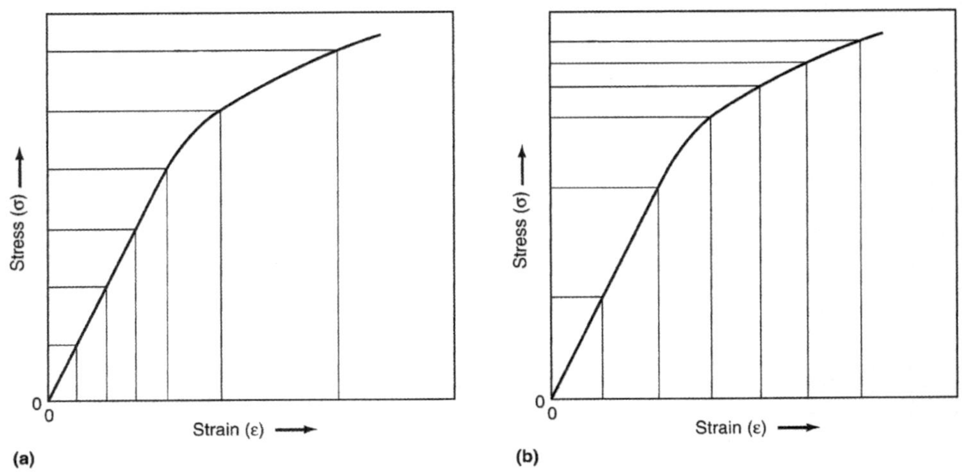

Fig. 26 Illustration of the differences between constant stress increments and constant strain increments. (a) Equal stress increments (increasing strain increments). (b) Equal strain increments (decreasing stress increments)

is being performed as part of a product specification, and the elongation is specified to meet a value of 3% or less, special techniques, which are described, are to be used to measure the final gage length. These measurements are discussed in a previous section, "Elongation," in this chapter.

Variability of Tensile Properties

Even carefully performed tests will exhibit variability because of the nonhomogenous nature of metallic materials. Figure 27 (Ref 10) shows the three-sigma distribution of the offset yield strength and tensile strength values that were obtained from multiple tests on a single aluminum alloy. Distribution curves are presented for the results from multiple tests of a single sheet and for the results from tests on a number of sheets from a number of lots of the same alloy. Because these data are plotted with the minus three-sigma value as zero, it appears there is a difference between the mean values; however, this appearance is due only to the way the data are presented. Figures 28(a) and (b) show lines of constant offset yield strength and constant tensile strength, respectively, for a 1 in. thick, quenched and tempered plate of an alloy steel. In this case, rectangular test pieces $1\frac{1}{2}$ in. wide were taken along the transverse direction (*T* orientation in Fig. 20) every 3 in. along each of the four test-piece centerlines shown. These data indicate that the yield and tensile strengths vary greatly within this relatively small sample and that the shape and location of the yield strength contour lines are not the same as the shape and location of the tensile strength lines.

Effect of Strain Concentrations. During testing, strain concentrations (often called stress concentrations) occur in the test piece where there is a change in the geometry. In particular,

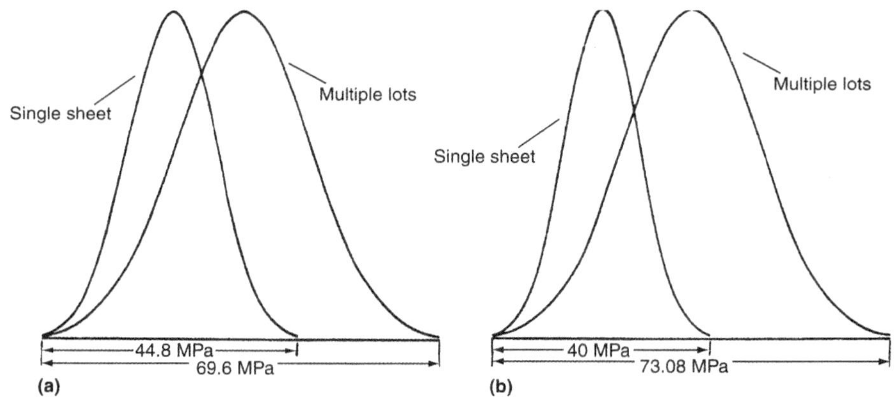

Fig. 27 Distribution of (a) yield and (b) tensile strengths for multiple tests on single sheet and on multiple lots of aluminum alloy 7075-T6. Source: Ref 10

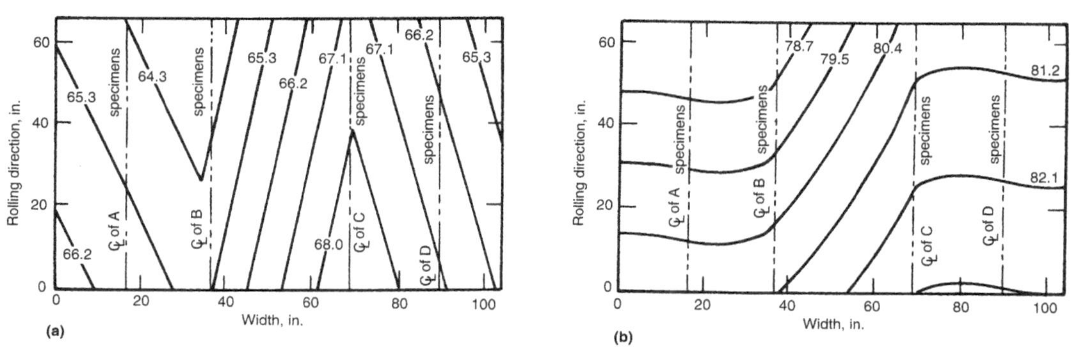

Fig. 28 Contour maps of (a) constant yield strength (0.5% elongation under load, ksi) and (b) constant tensile strength (ksi) for a plate of alloy steel

the transition radii between the reduced section and the grip ends are important, as previously noted in the section "Test-Piece Geometry." Most test methods specify a minimum value for these radii. However, because there is a change in geometry, there is still a strain concentration at the point of tangency between the radii and the reduced section. Figure 29(a) (Ref 11) shows a test piece of rubber with an abrupt change of section, which is a model of a tensile test piece in the transition region. Prior to applying the force at the ends of the model, a rectangular grid was placed on the test piece. When force is applied, it can be seen that the grid is severely distorted at the point of tangency but to a much lesser degree at the center of the model. The distortion is a visual measure of strain. The strain distribution across section n-n is plotted in Fig. 29(b). From the stress-strain curve for the material (Fig. 29c), the stresses on this section can be determined. It is apparent that the test piece will yield at the point of tangency prior to general yielding in the reduced section. The ratio between the nominal strain and actual, maximum strain is often referred to as the strain-concentration factor, or the stress-concentration factor if the actual stress is less than the elastic limit. This ratio is often abbreviated as k_t. Studies have shown that k_t is about 1.25 when the radii are ½ in., the width (or diameter) of the reduced section is 0.500 in., and the width (or diameter) of the grip end is ¾ in. That is, the actual strain or the actual elastic stress at the transition (if less than the yield of the material) is 25% greater than would be expected without consideration of the strain or stress concentration. The value of k_t decreases as the radii in-

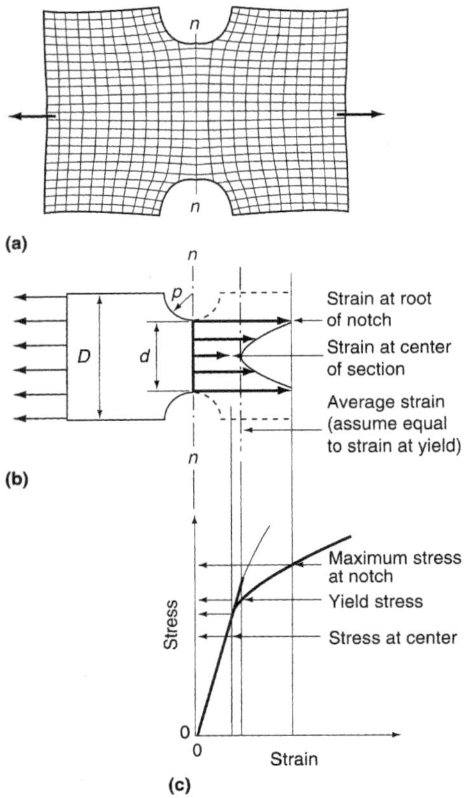

Fig. 29 Effect of strain concentrations on section n-n. (a) Strain distribution caused by an abrupt change in cross section (grid on sheet of rubber) (Ref 11). (b) Schematic of strain distribution on cross section (Ref 11). (c) Calculation of stresses at abrupt change in cross section n-n by graphical means

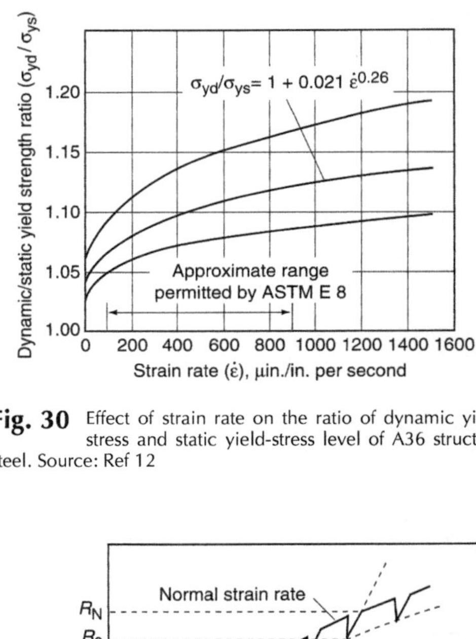

Fig. 30 Effect of strain rate on the ratio of dynamic yield-stress and static yield-stress level of A36 structural steel. Source: Ref 12

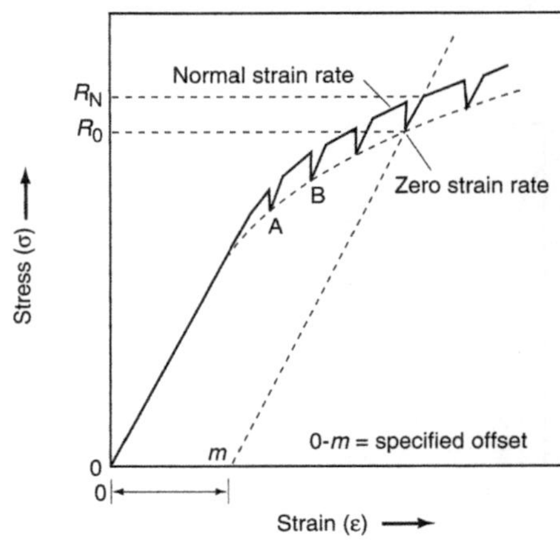

Fig. 31 Stress-strain curves for tests conducted at "normal" and "zero" strain rates

crease such that, for the above example, if the radii are 1.0 in., and k_t decreases to about 1.15.

Various techniques have been tried to minimize k_t, including the use of spirals instead of radii, but there will always be strain concentration in the transition region. This indicates that the yielding of the test piece will always initiate at this point of tangency and proceed toward midlength. For these reasons, it is extremely important that the radii be as large as feasible when testing materials with low ductility.

Strain concentrations can be caused by notches deliberately machined in the test piece, nicks from accidental causes, or shear burrs, machining marks, or gouges that occur during the preparation of the test piece or from many other causes.

Effect of Strain Rate. Although the mechanical response of different materials varies, the strength properties of most materials tend to increase at higher strain rates. For example, the variability in yield strength of ASTM A 36

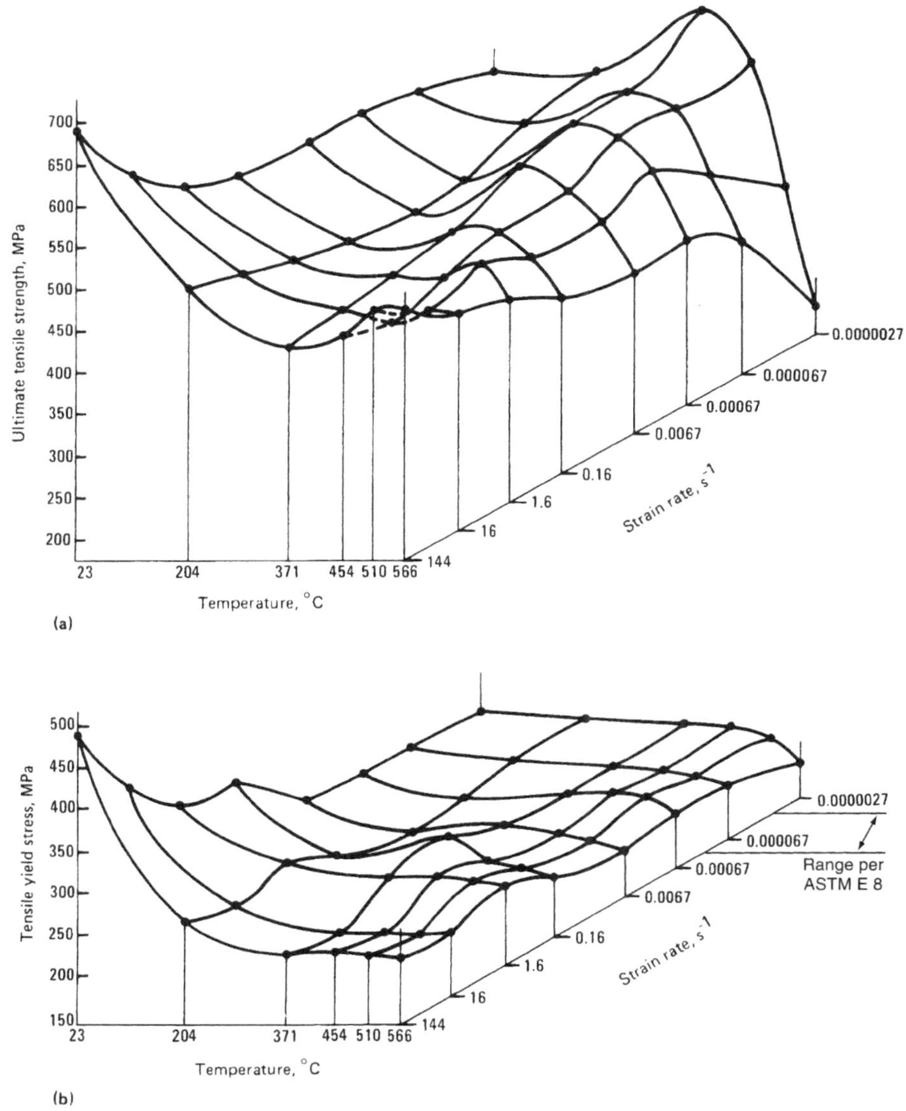

Fig. 32 Effect of temperature and strain rate on (a) tensile strength and (b) yield strength of 2¼ Cr-1 Mo Steel. Note: Strain-rate range permitted by ASTM Method E 8 when determining yield strength at room temperature is indicated. Source: Ref 13

structural steel over a limited range of strain rates is shown in Fig. 30 (Ref 12). A "zero-strain-rate" stress-strain curve (Fig. 31) is generated by applying forces to a test piece to obtain a small plastic strain and then maintaining that strain until the force ceases to decrease (Point A). Force is reapplied to the test piece to obtain another increment of plastic strain, which is maintained until the force ceases to decrease (Point B). This procedure is continued for several more cycles. The smooth curve fitted through Points A, B, and so on is the "zero-strain-rate" stress-strain curve, and the yield value is determined from this curve.

The effect of strain rate on strength depends on the material and the test temperature. Figure 32 (Ref 13) shows graphs of tensile strength and yield strength for a common heat-resistant low-alloy steel (2¼Cr-1Mo) over a wide range of temperatures and strain rates. In this figure, the strain rates were generally faster than those prescribed in ASTM E 8.

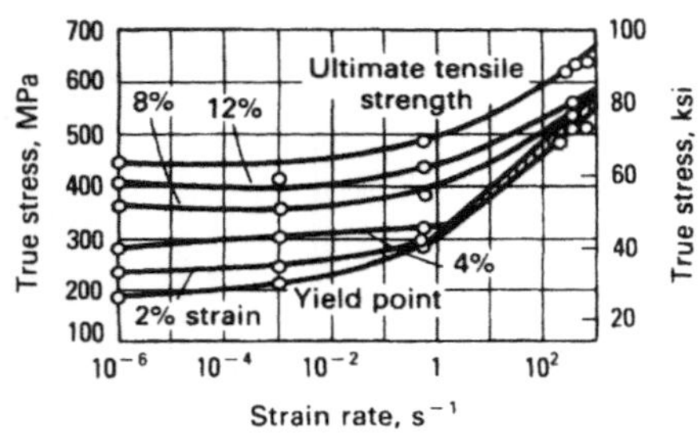

Fig. 33 True stresses at various strains vs. strain rate for a low-carbon steel at room temperature. The top line in the graph is tensile strength, and the other lines are yield points for the indicated level of strain. Source: Ref 14

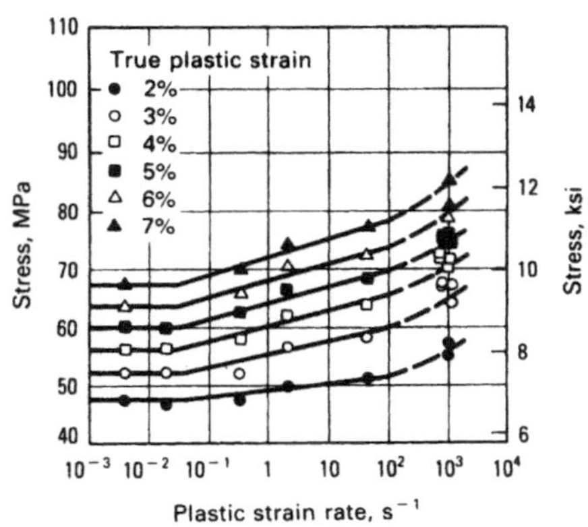

Fig. 34 Uniaxial stress/strain/strain rate data for aluminum 1060-O. Source: Ref 15

Another example of strain effects on strength is shown in Fig. 33 (Ref 14). This figure illustrates true yield stress at various strains for a low-carbon steel at room temperature. Between strain rates of 10^{-6} s^{-1} and 10^{-3} s^{-1} (a thousandfold increase), yield stress increases only by 10%. Above 1 s^{-1}, however, an equivalent rate increase doubles the yield stress. For the data in Fig. 33, at every level of strain the yield stress increases with increasing strain rate. However, a decrease in strain-hardening rate is exhibited at the higher deformation rates. For a low-carbon steel tested at elevated temperatures, the effects of strain rate on strength can become more complicated by various metallurgical factors such as dynamic strain aging in the "blue brittleness" region of some mild steels (Ref 14).

Structural aluminum is less strain-rate sensitive than steels. Figure 34 (Ref 15) shows data obtained for 1060-O aluminum. Between strain rates of 10^{-3} s^{-1} and 10^3 s^{-1} (a millionfold increase), the stress at 2% plastic strain increases by less than 20%.

ACKNOWLEDGMENT

This article was adapted from J.M. Holt, Uniaxial Tension Testing, *Mechanical Testing and Evaluation,* Volume 8, *ASM Handbook,* ASM International, 2000, p 124–142

REFERENCES

1. D. Lewis, Tensile Testing of Ceramics and Ceramic-Matrix Composites, *Tensile Testing,* P. Han, Ed., ASM International, 1992, p 147–182
2. R.J. Del Vecchio, Tensile Testing of Elastomers, *Tensile Testing,* P. Han, Ed., ASM International, 1992, p 135–146
3. N.E. Dowling, *Mechanical Behavior of Materials—Engineering Methods for Deformation, Fracture, and Fatigue,* 2nd ed., Prentice Hall, 1999, p 123
4. R.L. Brockenbough and B.G. Johnson, "Steel Design Manual," United States Steel Corporation, ADUSS 27 3400 03, 1974, p 2–3
5. P.M. Mumford, Test Methodology and Data Analysis, *Tensile Testing,* P. Han, Ed., ASM International, 1992, p 55
6. "Standard Test Method for Young's Modu-

lus, Tangent Modulus, and Chord Modulus," E 111, ASTM
7. *Making, Shaping, and Treating of Steel,* 10th ed., U.S. Steel, 1985, Fig. 50-12 and 50-13
8. "Standard Test Methods and Definitions for Mechanical Testing of Steel Products," A 370, Annex 6, *Annual Book of ASTM Standards,* ASTM, Vol 1.03
9. "Conversion of Elongation Values, Part 1: Carbon and Low-Alloy Steels," 2566/1, International Organization for Standardization, revised 1984
10. W.P. Goepfert, Statistical Aspects of Mechanical Property Assurance, *Reproducibility and Accuracy of Mechanical Tests,* STP 626, ASTM, 1977, p 136–144
11. F.B. Seely and J.O. Smith, *Resistance of Materials,* 4th ed., John Wiley & Sons, p 45
12. N.R.N. Rao et al., "Effect of Strain Rate on the Yield Stress of Structural Steel," Fritz Engineering Laboratory Report 249.23, 1964
13. R.L. Klueh and R.E. Oakes, Jr., High Strain-Rate Tensile Properties of 2¼ Cr-1 Mo Steel, *J. Eng. Mater. Technol.,* Oct 1976, p 361–367
14. M.J. Manjoine, Influence of Rate of Strain and Temperature on Yield Stresses of Mild Steel, *J. Appl. Mech.,* Vol 2, 1944, p A-211 to A-218
15. A.H. Jones, C.J. Maiden, S.J. Green, and H. Chin, Prediction of Elastic-Plastic Wave Profiles in Aluminum 1060-O under Uniaxial Strain Loading, *Mechanical Behavior of Materials under Dynamic Loads,* U.S. Lindholm, Ed., Springer-Verlag, 1968, p 254–269

SELECTED REFERENCES

- "Standard Method of Sharp-Notch Tension Testing of High-Strength Sheet Materials," E 338, ASTM
- "Standard Method of Sharp-Notch Tension Testing with Cylindrical Specimens," E 602, ASTM
- "Standard Methods and Definitions for Mechanical Testing of Steel Products," A 370, ASTM
- "Standard Methods of Tension Testing of Metallic Foil," E 345, ASTM
- "Standard Test Methods for Poisson's Ratio at Room Temperature," E 132, ASTM
- "Standard Test Methods for Static Determination of Young's Modulus of Metals at Low and Elevated Temperatures," E 231, ASTM
- "Standard Test Methods for Young's Modulus, Tangent Modulus, and Chord Modulus," E 111, ASTM
- "Standard Methods of Tension Testing of Metallic Materials," E 8, ASTM
- "Standard Methods of Tension Testing Wrought and Cast Aluminum- and Magnesium-Alloy Products," B 557, ASTM
- "Standard Recommended Practice for Elevated Temperature Tension Tests of Metallic Materials," E 21, ASTM
- "Standard Recommended Practice for Verification of Specimen Alignment Under Tensile Loading," E 1012, ASTM

CHAPTER 4

Tensile Testing Equipment and Strain Sensors

TENSILE-TESTING EQUIPMENT consists of several types of devices used to apply controlled tensile loads to test specimens (test pieces). The equipment is capable of varying the speed of load application and accurately measures the forces, strains, and elongations applied to the test piece.

Commercial tensile-testing equipment became available in the late 1800s. The earliest equipment used manual methods, such as hand cranks, to apply the load. In 1890, Tinius Olsen was granted a patent on the "Little Giant," a hand-cranked, 180 kN (40,000 lbf) capacity testing machine. In 1891, Olsen produced the first autographic machine capable of producing a stress-strain diagram (Ref 1). An example of an 1890 machine is shown in Fig. 1. Tensile testing equipment has evolved from purely mechanical machines to more advanced electromechanical and servohydraulic machines with advance electronics and microcomputers. Electronic circuitry and microprocessors have increased the reliability of experimental data, while reducing the time to analyze information. This transition has made it possible to determine rapidly and with great precision ultimate tensile strength and elongation, yield strength, modulus of elasticity, and other mechanical properties. Current equipment manufacturers also offer workstation configurations that automate mechanical testing.

Conventional test machines for measuring mechanical properties include tension testers, compression testers, or the more versatile universal testing machine (UTM) (Ref 2). UTMs have the capability to test material in tension, compression, or bending. The word *universal* refers to the variety of stress states that can be studied. UTMs can load material with a single, continuous (monotonic) pulse or in a cyclic manner. Other conventional test machines may be limited to either tensile loading or compressive loading, but not both. These machines have less versatility than UTM equipment, but are less expensive to purchase and maintain. The basic aspects of UTM equipment and testing generally apply to tension or compression testing machines as well.

This chapter reviews the current technology and examines force application systems, force measurement, strain measurement, important instrument considerations, gripping of test specimens, test diagnostics, and the use of computers for gathering and reducing data. The influence of the machine stiffness on the test results is also described, along with a general assessment of test accuracy, precision, and repeatability of modern equipment. A discussion of tensile test specimens can be found in Chapter 3, "Uniaxial Tensile Testing."

Fig. 1 Screw-driven balance-beam universal testing machine (1890 model)

Testing Machines

Although there are many types of test systems in current use, the most common are UTMs, which are designed to test specimens in tension, compression, or bending. The testing machines are designed to apply a force to a material to determine its strength and resistance to deformation. Regardless of the method of force application, testing machines are designed to drive a crosshead or platen at a controlled rate, thus applying a tensile or compressive load to a specimen. Such testing machines measure and indicate the applied force in pound-force (lbf), kilogram-force (kgf), or newtons (N). These customary force units are related by the following: 1 lbf = 4.448222 N; 1 kgf = 9.80665 N. All current testing machines are capable of indicating the applied force in either lbf or N (the use of kgf is not recommended).

The load-applying mechanism may be a hydraulic piston and cylinder with an associated hydraulic power supply, or the load may be administered via precision-cut machine screws driven by the necessary gears, reducers, and motor to provide a suitable travel speed. In some light-capacity machines (only a few hundred pounds maximum), the force is applied by an air piston and cylinder. Gear-driven systems obtain load capacities up to approximately 600 kN (1.35×10^5 lbf), while hydraulic systems can obtain forces up to approximately 4500 kN (1×10^6 lbf).

Whether the machine is a gear-driven system or hydraulic system, at some point the test machine reaches a maximum speed for loading the specimen. Gear driven test machines have a maximum crosshead speed limited by the speed of the electric motor in combination with the design of the gear box transmission. Crosshead speed of hydraulic machines is limited to the capacity of the hydraulic pump to deliver a steady pressure on the piston of the actuator or crosshead. Servohydraulic test machines offer a wider range of crosshead speeds; however, there are continuing advances in the speed control of screw-driven machines, which can be just as versatile as, or perhaps more versatile than, servohydraulic machines.

Conventional gear-driven systems are generally designed for speeds of about 0.001 to 500 mm/min (4×10^{-6} to 20 in./min), which is suitable for quasi-static testing. Servohydraulic systems are generally designed over a wider range of test speeds, such as:

- 1 μm/h test speeds for creep-fatigue, stress-corrosion, and stress-rupture testing
- 1 μm/min test speeds for fracture testing of brittle materials
- 10 m/s (400 in./s) test speeds for dynamic testing of components like bumpers or seat belts

Servohydraulic UTM systems may also be designed for cycle rates from 1 cycle/day to over 200 cycles/s. Gear-driven systems typically allow cycle rates between 1 cycle/h and 1 cycle/s.

Gear-driven (or screw-driven) machines are electromechanical devices that use a large actuator screw threaded through a moving crosshead (Fig. 2). The screw is turned in either direction by an electric motor through a gear reduction system. The screws are rotated by a variable-control motor and drive the moveable crosshead up or down. This motion can load the specimen in either tension or compression, depending on how the specimen is to be held and tested.

Screw-driven testing machines currently used are of either a one-, two-, or four-screw design. To eliminate twist in the specimen from the rotation of the screws in multiple-screw systems, one screw has a right-hand thread, and the other has a left-hand thread. For alignment and lateral stability, the screws are supported in bearings on each end. In some machines, loading crossheads are guided by columns or guideways to achieve alignment.

A range of crosshead speeds can be achieved by varying the speed of the electric motor and by changing the gear ratio. A closed-loop servodrive system ensures that the crosshead moves at a constant speed. The desired or user-selected speed and direction information is compared with a known reference signal, and the servomechanism provides positional control of the moving crosshead to reduce any error or difference. State-of-the-art systems use precision optical encoders mounted directly on preloaded twin ball screws. These types of systems are capable of measuring crosshead displacement to an accuracy of 0.125% or better with a resolution of 0.6 μm.

As noted previously, typical screw-driven machines are designed for speeds of 1 to 20 mm/min (0.0394–0.788 in./min) for quasi-static test applications; however, machines can be designed to obtain higher speeds, although the useful force available for application to the specimen decreases as the speed of the crosshead motion increases. Modern high-speed systems generally are useful in ranges up to 500 mm/min (20 in./min) (Ref 3). Nonetheless, top crosshead speeds of 1250 mm/min (50 in./min) can be attained in screw-driven machines, and servohydraulic machines can be driven up to 2.5×10^5 mm/min (10^4 in./min) or higher.

Due to the high forces involved, bearings and gears require particular attention to reduce friction and wear. Backlash, which is the free movement between the mechanical drive components, is particularly undesirable. Many instruments incorporate antibacklash preloading so that forces are translated evenly through the lead screw and crosshead. However, when the crosshead direction is constantly in one direction, antibacklash devices may be unnecessary.

Servohydraulic machines use a hydraulic pump and servohydraulic valves that move an actuator piston (Fig. 3). The actuator piston is attached to one end of the specimen. The motion of the actuator piston can be controlled in both directions to conduct tension, compression, or cyclic loading tests.

Servohydraulic test systems have the capability of testing at rates from as low as 45×10^{-11}

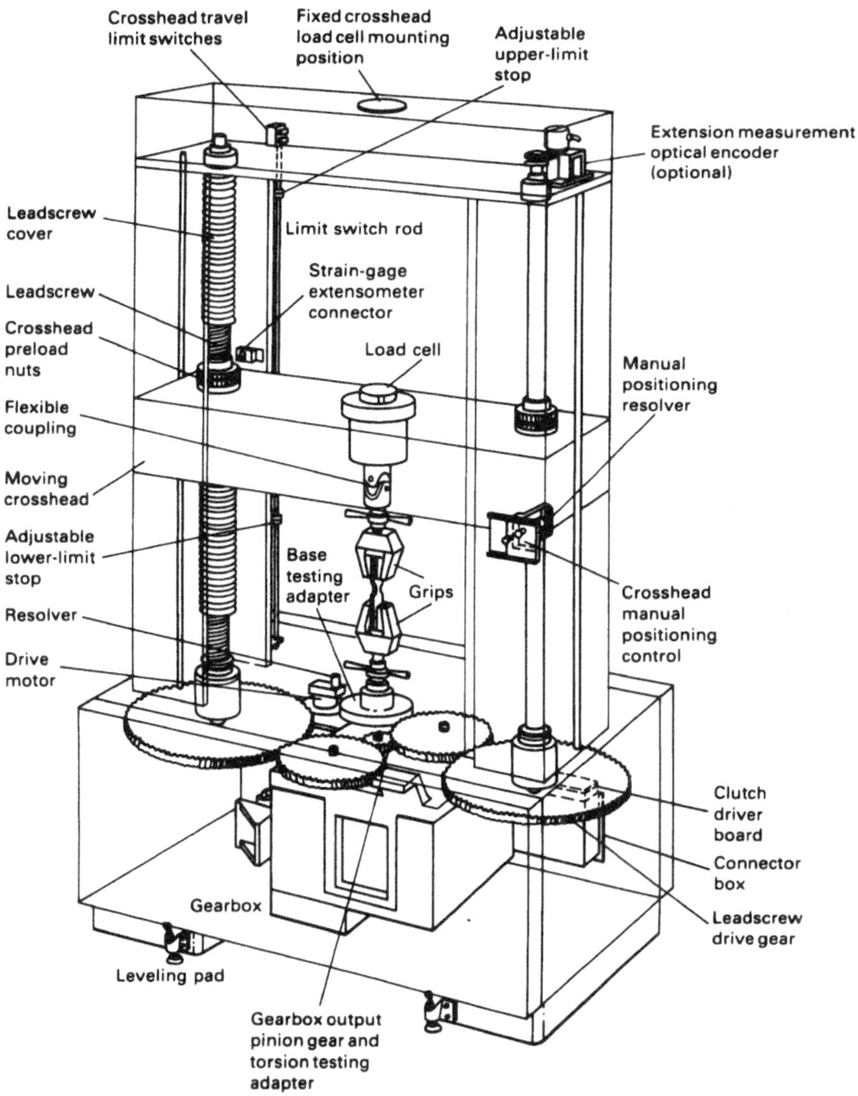

Fig. 2 Components of an electromechanical (screw-driven) testing machine. For the configuration shown, moving the lower (intermediate) head upward produces tension in the lower space between the crosshead and the base

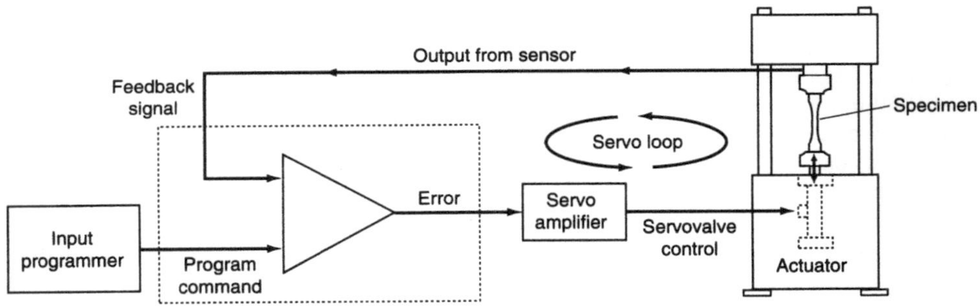

Fig. 3 Schematic of a basic servohydraulic, closed-loop testing machine

m/s (1.8×10^{-9} in./s) to 30 m/s (1200 in./s) or more. The actual useful rate for any particular system depends on the size of the actuator, the flow rating of the servovalve, and the noise level present in the system electronics. A typical servohydraulic UTM system is shown in Fig. 4.

Hydraulic actuators are available in a wide variety of force ranges. They are unique in their ability to economically provide forces of 4450 kN (1,000,000 lbf) or more. Screw-driven machines are limited in their ability to provide high forces due to problems associated with low machine stiffness and large and expensive loading screws, which are increasingly more difficult to produce as the force rating goes up.

Microprocessors for Testing and Data Reduction. Contemporary UTMs are controlled by microprocessor-based electronics. One class of controller is based on dedicated microprocessors for test machines (Fig. 4). Dedicated microprocessors are designed to perform specific tasks and have displays and input functions that are limited to those tasks. The dedicated microprocessor sends signals to the experimental apparatus and receives information from various sensors. The data received from sensors can be passed to oscilloscopes or computers for display and storage. The experimental results consist of time and voltage information that must be further reduced to analyze material behavior. Analysis of the data requires the conversion of test results, such as voltage, to specific quantities, such as displacement and load, based on known conversion factors.

The second class of controller is the personal computer (PC) designed with an electronic interface to the experimental apparatus, and the appropriate application software. The software takes the description of the test to be performed, including specimen geometry data, and establishes the requisite electronic signals. Once the test is underway, the computer controls the tests and collects, reduces, displays, and stores the data. The obvious advantage of the PC-based controller is reduced time to generate graphic results, or reports. The other advantage is the elimination of some procedural errors, or the reduction of the interfacing details between the operator and the experimental apparatus. Some systems are designed with both types of controllers. Having both types of controllers provides maximum flexibility in data gathering with a minimal amount of time required for reducing data when conducting standard experiments.

Principles of Operation

The operation of a universal testing machine can be understood in terms of the main elements for any stress analysis, which include material response, specimen geometry, and load or boundary condition.

Material response, or material characterization, is studied by adopting standards for the

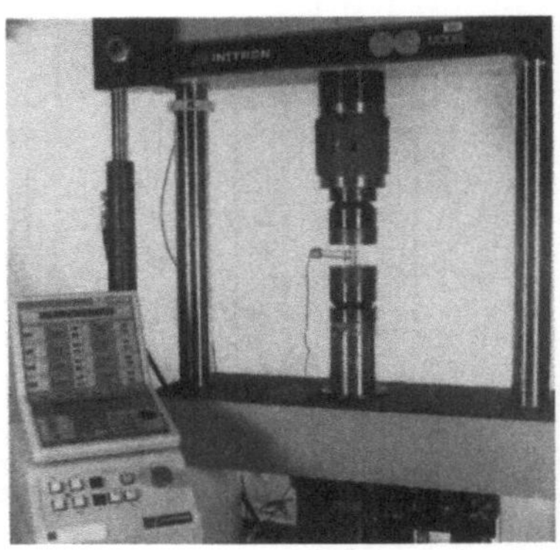

Fig. 4 Servohydraulic testing machine and load frame with a dedicated microprocessor-based controller

other two elements. Specimen geometries are described in the section "Tensile Testing Requirements and Standards" at the end of this chapter. This section briefly describes load condition factors, such as strain rate, machine rigidity, and various testing modes by load control, speed control, strain control, and strain-rate control.

Strain Rate

Strain rate, or the rate at which a specimen is deformed, is a key test variable that is controlled within prescribed limits, depending on the type of test being performed. Table 1 summarizes the general strain-rate ranges that are required for various types of property tests. Conventional (quasi-static) tensile tests require strain rates between 10^{-5} and 10^{-1} s^{-1}.

A typical mechanical test on metallic materials is performed at a strain rate of approximately 10^{-3} s^{-1}, which yields a strain of 0.5 in 500 s. Conventional equipment and techniques generally can be extended to strain rates as high as 0.1 s^{-1} without difficulty. Tests at higher strain rates necessitate additional considerations of machine stiffness and strain measurement techniques. In terms of machine capability, servohydraulic load frames equipped with high-capacity valves can be used to generate strain rates as high as 200 s^{-1}. These tests are complicated by load and strain measurement and data acquisition.

If the crosshead speed is too high, inertia effects can become important in the analysis of the specimen stress state. Under conditions of high crosshead speed, errors in the load cell output and crosshead position data may become unacceptably large. A potential exists to damage load cells and extensometers under rapid loading. The damage occurs when the specimen fractures and the load is instantaneously removed from the specimen and the load frame.

At strain rates greater than 200 s^{-1}, the required crosshead speeds exceed the speeds easily obtained with screw-driven or hydraulic machines. Specialized high strain rate methods are discussed in more detail in Chapter 15, "High Strain Rate Testing."

Determination of Strain Rates for Quasi-Static Tensile Tests

Strength properties for most materials tend to increase at higher rates of deformation. In order to quantify the effect of deformation rate on strength and other properties, a specific definition of strain rate is required. During a conventional (quasi-static) tension test, for example, ASTM E 8 "Tension Testing of Metallic Materials" prescribes an upper limit of deformation rate as determined quantitatively during the test by one of the following methods (listed in decreasing order of precision):

- Rate of straining
- Rate of stressing (when loading is below the proportional limit)
- Rate of crosshead separation during the tests
- Elapsed time
- Free-running crosshead speed

For some materials, the free-running crosshead speed, which is the least accurate, may be adequate, while for other materials, one of the remaining methods with higher precision may be necessary in order to obtain test values within acceptable limits. When loading is below the proportional limit, the deformation rate can be specified by the "loading rate" units of stress per unit of time such that:

$$\dot{\sigma} = E\dot{\varepsilon}$$

where, according to Hooke's law, $\dot{\sigma}$ is stress. E is the modulus of elasticity, $\dot{\varepsilon}$ is strain, and the superposed dots denote time derivatives.

ASTM E 8 specifies that the test speed must be low enough to permit accurate determination of loads and strains. When the rate of stressing is stipulated, ASTM E 8 requires that it not exceed 690 MPa/min (100 ksi/min). This corresponds to an elastic strain rate of about 5×10^{-5} s^{-1} for steel or 15×10^{-5} s^{-1} for aluminum. When the rate of straining is stipulated, ASTM E 8 prescribes that after the yield point has been passed, the rate can be increased to about 1000×10^{-5} s^{-1}; presumably, the stress rate limitation must be applied until the yield point is passed. Lower limits are also given in ASTM E 8.

Table 1 Strain rate ranges for different tests

Type of test	Strain rate range, s^{-1}
Creep tests	10^{-8} to 10^{-5}
Pseudostatic tensile or compression tests	10^{-5} to 10^{-1}
Dynamic tensile or compression tests	10^{-1} to 10^{2}
Impact bar tests involving wave propagation effects	10^{2} to 10^{4}

Source: Ref 4

In ASTM standard E 345, "Tension Testing of Metallic Foil," the same upper limit on the rate of stressing is recommended. In addition, a lower limit of 7 MPa/min (1 ksi/min) is given. ASTM E 345 further specifies that when the yield strength is to be determined, the strain rate must be in the range from approximately 3×10^{-5} to 15×10^{-5} s^{-1}.

Inertia Effects

A fundamental difference between a high strain rate tensile test and a quasi-static tensile test is that inertia and wave propagation effects are present at high rates. An analysis of results from a high strain rate test thus requires consideration of the effect of stress wave propagation along the length of the test specimen in order to determine how fast a uniaxial test can be run to obtain valid stress-strain data.

For high loading rates, the strain in the specimen may not be uniform. Figure 5 illustrates an elemental length dx_0 of a tensile test specimen whose initial cross-sectional area is A_0 and whose initial location is prescribed by the coordinate x. Neglecting gravity, no forces act on this element in its initial configuration. After the test has begun, the element is shown displaced by a distance u, deformed to new dimensions dx and A, and subjected to forces F and $F + dF$. The difference, dF, between these end-face forces causes the motion of the element that is manifested by the displacement, u. This motion is governed by Newton's second law, force equals mass times acceleration:

$$dF = \rho_0 A_0 dx_0 \left(\frac{d^2u}{dt^2}\right) \qquad \text{(Eq 1)}$$

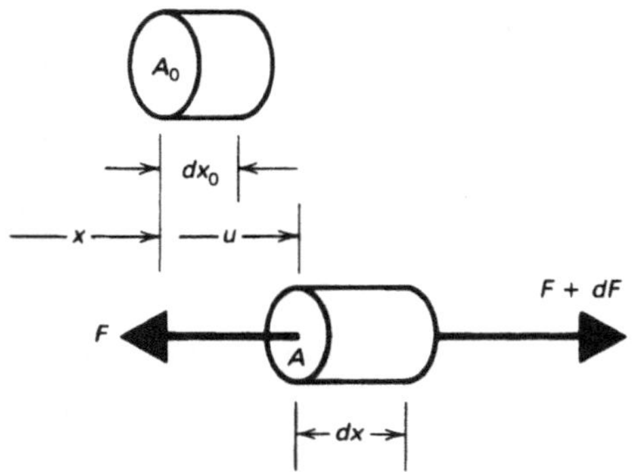

Fig. 5 The deformation of an elemental length, dx_0, of a tensile test specimen of initial cross-sectional area, A_0, by a stress wave. The displacement of the element is u; the differential length of the element as a function of time is dx; the forces acting on the faces of the element are given by F and $F + dF$.

where $\rho_0 A_0 dx_0$ is the mass of the element, $A_0 dx_0$ is the volume, ρ is the density of the material, and (d^2u/dt^2) is its acceleration. Tests that are conducted very slowly involve extremely small accelerations. Thus, Eq 1 shows that the variation of force dF along the specimen length is negligible.

However, for tests of increasingly shorter durations, the acceleration term on the right side of Eq 1 becomes increasingly significant. This produces an increasing variation of axial force along the length of the specimen. As the force becomes more nonuniform, so must the stress. Consequently, the strain and strain rate will also vary with axial position in the specimen. When these effects become pronounced, the concept of average values of stress, strain, and strain rate become meaningless, and the test results must be analyzed in terms of the propagation of waves through the specimen. This is shown in Table 1 as beginning near strain rates of 10^2 s^{-1}.

In an intermediate range of strain rates (denoted as dynamic tests in Table 1), an effect known as "ringing" of the load-measuring device obscures the interpretation of test data. An example of this condition is shown in Fig. 6, which is a tracing of load cell force versus time during a dynamic tensile test of a 2024-T4 aluminum specimen. Calculation showed that the oscillations apparent in the figure are consistent with vibrations at the approximate natural frequency of the load cell used for this test (Ref 5, 6).

In many machines currently available for dynamic testing, electronic signal processing is used to filter out such vibrations, thus making the instrumentation records appear much smoother than the actual load cell signal. However, there is still a great deal of uncertainty in the interpretation of dynamic test data. Conse-

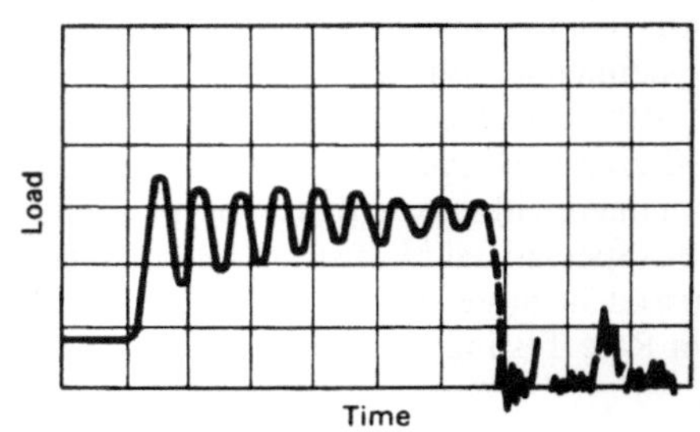

Fig. 6 Oscilloscope record of load cell force versus time during a dynamic tensile test depicting the phenomenon of ringing. The uncontrolled oscillations result when the loading rate is near the resonant frequency of the load cell. The scales are arbitrary. Source: Ref 5

quently, the average value of the high-frequency vibrations associated with the load cell can be expected to differ from the force in the specimen. This difference is caused by vibrations near the natural frequency of the testing machine, which are so low that the entire test can occur in less than 1⁄10 of a cycle. Hence, these low-frequency vibrations usually are impossible to detect in a test record, but can produce significant errors in the analysis of test results. The ringing frequency for typical load cells ranges from 2400 to 3600 Hz.

Machine Stiffness

The most common misconception relating to strain rate effects is that the testing machine is much stiffer than the specimen. Such an assumption leads to the concept of deformation of the specimen by an essentially rigid machine. However, for most tests the opposite is true: the conventional tensile specimen is much stiffer than most testing machines. As shown in Fig. 7, for example, if crosshead displacement is defined as the relative displacement, Δ, that would occur under conditions of zero load, then with a specimen gripped in a testing machine and the driving mechanism engaged, the crosshead displacement equals the deformation in the gage length of the specimen plus elastic deflections in components such as the machine frame, load cell, grips, and specimen ends. Before yielding, the gage length deformation is a small fraction of the crosshead displacement.

After the onset of gross plastic yielding of the specimen, conditions change. During this phase of deformation, the load varies slowly as the material strain hardens. Thus, the elastic deflections in the machine change slowly, and most of the relative crosshead displacement produces plastic deformation in the specimen. Qualitatively, in a test at approximately constant crosshead speed, the initial elastic strain rate in the specimen will be small, but the specimen strain rate will increase when plastic flow occurs.

Quantitatively, this effect can be estimated as follows. Consider a specimen having an initial cross-sectional area A_0 and modulus of elasticity E gripped in a testing machine so that its axially stressed gage length initially is L_0. (This discussion is limited to the range of testing speeds where wave propagation effects are negligible. This restriction implies that the load is uniform throughout the gage length of the specimen.) Denote the stiffness of the machine, grips, and so on, by K and the crosshead displacement rate (nominal crosshead speed) by S. The ratio S/L_0 is sometimes called the nominal rate of strain, but because it is often substantially different from the rate of strain in the specimen, the term specific crosshead rate is preferred (Ref 8).

Let loading begin at time t equal to zero. At any moment thereafter, the displacement of the crosshead must equal the elastic deflection of the machine plus the elastic and plastic deflections of the specimen. Letting s denote the engineering stress in the specimen, the machine deflection is then sA_0/K. It is reasonable to assume that Hooke's law adequately describes the elastic deformation of the specimen at ordinary stress levels. Thus, the elastic strain e_e is s/E.

Denoting the average plastic strain in the specimen by e_p, the above displacement balance can be expressed as:

$$\int_0^t S\,dt = s\left(\frac{A_0}{K} + \frac{L_0}{E}\right) + e_pL_0 \quad \text{(Eq 2)}$$

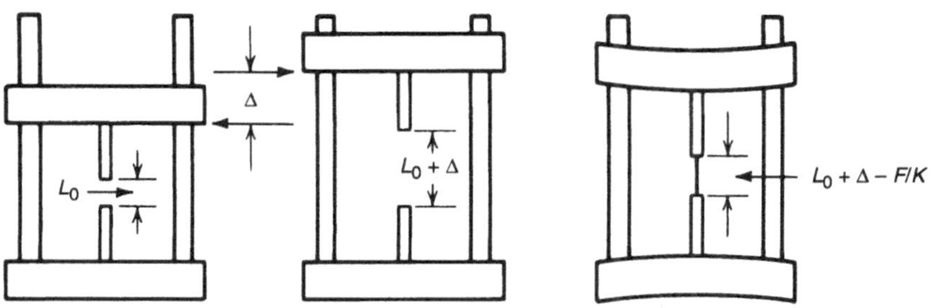

Fig. 7 Schematic illustrating crosshead displacement and elastic deflection in a tensile testing machine. Δ is the displacement of the crosshead relative to the zero load displacement; L_0 is the initial gage length of the specimen; K is the composite stiffness of the grips, loading frame, load cell, specimen ends, etc.; F is the force acting on the specimen. The development of Eq 2 through 12 describes the effects of testing machine stiffness on tensile properties. Source: Ref 7

Differentiating Eq 2 with respect to time and dividing by L_0 gives:

$$\frac{S}{L_0} = \left(\frac{\dot{s}}{E}\right)\left(\frac{A_0E}{KL_0} + 1\right) + \dot{e}_p \qquad \text{(Eq 3)}$$

The strain rate in the specimen is the sum of the elastic and plastic strain rates:

$$\dot{e} = \dot{e}_e + \dot{e}_p = \left(\frac{\dot{s}}{E}\right) + \dot{e}_p \qquad \text{(Eq 4)}$$

Using Eq 3 to eliminate the stress rate from Eq 4 yields:

$$\dot{e} = \frac{\left(\frac{SK}{A_0E} + \dot{e}_p\right)}{\left(\frac{KL_0}{A_0E} + 1\right)} \qquad \text{(Eq 5)}$$

Thus, it is seen that the specimen strain rate usually will differ from the specific crosshead rate by an amount dependent on the rate of plastic deformation and the relative stiffnesses of the specimen (A_0E/L_0) and the machine, K.

Determination of Testing Machine Stiffness

Machine stiffness is the amount of deflection in the load frame and the grips for each unit of load applied to the specimen. This deflection not only encompasses elastic deflection of the load frame, but includes any motion in the grip mechanism, or at any interface (threads, etc.) in the system. These deflections are substantial during the initial loading of the specimen, that is, through the elastic regime. This means that the initial crosshead speed (specified by the operator) is not an accurate measure of specimen displacement (strain). If the strain in the elastic regime is not accurately known, then extremely large errors may result in the calculation of Young's modulus (E, the ratio of stress versus strain in the elastic regime). In the analysis by Hockett and Gillis (Ref 9), the machine stiffness K is accounted for in the following equation:

$$K = \left(\frac{S}{\dot{P}_0} - \frac{L_0}{A_0E}\right)^{-1} \qquad \text{(Eq 6)}$$

where L_0 is initial specimen gage length, S is crosshead speed of the testing machine, A_0 is initial cross-sectional area of the specimen, $\dot{P}_0$ is specimen load rate ($dF/dt = A_0\dot{s}$), and E is Young's modulus of the specimen material.

Research in this area showed that a significant amount of scatter was found in the measurement of machine stiffness. This variability can be attributed to relatively small differences in test conditions. For characterization of the elastic response of a material and for a precise measure of yield point, the influence of machine stiffness requires that an extensometer, or a bonded strain gage, be used. After yielding of the specimen material, the change of machine deflection is very small because the load changes slowly. If the purpose of the experiment is to study large strain behavior, then the error associated with the use of the crosshead displacement is small relative to other forms of experimental uncertainties.

Control Modes

During a test, control circuits and servomechanisms monitor and control the key experimental conditions, such as force, specimen deformation, and the position of the moveable crosshead. These are the key boundary conditions, which are analyzed to provide mechanical property data. These boundary conditions on the specimen can also be controlled in different ways, such as constant load control, constant strain control, and constant crosshead speed control. Constant crosshead speed is the most common method for tensile tests.

Constant Load Rate Testing. With appropriate modules on a UTM system, a constant load rate test can be accomplished easily. In this configuration, a load-control module allows the machine with the constant rate of extension to function as a constant load rate device. This is accomplished by a feedback signal from a load cell, which generates a signal that automatically adjusts to the motion controller of the crosshead. Usually, the servomechanism system response is particularly critical when materials are loaded through the yield point.

Constant Strain Rate Testing. Commercial systems have been developed to control the experiment based on a constant rate of straining in the specimen. These systems rely on extensometers measuring the change in gage length to provide data on strain as a function of time. The resulting signal is processed to determine the current strain rate and is used to adjust the crosshead displacement rate throughout the test. Again, servomechanism response time is particularly critical when materials are taken through yield.

To maintain a constant average strain rate during a test, the crosshead speed must be adjusted as plastic flow occurs so that the sum $(SK/A_0E + e_p)$ remains constant. For most metallic materials at the beginning of a test, the plastic strain rate is ostensibly zero, and from Eq 5 the initial strain rate is:

$$\dot{e}_0 = \frac{\left(\frac{S_0}{L_0}\right)}{1 + \left(\frac{A_0E}{KL_0}\right)} \quad \text{(Eq 7)}$$

where S_0 is the crosshead speed at the beginning of the test. For materials that have a definite yield, $\dot{s} = 0$ at the yield point. Therefore, from Eq 3 and 4, the yield point strain rate is:

$$\dot{e}_1 = \left(\frac{S_1}{L_0}\right) \quad \text{(Eq 8)}$$

where S_1 is the crosshead speed at the yield point. Equating these two values of strain rate shows that the crosshead speed must be reduced from its initial value to its yield-point value by a factor of:

$$\frac{S_0}{S_1} = \left(1 + \frac{A_0E}{KL_0}\right) \quad \text{(Eq 9)}$$

For particular measured values of machine stiffness given in Table 2, this factor for a standard 12.8 mm (0.505 in.) diameter steel specimen is typically greater than 20 and can be as high as 100. Only for specially designed machines will the relative stiffness of the machine exceed that of the specimen. Even for wire-like specimens, the correspondingly delicate gripping arrangement will ensure that the machine stiffness is less than that of the specimen. Thus, large changes in crosshead speed usually are required to maintain a constant strain rate from the beginning of the test through the yield point.

Furthermore, for many materials, the onset of yielding is quite rapid, so that this large change in speed must be accomplished quickly. Making the necessary changes in speed generally requires not only special strain-sensing equipment, but also a driving unit that is capable of extremely fast response. The need for fast response in the driving system eliminates the use of screw-driven machines for constant strain-rate testing. Servohydraulic machines may be capable of conducting tests at constant strain rate through the yield point of a material.

Equation 9 indicates the magnitude of speed changes required only for tests in which there is no yield drop. For materials having upper and lower yield points, the direction of crosshead motion may have to be reversed after initial yielding to maintain a constant strain rate. This reversal may be necessary, because plastic strains beyond the upper yield point can be imposed at a strain rate greater than the desired rate by recovery of elastic deflections of the machine as the load decreases.

Another important test feature related to the speed change capability of the testing machine is the rate at which the crosshead can accelerate from zero to the prescribed test speed at the beginning of the test. For a slow test this may not be critical, but for a high-speed test, the yield point could be passed before the crosshead achieves full testing speed. Thus, the crosshead may still be accelerating when it should be decelerating, and accurate information concerning the strain rate will not be obtained. With the advent of closed-loop servohydraulic machines and electromagnetic shakers, the speed at which the ram (crosshead) responds is two orders of magnitude greater than for screw-driven machines.

Tests at Constant Crosshead Speeds. Machines with a constant rate of extension are the most common type of screw-driven testers and are characterized by a constant rate of crosshead travel regardless of applied loads. They permit testing without speed variations that might alter test results; this is particularly important when testing rate-sensitive materials such as polymers, which exhibit different ultimate strengths and elongations when tested at different speeds.

For a gear-driven system, applying the boundary condition is as simple as engaging the electric motor with a gear box transmission. At this point, the crosshead displacement will be whatever speed and direction was selected. More sophisticated systems use a command signal that

Table 2 Experimental values of testing machine stiffness

Machine stiffness		
kg/mm	lb/In.	Source
740	41,500	Ref 10
460	26,000	Ref 11
1800	100,000	Ref 12
1390–2970	77,900–166,500	Ref 13

is compared with a feedback signal from a transducer monitoring the position of the crosshead. Using this feedback circuit, the desired boundary condition can be achieved.

Tensile tests usually can be carried out at a constant crosshead speed on a conventional testing machine, provided the machine has an adequate speed controller and the driving mechanism is sufficiently powerful to be insensitive to changes in the loading rate. Because special accessory equipment is not required, such tests are relatively simple to perform. Also, constant crosshead speed tests typically provide as good a comparison among materials and as adequate a measure of strain-rate sensitivity as constant strain-rate tests.

Two of the most significant test quantities—yield strength and ultimate tensile strength—frequently can be correlated with initial strain rate and specific crosshead rate, respectively. The strain rate up to the proportional limit equals the initial strain rate. Thus, for materials that yield sharply, the time-average strain rate from the beginning of the test to yield is only slightly greater than the initial strain rate:

$$\dot{e}_0 = \left(\frac{\frac{S}{L_0}}{1 + \frac{A_0E}{KL_0}} \right) \quad \text{(Eq 10)}$$

even though the instantaneous strain rate at yield is the specific crosshead rate:

$$\dot{e}_1 = \left(\frac{S}{L_0} \right) \quad \text{(Eq 11)}$$

However, beyond the yield point, the stress rate is small so that the strain rate remains close to the specific crosshead rate (Eq 11). Thus, ductile materials, for which a rather long time will elapse before reaching ultimate strength, have a time-average strain rate from the beginning of the test to ultimate that is only slightly less than the specific crosshead rate. Also, because the load rate is zero at ultimate as well as at yield, the instantaneous strain rate at ultimate equals the specific crosshead rate.

During a test at constant crosshead speed, the variation of strain rate from initial to yield-point values is precisely the inverse of the crosshead speed change required to maintain a constant strain rate (Eq 9):

$$\frac{\dot{e}_1}{\dot{e}_0} = \left(1 + \frac{A_0E}{KL_0} \right) \quad \text{(Eq 12)}$$

Consequently, in an ordinary tensile test, the yield strength and ultimate tensile strength may be determined at two different strain rates, which can vary by a factor of 20 to 100, depending on machine stiffness. If a yield drop occurs, elastic recovery of machine deflections will impose a strain rate even greater than the specific crosshead rate given by Eq 12.

A point of interest from the analysis involves testing of different sized specimens at about the same initial strain rate. Assuming that these tests are to be made on one machine under conditions for which K remains substantially constant, the crosshead speed must be adjusted to ensure that specimens of different lengths, diameters, or materials will experience the same initial strain rate. In the typical case where the specimen is much stiffer than the machine, $(1 + A_0E/KL_0)$ in Eq 10 can be approximated simply by (A_0E/KL_0), so that the initial strain rate is approximately $\dot{e}_0 = SK/A_0E$. Thus, specimens of various lengths, tested at the same crosshead speed, will generally experience nearly the same initial strain rate. However, changing either the specimen cross section or material necessitates a corresponding change in crosshead speed to obtain the same initial rate.

A change in specimen length has substantially the same effect on both the specific crosshead rate (S/L_0) and the stiffness ratio of specimen to machine (A_0E/KL_0) and, therefore, has no net effect. For example, an increase in specimen length tends to decrease the strain rate by distributing the crosshead displacement over the longer length; however, at the same time, the increase in length reduces the stiffness of the specimen so that more of the crosshead displacement goes into deformation of the specimen and less into deflection of the machine. These two effects are almost exactly equal in magnitude. Thus, no change in initial strain rate is expected for specimens of different lengths tested at the same crosshead speed.

Load-Measurement Systems

Prior to the development of load cells, testing machine manufacturers used several types of devices for the measurement of force. Early systems, some of which are still in use, employ a graduated balanced beam similar to platform-scale weighing systems. Subsequent systems have used Bourdon tube hydraulic test gages, Bourdon tubes with various support and assist devices, and load cells of several types. One of the most common load-measuring systems, prior to the development of load cells, was the displacement pendulum, which measured load by the movement of the balance displacement pendulum. The pendulum measuring system was used widely, because it is applicable to both hydraulic and screw-driven machines and has a high degree of reliability and stability. Many machines of this design are still in use, and they are still manufactured in Europe, India, South America, and Asia. Another widely used testing system was the Emery-Tate oil-pneumatic system, which accurately senses the hydraulic pressure in a closed, flat capsule.

Load Cells. Current testing machines use strain-gage load cells and pressure transducers. In a load cell, strain gages are mounted on precision-machined alloy-steel elements, hermetically sealed in a case with the necessary electrical outlets, and arranged for tensile and/or compressive loading. The load cell can be mounted so that the specimen is in direct contact, or the cell can be indirectly loaded through the machine crosshead, table, or columns of the load frame. The load cell and the load cell circuit are calibrated to provide a specific voltage as an output signal when a certain force is detected. In pressure transducers, which are variations of strain-gage load cells, the strain-gaged member is activated by the hydraulic pressure of the system.

Strain gages, strain-gage load cells, and pressure transducers are manufactured to several degrees of accuracy; however, when used as the load-measuring mechanism of a testing machine, the mechanisms must conform to ASTM E 4, as well as to the manufacturer's quality standards. Load cells are rated by the maximum force in their operating range, and the deflection of the load cell must be maintained within the elastic regime of the material from which the load cell was constructed. Because the load cell operates within its elastic range, both tensile and compressive forces can be monitored.

Electronics provide a wide range of signal processing capability to optimize the resolution of the output signal from the load cell. Temperature-compensating gages reduce measurement errors from changes in ambient temperature. A prior knowledge of the mechanical properties of the material being studied is also useful to obtain full optimization of these signals.

Within individual load cells, mechanical stops can be incorporated to minimize possible damage that could be caused by accidental overloads. Also, guidance and supports can be included to prevent the deleterious effects of side loading and to give desired rigidity and ruggedness. This is important in tensile testing of metals because of the elastic recoil that can occur when a stiff specimen fails.

Calibration of load-measuring devices refers to the procedure of determining the magnitude of error in the indicated loads. Only load-indicating mechanisms that comply with standard calibration methods (e.g., ASTM E 74) should be used for the load calibration and verification of universal testing machines (see the section "Force Verification of Universal Testing Machines" later in this chapter).

Calibration of load-measuring devices for mechanical test machines is covered in specifications of several standards organizations such as:

Specification number	Specification title
ASTM E 74	Standard Practice for Calibration of Force-Measuring Instruments for Verifying the Force Indication of Testing Machines
EN 10002-3	Part 3: Calibration of Force-Proving Instruments Used for the Verification of Testing Machines
ISO 376	Metallic Materials—Calibration of Force-Proving Instruments Used for the Verification of Testing Machines
BS EN 10002-3	Calibration of Force-Proving Instruments Used for the Verification of Uniaxial Testing Machines

To ensure valid load verification, calibration procedures should be performed by skilled personnel who are knowledgeable about testing machines and related instruments and the proper use of calibration standards.

Load verification of load-weighing systems can be accomplished using methods based on the use of standard weights, standard weights and lever balances, and elastic calibration devices. Of these calibration methods, elastic calibration devices have the fewest inherent problems and are widely used. The two main types of elastic load-calibration devices are elastic proving rings and strain-gage load cells, as briefly described below.

The elastic proving ring (Fig. 8a, b) is a forged steel ring that is precisely machined to a fine finish and closely maintained tolerances.

This device has a uniform and repeatable deflection throughout its loaded range. Elastic proving rings usually are designed to be used only in compression, but special rings are designed to be used in tension or compression.

As the term "elastic device" implies, the ring is used well within its elastic range, and the deflection is read by a precise micrometer. Proving rings are available with capacities ranging from 4.5 to over 5000 kN (1000 to 1.2×10^6 lbf). Their usable range is from 10 to 100% of load capacity, based on compliance with the ASTM E 74 verification procedure.

Proving rings vary in weight from about 2 kg (5 lb) to hundreds of kilograms (or several hundred pounds). They are portable and easy to use. After initial certification, they should be recalibrated and recertified at intervals not exceeding 2 years.

Proving rings are not load rings. Although the two devices are of similar design and construction, only proving rings that use a precise micrometer for measuring deflection can be used for calibration. Load rings employ a dial indicator to measure deflection and usually do not comply with the requirements of ASTM E 74.

Calibration strain-gage load cells are precisely machined high-alloy steel elements designed to have a positive and predetermined uniform deflection under load. The steel load cell element contains one or more reduced sections, onto which wire or foil strain gages are attached to form a balanced circuit containing a temperature-compensating element.

Strain-gage load cells used for calibration purposes are either compression or tension-compression types and have built-in capacities ranging from about 0.4 to 4000 kN (100 to 1,000,000 lbf). Their usable range is typically from 5 to 100% of capacity load, and their accuracy is ±0.05%, based on compliance with applicable calibration procedures, such as ASTM E 74. Figure 9 illustrates a load cell system used to calibrate a UTM. This particular system incorporates a digital load indicator unit.

Comparison of Elastic Calibration Devices. The deflection of a proving ring is measured in divisions that are assigned a value in lbf, kgf, or N. The force is then calculated in the desired units. Although the deflection of a load cell is given numerically and a force value can be assigned with a load cell reading, electric circuits can provide direct readout in lbf, kgf, or N. Thus, certified load cells are more practical and convenient to use and minimize errors in calculation.

In small capacities (5 to 20 kN, or 1000 to 5000 lbf), proving rings and load cells are of

(a) (b)

Fig. 8 Proving rings. (a) Elastic proving ring with precision micrometer for deflection/load readout. (b) Load calibration of 120,000 lbf screw-driven testing machine with a proving ring similar size and weight (2 to 5 kg, or 4 to 10 lb). In large capacities (2000 to 2700 kN, or 400,000 to 600,000 lbf), load cells are about one half the size and weight of proving rings. Proving rings are a single-piece, self-contained unit. A load cell calibration kit consists of two parts: the load cell and the display indicator (Fig. 9). Although the display indicator is designed to be used with a load cell of any capacity, it can only be used with load cells that have been verified with it as a system.

Although both proving rings and load cells are portable, the lighter weight and smaller size of high-capacity load cells enhance their suitability for general use. Load cells and their display indicators require a longer setup time: however, their direct readout feature reduces the overall calibration and reporting time. After initial certification, the load cell should be recalibrated after one year and thereafter at intervals not exceeding two years.

Both types of calibration devices are certified in accordance with the provisions of calibration standards. In the United States, devices are certified in accordance with ASTM E 74 and the verification values determined by the National Institute of Standards and Technology (NIST). NIST maintains a 1,000,000 lbf deadweight calibrator that is kept in a temperature- and humidity-controlled environment. This force-calibrating machine incorporates twenty 50,000 lb stainless steel weights, each accurate to within ±0.25 lb. This machine, and six others of smaller capacities, are used to calibrate elastic calibrating devices, which in turn are employed to accurately calibrate other testing equipment.

Elastic calibrating devices for verification of testing machines are calibrated to primary standards, which are weights. The masses of the weights used are determined to 0.005% of their values.

Strain-Measurement Systems

Deformation of the specimen can be measured in several ways, depending on the size of specimen, environmental conditions, and measurement requirements for accuracy and precision of anticipated strain levels. A simple method is to use the velocity of the crosshead while tracking the load as a function of time. For the load and time data pair, the stress in the specimen and the amount of deformation, or strain, can be calculated. When the displacement of the platen is assumed to be equal to the specimen displacement, an error is introduced by the fact that the entire load frame has been deflected under the stress state. This effect is related to the concept of machine stiffness, as previously discussed.

Extensometry

The elongation of a specimen during load application can be measured directly with various types of devices, such as clip-on extensometers (Fig. 10), directly-mounted strain gages (Fig. 11), and various optical devices. These devices are used extensively and can provide a high degree of deformation- (strain-) measurement accuracy. Other more advanced instrumentations, such as laser interferometry and video extensometers, are also available.

Various types of extensometers and strain gages are described below. Selection of a device for strain measurement depends on various factors:

- The useable range and accuracy of the gage
- Techniques for mounting the gage
- Specimen size
- Environmental test conditions
- Electronic circuit configuration and analysis for signal processing

The last item should include the calibration of the extensometer device over its full operating

Fig. 9 Load cell and digital load indicator used to calibrate a 200,000 lbf hydraulic testing machine

range. In addition, one challenge of working with clip-on extensometers is to ensure proper attachment to the specimen. If the extensometer slips as the specimen deforms, the resulting signal will give a false reading.

Clip-on extensometers can be attached to a test specimen to measure elongation or strain as the load is applied. This is particularly important for metals and similar materials that exhibit high stiffness. As shown in Fig. 12, typical extensometers have fixed gage lengths such as 25 or 50 mm (1 or 2 in.). They are also classified by maximum percent elongation so that a typical 25 mm (1 in.) gage length unit would have different models for 10, 50, or 100% maximum strain. Extensometers are used to measure axial strain in specimens. There also are transverse strain-measuring devices that indicate the reduction in width or diameter as the specimen is tested.

The two basic types of clip-on extensometers are linear variable differential transformer (LVDT) devices and strain-gage devices. These two types are described along with a description of earlier dial-type extensometers.

Early extensometers were held to the specimen with center points matching the specimen gage-length punch marks, and elongation was indicated between the points by a dial indicator. Because of mechanical problems associated with these early devices, most dial extensometers use knife edges and leaf-spring pressure for specimen attachment. An extensometer using a dial indicator to measure elongation is shown in Fig. 13. The dial indicator usually is marked off in 0.0025 mm (0.0001 in.) increments and measures the total extension between the gage points. This value divided by the gage length gives strain in mm/mm, or in./in.

LVDT extensometers employ an LVDT with a core, which moves from specimen deformation and produces an electrical signal proportional to amount of core movement (Fig. 14). LVDT extensometers are small, lightweight, and easy to use. Knife edges provide an exact point of contact and are mechanically set to the exact gage length. Unless the test report specifies total elongation, center punch marks or scribed lines are not required to define the gage length. They are available with gage lengths ranging from 10 to 2500 mm (0.4 to 100 in.) and can be fitted with breakaway features (Fig. 15), sheet metal clamps, low-pressure clamping arrangements (film clamps, as shown in Fig. 16), and other devices. Thus, they can be used on small specimens—such as thread, yarn, and foil—and on large test specimens—such as reinforcing bars, heavy steel plate, and tubing up to 75 mm (3 in.) in diameter.

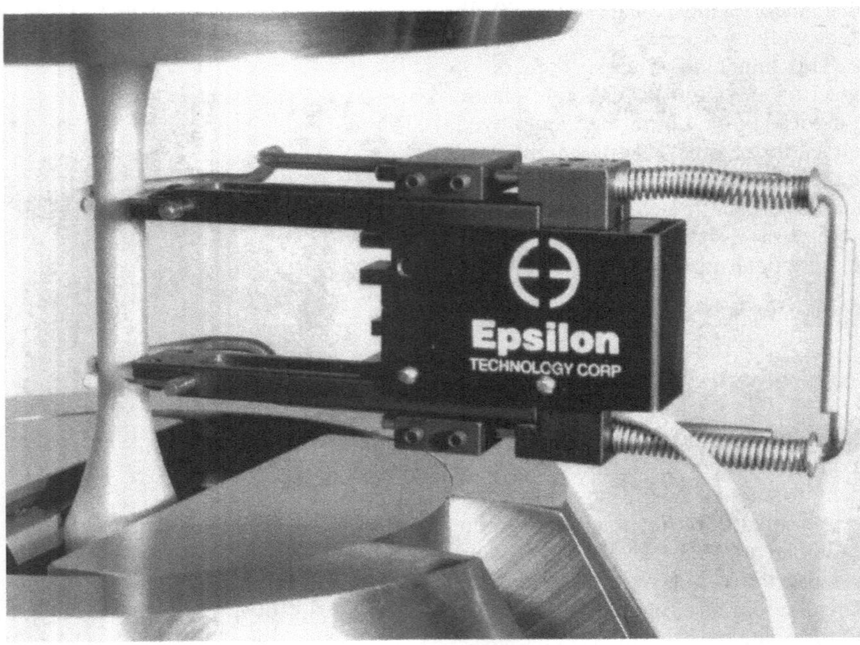

Fig. 10 Test specimen with an extensometer attached to measure specimen deformation. Courtesy of Epsilon Technology Corporation

Modifications of the LVDT extensometer also permit linear measurements at temperatures ranging from −75 to 1205 °C (−100 to 2200 °F). Accurate measurements can also be made in a vacuum. For standard instruments, the working temperature range is approximately −75 to 120 °C (−100 to 250 °F). However, by substituting an elevated-temperature transformer coil, the usable range of the instrument can be extended to −130 to 260 °C (−200 to 500 °F).

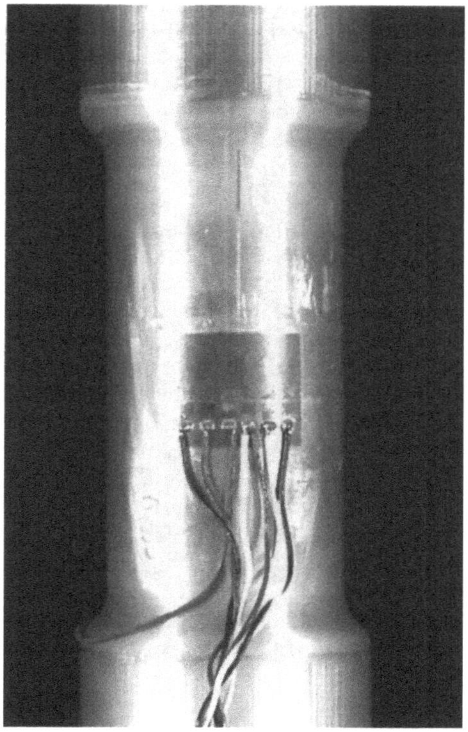

Fig. 11 Strain gages mounted directly to a specimen

Strain-gage extensometers, which use strain gages rather than LVDTs, are also common and are lighter in weight and smaller in size, but strain gages are somewhat more fragile than LVDTs. The strain gage usually is mounted on a pivoting beam, which is an integral part of the extensometer. The beam is deflected by the movement of the extensometer knife edge when the specimen is stressed. The strain gage attached to the beam is an electrically conductive small-sized grid that changes its resistance when deformed in tension, compression, bending, or torsion. Thus, strain gages can be used to supply the information necessary to calculate strain, stress, angular torsion, and pressure.

Strain gages have been improved and refined, and their use has become widespread. Basic types include wire gages, foil gages, and capacitive gages. Wire and foil bonded resistance strain gages are used for measuring stress and strain and for calibration of load cells, pressure transducers, and extensometers. These gages typically measure 9.5 to 13 mm (3/8 to 1/2 in.) in width and 13 to 19 mm (1/2 to 3/4 in.) in length

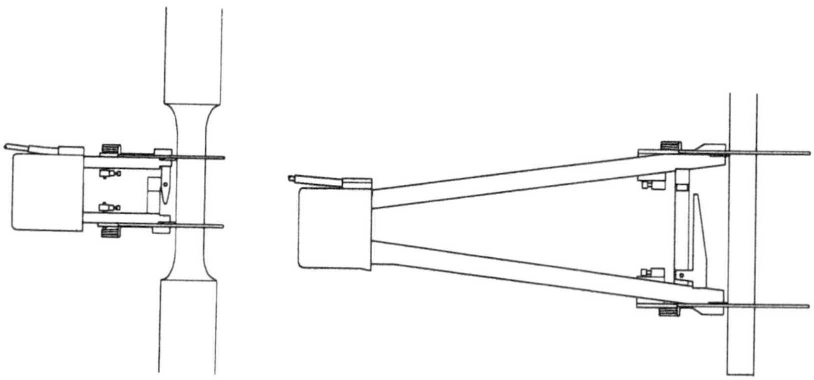

(a) (b)

Fig. 12 Typical clip-on extensometers. (a) Extensometer with 25 mm (1 in.) gage length and ±3.75 mm (±0.150 in.) travel suitable for static and dynamic applications with a variety of specimen geometries, dimensions, and materials. (b) Extensometer with 50 mm (2 in.) gage length and 25 mm (1 in.) travel suitable for large specimens and materials with long elongation patterns and are adhesively bonded to a metal element (Fig. 17).

Operation of strain-gage extensometers is based on gages that are bonded to a metallic element and connected to a bridge circuit. Deflection of the element, due to specimen strain, changes the gage's resistance that produces an output signal from a bridge circuit. This signal is amplified and processed by signal conditioners before being displayed on a digital readout, chart recorder, or computer. The circuitry in the strain-measuring system allows multiple ranges of sensitivity, so one transducer can be used over broad ranges. The magnification ratio, which is the ratio of output to extensometer deflection, can be as high as 10,000 to 1.

Strain Gages Mounted Directly to the Test Specimen. For some strain measurements, strain gages are mounted on the part being tested (Fig. 11). When used in this manner, they differ from extensometers in that they measure average unit elongation over nominal gage length rather than total elongation between definite gage points. For some testing applications, strain gages are used in conjunction with extensometers (Fig. 17).

In conventional use, wire or foil strain gages, when mounted on structures and parts for stress analysis, are discarded with the tested item. Thus, strain gages are seldom used in production testing of standard tension specimens. Foil strain gages currently are the most widely used, due to the ease of their attachment.

Averaging Extensometers. Typically extensometers are either nonaveraging or averaging types. A nonaveraging extensometer has one fixed nonmovable knife edge or center point and one movable knife edge or center point on the same side of the specimen. This arrangement results in extension measurements that are taken on one side of the specimen only; such measurements do not take into account that elongation may be slightly different on the other side.

For most specimens, notably those with machined rounds or reduced gage length flats, there is no significant difference in elongation between the two sides. However, for as-cast specimens, high-modulus materials, some forged parts, and specimens made from tubing, a dif-

Fig. 13 Dial-type extensometer, 50 mm (2 in.) gage length

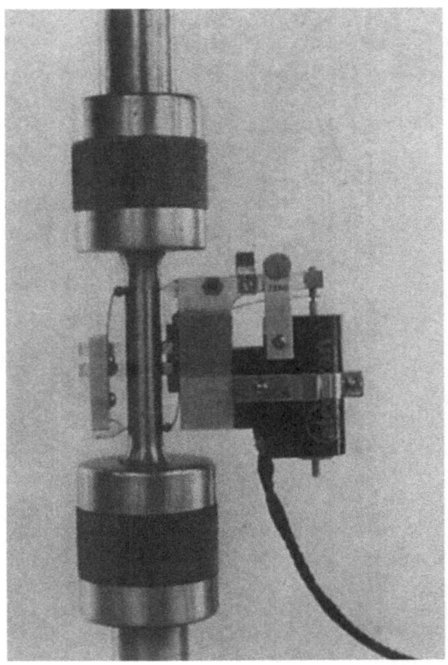

Fig. 14 Averaging LVDT extensometer (50 mm, or 2 in. gage length) mounted on a threaded tension specimen

ference in elongation sometimes exists on opposite sides of the specimen when subjected to a tensile load. This is due to part configuration and/or internal stress. Misalignment of grips also contributes to elongation measurement variations in the specimen. For these situations, averaging extensometers are used. Averaging extensometers use dual-measuring elements that measure elongation on both sides of a sample (Fig. 18); the measurements are then averaged to obtain a mean strain.

Optical Systems. Lasers and other systems can also be used to obtain linear strain measurements. Optical extensometers are particularly useful with materials such as rubber, thin films, plastics, and other materials where the weight of a conventional extensometer would distort the workpiece and affect the readings obtained. In the past, such strain-measuring systems were expensive, and their principal use has been primarily in research and development work. However, these optical techniques are becoming more accessible for commercial testing machines. For example, bench-top UTM systems with a laser extensometer are available (Fig. 19). This laser extensometer allows accurate measurement of strain in thin films, which would not otherwise be practical by mechanical attachment of extensometer devices. Optical systems also allow noncontact measurement from environmental test chambers.

Calibration, Classification, and Verification of Extensometers. All types of extensometers for materials testing must be verified, classified, and calibrated in accordance with applicable standards. Calibration of extensometers refers to the procedure of determining the magnitude of error in strain measurements. Verification is a calibration to ascertain whether the errors are within a predetermined range. Verification also implies certification that an extensometer meets stated accuracy requirements, which are defined by classifications such as those in ASTM E 83 (Table 3).

Several calibration devices can be used, including an interferometer, calibrated standard gage blocks and an indicator, and a micrometer

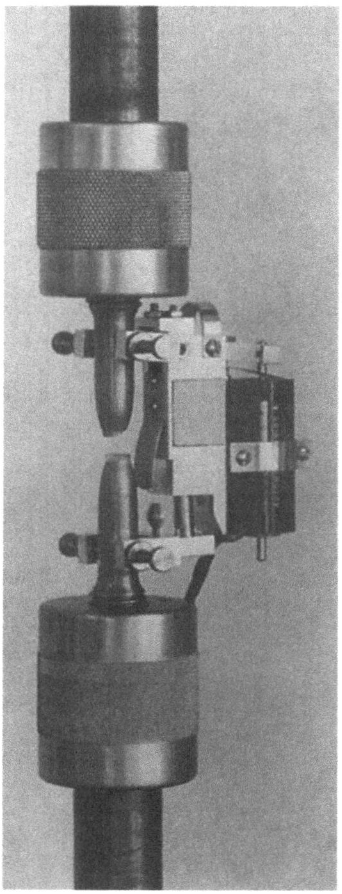

Fig. 15 Breakaway-type LVDT extensometer (50 mm, or 2 in. gage length) that can remain on the specimen through rupture

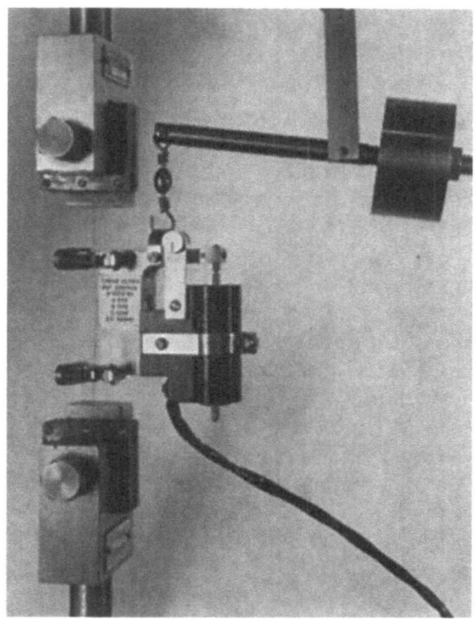

Fig. 16 Averaging LVDT extensometer (50 mm, or 2 in. gage length) mounted on a 0.127 mm (0.005 in.) wire specimen. The extensometer is fitted with a low-pressure clamping arrangement (film clamps) and is supported by a counterbalance device.

screw. Applicable standards for extensometer calibration or verification include:

Specification number	Specification title
DIN EN 10002-4	Part 4: Verification of Extensometers Used in Uniaxial Testing, Tensile
ISO 9513	Metallic Materials—Verification of Extensometers Used in Uniaxial Testing
BS EN 10002-4	Verification of Extensometers Used in Uniaxial Testing
ASTM E 83	Standard Practice for Verification and Classification of Extensometers
BS 3846	Methods for Calibration and Grading of Extensometers for Testing of Materials

Verification and classification of extensometers are applicable to instruments of both the averaging and nonaveraging type.

Procedures for the verification and classification of extensometers can be found in ASTM E 83. It establishes six classes of extensometers (Table 3), which are based on allowable error deviations, as discussed later in this article. This standard also establishes a verification procedure to ascertain compliance of an instrument to a particular classification. In addition, it stipulates that a certified calibration apparatus must be used for all applied displacements and that the accuracy of the apparatus must be five times more precise than allowable classification errors. Ten displacement readings are required for verification of a classification.

Class A extensometers, if available, would be used for determining precise values of the modulus of elasticity and for precise measurements of permanent set or very slight deviations from Hooke's law. Currently, however, there are no commercially available extensometers manufactured that are certified to comply with class A requirements.

Class B-1 extensometers are frequently used to determine values of the modulus of elasticity and to measure permanent set or deviations from Hooke's law. They are also used for determining values such as the yield strength of metallic materials.

Class B-2 extensometers are used for determining the yield strength of metallic materials.

All LVDT and strain-gage extensometers can comply with class B-1 or class B-2 requirements if their measuring ranges do not exceed 0.5 mm (0.02 in.). Instruments with measuring ranges of over 0.5 mm (0.02 in.) can be class C instruments.

Most electrical differential transformer extensometers of 500-strain magnification and higher can conform to class B-1 requirements throughout their measuring range. Extensometers of less

Fig. 17 Test specimen with bonded resistance strain gages and a 25 mm (1 in.) gage length extensometer mounted on the reduced section

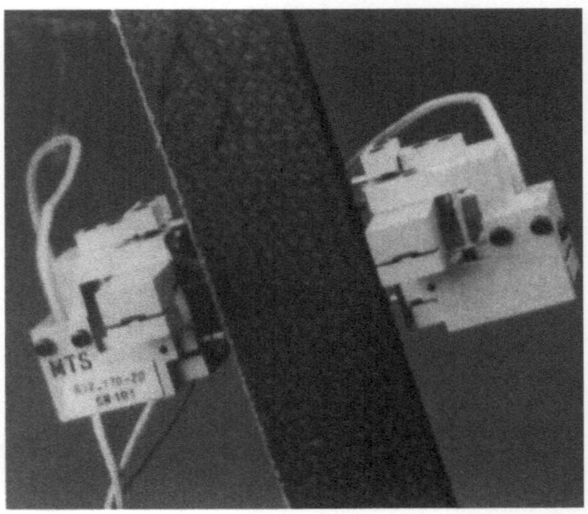

Fig. 18 Averaging extensometer with dual measuring elements mounted on a specimen. Source: Ref 3

than 500-strain magnification can comply only with class B-1 requirements in their lower (40%) measuring range and are basically class B-2 instruments.

Dial Extensometers. Although all dial instruments usually are considered class C instruments, the majority (up to a gage length of 200 mm, or 8 in.) are class B-1 and class B-2 in their initial 40% measuring range, and class C throughout the remainder of the range. Dial instruments are used universally for determining yield strength by the extension-under-load method and yield strength of 0.1% offset and greater.

Class C and D Extensometers. Extensometers with a gage length of 610 mm (24 in.) begin in class C, although their overall measuring range must be considered as class D.

Gripping Techniques

The use of proper grips and faces for testing materials in tension is critical in obtaining meaningful results. Trial and error often will solve a particular gripping problem. Tensile testing of most flat or round specimens can be

Fig. 19 Bench-top UTM with laser extensometer. Courtesy of Tinius Olsen Testing Machine Company, Inc.

Table 3 Classification of extensometer systems

	Error of strain not to exceed the greater of(a):		Error of gage length not to exceed the greater of:	
Classification	Fixed error, in./in.	Variable error, % of strain	Fixed error, in.	Variable error, % of gage length
Class A	0.00002	±0.1	±0.001	±0.1
Class B-1	0.0001	±0.5	±0.0025	±0.25
Class B-2	0.0002	±0.5	±0.005	±0.5
Class C	0.001	±1	±0.01	±1
Class D	0.01	±1	±0.01	±1
Class E	0.1	±1	±0.01	±1

(a) Strain of extensometer system—ratio of applied extension to the gage length. Source: ASTM E 83 accommodated with wedge-type grips (Fig. 20). Wire and other forms may require different grips, such as capstan or snubber types. The load capacities of grips range from under 4.5 kgf (10 lbf) to 45,000 kgf (100,000 lbf) or more. ASTM E 8 describes the various types of gripping devices used to transmit the measured load applied by the test machine to the tensile test specimen. Additional information on gripping devices can also be found in Chapter 3, "Uniaxial Tensile Testing."

Screw-action grips, or mechanical grips, are low in cost and are available with load capacities of up to 450 kgf (1000 lbf). This type of grip, which is normally used for testing flat specimens, can be equipped with interchangeable grip faces that have a variety of surfaces. Faces are adjustable to compensate for different specimen thicknesses.

Wedge-type grips (Fig. 20) are self-tightening and are built with capacities of up to 45,000 kgf (100,000 lbf) or more. Some units can be tightened without altering the vertical position of the faces, making it possible to preselect the exact point at which the specimen will be held. The wedge-action design works well on hard-to-hold specimens and prevents the introduction of large compressive forces that cause specimen buckling.

Pneumatic-action grips are available in various designs with capacities of up to 90 kgf (200 lbf). This type of grip clamps the specimen by lever arms that are actuated by compressed air cylinders built into the grip bodies. A constant force maintained on the specimen compensates for decrease of force due to creep of the specimen in the grip. Another advantage of this design is the ability to optimize gripping force by adjusting the air pressure, which makes it possible to minimize specimen breaks at the grip faces.

Buttonhead grips enable the rapid insertion of threaded-end or mechanical-end specimens. They can be manually or pneumatically operated, as required by the type of material or test conditions.

Alignment. Whether the specimen is threaded into the crossheads, held by grips, or is in direct contact with platens, the specimen must be well aligned with the load cell. Any misalignment will cause a deviation from uniaxial stress in the material studied.

Environmental Chambers

Elevated- and low-temperature tensile tests are conducted with basically the same specimens and procedures as those used for room-temperature tensile tests. However, the specimens must be heated or cooled in an appropriate environmental chamber (Fig. 21). Also, the test fixtures must be sufficiently strong and corrosion resistant, and the strain-measuring system must be usable at the test temperature.

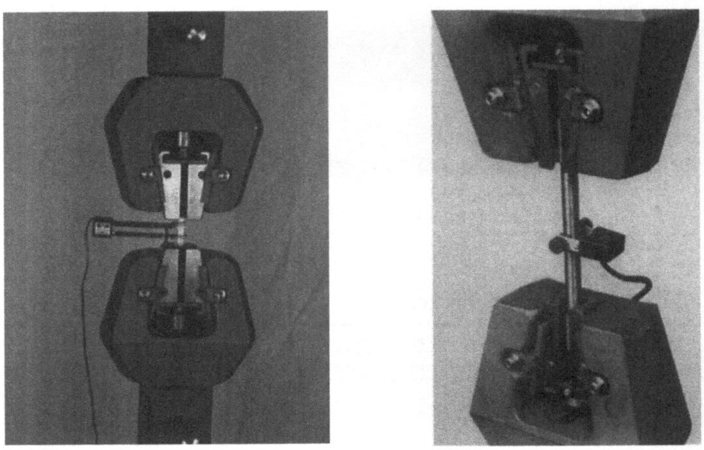

(a) (b)

Fig. 20 Test setup using wedge grips on (a) a flat specimen with axial extensometer and (b) a round specimen with diametral extensometer

Strain gages are generally adequate between cryogenic temperatures and about 600 °C (1100 °F), but at higher temperatures, other devices must be used. Rod and tube extensometers, which are manufactured from a variety of materials, are most commonly used. When testing is done below room temperature, Teflon is suitable. Nickel-base superalloys are adequate for testing in air at up to 1100 °C (2010 °F). Above 1100 °C, ceramics are used in reactive atmospheres, whereas refractory metals are adequate for inert environments.

Environmental chambers contain automated systems for temperature control and can also simulate vacuum and high-humidity environments. More detailed information on environmental chambers can be found in Chapter 13, "Hot Tensile Testing," and Chapter 14, "Tensile Testing at Low Temperatures."

Force Verification of Universal Testing Machines

The calibration and verification of UTM systems refer to two different methods that are not synonymous. Calibration of testing machines refers to the procedure of determining the magnitude of error in the indicated loads. Verification is a calibration to ascertain whether the errors are within a predetermined range. Verification also implies certification that a machine meets stated accuracy requirements. Valid verification requires device calibration by skilled personnel who are knowledgeable about testing machines, related instruments, and the proper use of device calibration standards (such as ASTM E 74 for load indicators and ASTM E 83 for extensometer devices). After verification is

Fig. 21 Tensile-testing apparatus with environmental chamber for testing at up to 540 °C (1000 °F). Source: Ref 3

performed, the calibrator or agency must issue reports and certificates attesting to compliance of the equipment with the verification requirements, including the loading range(s) for which the system may be used.

Force Verification. For the load verification to be valid, the weighing system(s) and associated instrumentation and data systems must be verified annually. In no case should the time interval between verifications exceed 18 months. Testing systems and their loading ranges should be verified immediately after relocation of equipment, after repairs or parts replacement (mechanical or electric/electronic) that could affect the accuracy of the load-measuring system(s), or whenever the accuracy of indicated loads is suspect, regardless of when the last verification was made.

Force verification standards for mechanical testing machines include specifications from various standards organizations such as:

Specification number	Specification title
EN 10002-2	Metallic Materials—Tensile Testing—Part 2: Verification of the Force Measurements
DIN EN 10002-2	Part 2: Verification of the Force-Measuring System of Tensile Testing Machines
BS 1610	Materials Testing Machines and Force Verification Equipment
BS EN 10002-2	Verification of the Force Measuring System of the Tensile Testing Machine
ASTM E 4	Standard Practices for Force Verification of Testing Machines

To comply with ASTM E 4, one or a combination of the three allowable verification methods must be used in the determination of the loading range or multiple loading ranges of the testing system. These methods are based on the use of:

- Standard weights
- Standard weights and lever balances
- Elastic calibration devices

For each loading range, at least five (preferably more) verification load levels must be selected. The difference between any two successive test loads must not be larger than one third of the difference between the maximum and minimum test loads. The maximum can be the full capacity of an individual range. For example, acceptable test load levels could be 10, 25, 50, 75, and 100%, or 10, 20, 40, 70, and 100%, of the stated machine range.

Regardless of the load verification method used at each of the test levels, the values indicated by the load-measuring system(s) of the testing machine must be accurate to within ±1% of the loads indicated by the calibration standard. If all five or more of the successive test load deviations are within the ±1% required in ASTM E 4, the loading ranges may be established and reported to include all of the values. If any deviations are larger than ±1%, the system should be corrected or repaired immediately. For determining accuracy of values at various test loads (or the deviation from the indicated load of the standard), ASTM E 74 specifies the required calibration accuracy tolerances of the three allowable types of verification methods.

For determining material properties, the testing machine loads should be as accurate as possible. In addition, deformations resulting from load applications should be measured as precisely as possible. This is particularly important because the relationship of load to deformation, which may be, for example, extension or compression, is the main factor in determining material properties.

As described previously, load accuracy may be ensured by following the ASTM E 4 procedure. In a similar manner, the methods contained in ASTM E 83, if followed precisely, will ensure that the devices or instruments used for deformation (strain) measurements will operate satisfactorily.

Manufacturers of testing machines calibrate before shipping and certify conformation to the manufacturer's guarantee of accuracy and any applicable standards, such as ASTM E 4. Subsequent calibrations can be made by the manufacturer or another organization with recognized equipment that is properly maintained and recertified periodically.

Example: Calibrating a 60,000 lbf Capacity Testing Machine. A 60,000 lbf capacity dial-type UTM of either hydraulic or screw-driven design will have the following typical scale ranges:

- 0 to 60,000 lbf reading by 50 lbf divisions
- 0 to 30,000 lbf reading by 25 lbf divisions
- 0 to 12,000 lbf reading by 10 lbf divisions
- 0 to 1200 lbf reading by 1 lbf divisions

As discussed previously, the ASTM required accuracy is ±1% of the indicated load above 10% of each scale range. Most manufacturers produce equipment to an accuracy of ±0.5% of the indicated load or ± one division, whichever is greater.

According to ASTM specifications, the 60,000 lbf scale range must be within 1% at 60,000 lbf (±600 lbf) and at 6000 lbf (±60 lbf). In both cases, the increment division is 50 lbf. Although the initial calibration by the manufacturer is to closer tolerance than ASTM E 4, subsequent recalibrations are usually to the ±1% requirement. In the low range, the machine must be accurate (±1%) from 120 to 1200 lbf. Thus, the machine must be verified from 120 to 60,000 lbf.

If proving rings are used in calibration, a 60,000 lbf capacity proving ring is usable down to a 6000 lbf load level. A 6000 lbf capacity proving ring is usable down to a 600 lbf load level, and a 1000 lbf capacity proving ring is usable down to a 100 lbf load level.

If calibrating load cells are used, a 60,000 lbf capacity load cell is usable down to a 3000 lbf load level, a 6000 lbf capacity load cell is usable to a 3000 lbf load level, and a 600 lbf capacity load cell is usable down to a 120 lbf load level.

Before use, proving rings and load cells must be removed from their cases and allowed to stabilize to ambient (surrounding) temperature. Upon stabilization, either type of unit is placed on the table of the testing machine. At this stage, proving rings are ready to operate, but load cells must be connected to an appropriate power source and again be allowed to stabilize, generally for 5 to 15 min.

Each system is set to zero, loaded to the full capacity of the machine or elastic device, then unloaded to zero for checking. Loading to full capacity and unloading must be repeated until a stable zero is obtained, after which the load verification readings are made at the selected test load levels.

For the highest load range of 60,000 lbf, loads are applied to the calibrating device from its minimum lower limit (6000 lbf for proving rings and 3000 lbf for load cells) to its maximum 60,000 lbf in a minimum of five steps, or test load levels, as discussed previously in the section "Force Verification of Universal Testing Machines" in this chapter. In the verification loading procedure for proving rings, a "set-the-load" method usually is used. The test load is determined, and the nominal load is preset on the proving ring. The machine load readout is read when the nominal load on the proving ring is achieved. For load cells, a "follow-the-load" method can be used, wherein the load on the display indicator is followed until the load reaches the nominal load, which is the preselected load level on the readout of the testing machine.

In both methods, the load of the testing machine and the load of the calibration device are recorded. The error, E, and the percent error, E_p, can be calculated as:

$$E = A - B$$

$$E_p = \frac{(A - B)}{B} \times 100 \qquad \text{(Eq 13)}$$

where A is the load indicated by the machine being verified in lbf, kgf, or N, and B is the correct value of the applied load (lbf, kgf, or N), as determined by the calibration device.

This procedure is repeated until each scale range of the testing machine has been calibrated from minimum to maximum capacity. The necessary reports and certificates are then prepared, with the loading range(s) indicated clearly as required by ASTM E 4. Figures 8(b) and 9 illustrate UTMs being calibrated with elastic proving rings and calibration load cells.

Tensile Testing Requirements and Standards

Tensile testing requirements are specified in various standards for a wide variety of different materials and products. Table 4 lists various tensile testing specifications from several standards organizations. These specifications define requirements for the test apparatus, test specimens, and test procedures.

Standard tensile tests are conducted using a threaded tensile specimen geometry, like the standard ASTM geometry (Fig. 22) of ASTM E 8. To load the specimen in tension, the threaded specimen is screwed into grips attached to each crosshead. The boundary condition, or load, is applied by moving the crossheads away from one another.

For a variety of reasons, it is not always possible to fabricate a specimen as shown in Fig. 22. For thin plate or sheet materials, a flat, or dog-bone, specimen geometry is used. The dog-bone specimen is held in place by wedge shaped grips. The holding capacity of the grips provides a practical limit to the strength of material that a machine can test. Other specimen geometries can be tested, with certain cautions, and formulas for critical dimensions are given in ASTM E 8 and in Chapter 3, "Uniaxial Tensile Testing."

Accuracy, Repeatability, and Precision of Tensile Tests. Accuracy and precision of test results can only be quantified when known quantities are measured. One difficulty of assessing data is that no agreed-upon "material standard" exists as reference material with known properties for strength and elongation. Tests of the "standard material" would reveal the system accuracy, and repeated experiments would quantify its precision and repeatability.

A variety of factors influence accuracy, precision, and repeatability of test results. Sources for errors in tensile testing are mentioned in the appendix of ASTM E 8. Errors can be grouped into three broad categories:

- *Instrumental errors:* These can involve machine stiffness, accuracy and resolution of the load cell output, alignment of the specimen, gripping of the specimen, and accuracy of the extensometer.
- *Testing errors:* These can involve initial measurement of specimen geometry, electronic zeroing, and establishing a preload stress level in the specimen.
- *Material factors:* These describe the relationship between the material intended to be studied and that being tested. For example, does the material in the specimen represent the parent material, and is it homogenous? Other material factors would include specimen preparation, specimen geometry, and material strain-rate sensitivity.

The ASTM committee for tensile testing reported on a round robin set of experiments to assess repeatability and to judge precision of standard quantities. In this series (see appendix of ASTM E 8) six specimens of six materials were tested at six different laboratories. The comparison of measurements within a laboratory and between laboratories is given

Table 4 Tensile testing standards for various materials and product forms

Specification number	Specification title
ASTM A 770	Standard Specification for Through-Thickness Tension Testing of Steel Plates for Special Applications
ASTM A 931	Standard Test Method for Tension Testing of Wire Ropes and Strand
ASTM B 557	Standard Test Methods of Tension Testing Wrought and Cast Aluminum- and Magnesium-Alloy Products
ASTM B 557M	Standard Test Methods of Tension Testing Wrought and Cast Aluminum- and Magnesium-Alloy Products [Metric]
ASTM C 565	Standard Test Methods for Tension Testing of Carbon and Graphite Mechanical Materials
ASTM C 1275	Standard Test Method for Monotonic Tensile Strength Testing of Continuous Fiber-Reinforced Advanced Ceramics with Solid Rectangular Cross-Section Specimens at Ambient Temperature
ASTM C 1359	Standard Test Method for Monotonic Tensile Strength Testing of Continuous Fiber-Reinforced Advanced Ceramics with Solid Rectangular Cross-Section Specimens at Elevated Temperatures
ASTM D 76	Standard Specification for Tensile Testing Machines for Textiles
ASTM E 8	Standard Test Methods for Tension Testing of Metallic Materials
ASTM E 8M	Standard Test Methods for Tension Testing of Metallic Materials [Metric]
ASTM E 338	Standard Test Method of Sharp-Notch Tension Testing of High-Strength Sheet Materials
ASTM E 345	Standard Test Methods of Tension Testing of Metallic Foil
ASTM E 602	Standard Method for Sharp-Notch Tension Testing with Cylindrical Specimens
ASTM E 740	Standard Practice for Fracture Testing with Surface-Crack Tension Specimens
ASTM E 1450	Standard Test Method for Tension Testing of Structural Alloys in Liquid Helium
ASTM F 1501	Standard Test Method for Tension Testing of Calcium Phosphate Coatings
ASTM F 152	Standard Test Methods for Tension Testing of Nonmetallic Gasket Materials
ASTM F 19	Standard Test Method for Tension and Vacuum Testing Metallized Ceramic Seals
ASTM F 1147	Standard Test Method for Tension Testing of Porous Metal Coatings
BS EN 10002	Tensile Testing of Metallic Materials
BS 18	Method for Tensile Testing of Metals (Including Aerospace Materials)
BS 4759	Method for Determination of K-Values of a Tensile Testing System
BS 3688-1	Tensile Testing
BS 3500-6	Tensile Stress Relaxation Testing
BS 3500-3	Tensile Creep Testing
BS 3500-1	Tensile Rupture Testing
BS 1687	Medium-Sensitivity Tensile Creep Testing
BS 1686	Long-Period, High-Sensitivity, Tensile Creep Testing
DIN 53455	Tensile Testing: Testing of Plastics
DIN 53328	Testing of Leather, Tensile Test
DIN 50149	Tensile Test, Testing of Malleable Cast Iron
EN 10002-1	Metallic Materials—Tensile Testing—Part 1: Method of Test at Ambient Temperature
ISO 204	Metallic Materials—Uninterrupted Uniaxial Creep Testing Intension—Method of Test
ISO 783	Metallic Materials—Tensile Testing at Elevated Temperature
ISO 6892	Metallic Materials—Tensile Testing at Ambient Temperature
JIS B 7721	Tensile Testing Machines
JIS K 7113	Testing Methods for Tensile Properties of Plastics (English Version)

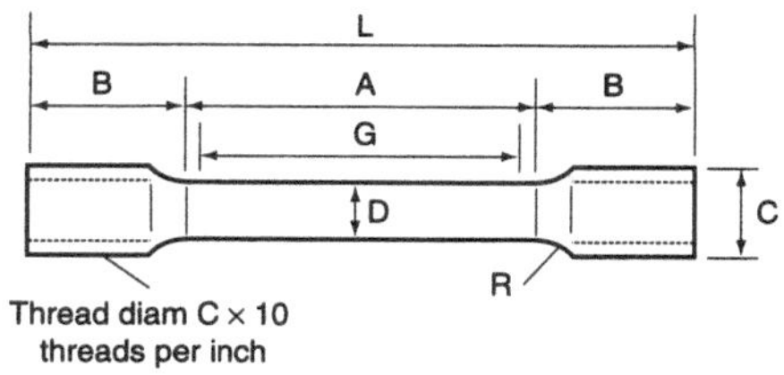

Abbreviation	Dimension	Measurement in.	Measurement mm
G	Gage length	2.4606	62.5 ±0.1
D	Diameter	0.4920	12.5 ±0.2
R	Radius of fillet	0.3937	10.0
A	Length of reduced section	2.953	75
L	Approximate overall length	5.7086	145.0
B	Length of end section	1.378	35.0
C	Diameter of end section	0.787	20

Fig. 22 Standard ASTM geometry for threaded tensile specimens. Dimensions for the specimen are taken from ASTM E 8M (metric units), or ASTM E 8 (English units).

Table 5 Results of round-robin tensile testing

Property	Coefficient of variation, % Within laboratory	Coefficient of variation, % Between laboratory
Tensile strength	0.91	1.30
0.02% yield strength	2.67	4.46
0.2% yield strength	1.35	2.32
Elongation in 5*D*	2.97	6.36
Reduction in area	2.80	4.59

Source: ASTM E 8

in Table 5. The data show the highest level of reproducibility in the strength measurements; the lowest reproducibility is found in elongation and reduction of area. Within-laboratory results were always more reproducible than those between laboratories.

ACKNOWLEDGMENTS

This chapter was adapted from:

- J.W. House and P.P. Gillis, Testing Machines and Strain Sensors, *Mechanical Testing and Evaluation,* Vol 8, *ASM Handbook,* ASM International, 2000, p 79–92
- M.A. Bishop, J.J. Martin, and K. Hendry, Chapter 2, Tensile-Testing Equipment, *Tensile Testing,* P. Han, Ed., ASM International, 1992, p 25–48
- P.P. Gillis and T.S. Gross, Effect of Strain Rate on Flow Properties, *Mechanical Testing,* Vol 8, *Metals Handbook,* 9th ed., American Society for Metals, 1985, p 38–46

REFERENCES

1. R.C. Anderson, *Inspection of Metals: Destructive Testing,* ASM International, 1988, p 83–119
2. H.E. Davis, G.E. Troxell, and G.F.W. Hauck, *The Testing of Engineering Materials,* 4th ed., McGraw-Hill, 1982, p 80–124
3. P. Han, Ed., *Tensile Testing,* ASM International, 1992, p 28
4. G.E. Dieter, *Mechanical Metallurgy,* McGraw-Hill, 2nd ed., 1976, p 349
5. D.J. Shippy, P.P. Gillis, and K.G. Hoge, Computer Simulation of a High Speed Tension Test, *J. Appl. Polym. Sci.,* Applied Polymer Symposia (No. 5), 1967, p 311–325
6. P.P. Gillis and D.J. Shippy, Vibration Analysis of a High Speed Tension Test, *J. Appl. Polym. Sci.,* Applied Polymer Symposia (No. 12), 1969, p 165–179
7. M.A. Hamstad and P.P. Gillis, Effective Strain Rates in Low-Speed Uniaxial Tension Tests, *Mater. Res. Stand.,* Vol 6 (No. 11), 1966, p 569–573
8. P. Gillis and J.J. Gilman, Dynamical Dislocation Theories of Crystal Plasticity, *J. Appl. Phys.,* Vol 36, 1965, p 3375–3386
9. J.E. Hockett and P.P. Gillis, Mechanical Testing Machine Stiffness, Parts I and II, *Int. J. Mech. Sci.,* Vol 13, 1971, p 251–275
10. W.G. Johnston, Yield Points and Delay Times in Single Crystals, *J. Appl. Phys.,* Vol 33, 1962, p 2716
11. H.G. Baron, Stress-Strain Curves of Some Metals and Alloys at Low Temperatures and High Rates of Strain, *J. Iron Steel Inst. (Brit.),* Vol 182, 1956, p 354
12. J. Miklowitz, The Initiation and Propagation of the Plastic Zone in a Tension Bar of Mild Steel as Influenced by the Speed of Stretching and Rigidity of the Testing Machine, *J. Appl. Mech. (Trans. ASME),* Vol 14, 1947, p A-31
13. M.A. Hamstead, "The Effect of Strain Rate and Specimen Dimensions on the Yield Point of Mild Steel," Lawrence Radiation Laboratory Report UCRL-14619, April 1966

CHAPTER 5

Tensile Testing for Design

DESIGN is the ultimate function of engineering in the development of products and processes, and an integral aspect of design is the use of mechanical properties derived from mechanical testing. The basic objective of product design is to specify the materials and geometric details of a part, component, and assembly so that a system meets its performance requirements. For example, minimum performance of a mechanical system involves transmission of the required loads without failure for the prescribed product lifetime under anticipated environmental (thermal, chemical, electromagnetic, radiation, etc.) conditions. Optimum performance requirements may also include additional criteria such as minimum weight, minimum life cycle cost, environmental responsibility, human factors, and product safety and reliability.

This chapter introduces the basic concepts of mechanical design and its general relation with the properties derived from tensile testing. Product design and the selection of materials are key applications of mechanical property data derived from testing. Although existing and feasible product shapes are of infinite variety and these shapes may be subjected to an endless array of complex load configurations, a few basic stress conditions describe the essential mechanical behavior features of each segment or component of the product. These stress conditions include the following:

- Axial tension or compression
- Bending, shear, and torsion
- Internal or external pressure
- Stress concentrations and localized contact loads

Mechanical testing under these basic stress conditions using the expected product load/time profile (static, impact, cyclic) and within the expected product environment (thermal, chemical, electromagnetic, radiation, etc.) provides the design data required for most applications.

In conducting mechanical tests, it is also very important to recognize that the material may contain flaws and that its microstructure (and properties) may be directional (as in composites) and heterogeneous or dependent on location (as in carburized steel). To provide accurate material characteristics for design, one must take care to ensure that the geometric relationships between the microstructure and the stresses in the test specimens are the same as those in the product to be designed.

It is also important to consider the complexity of materials selection for a combination of properties such as strength, toughness, weight, cost, and so on. This chapter briefly describes design criteria for some basic property combinations such as strength, weight, and costs. More detailed information on various performance indices in design, based on the methodology of Ashby, can be found in the article "Material Property Charts" in *Materials Selection and Design,* Volume 20 of *ASM Handbook.* The materials selection method developed by Ashby is also available as an interactive electronic product (Ref 1).

Product Design

Design involves the application of physical principles and experience-based knowledge to develop a predictive model of the product. The model may be a prototype, a simplified mathematical model, or a complex finite element model. Regardless of the level of sophistication of the model, reaching the product design objectives of material and geometry specifications for successful product performance requires accurate material parameters (Ref 2).

Modern design methods help manage the complex interactions between product geometry, material microstructure, loading, and environment. In particular, engineering mechanics (from simple equilibrium equations to complex finite element methods) extrapolates the results of basic mechanical testing of simple shapes under representative environments to predict the behavior of actual product geometries under real service environments.

In the following sections, a simple tie bar is used to illustrate the application of mechanical property data to material selection and design and to highlight the general implications for mechanical testing. Material subjected to the basic stress conditions is considered in order to establish design approaches and mechanical test methods, first in static loading and then in dynamic loading and aggressive environments. More detailed reference books on mechanical design and engineering methods are also listed in the "Selected References" at the end of this chapter.

Design for Strength in Tension

Figure 1 shows an axial tensile load applied to a tie bar representing, for example, a boom crane support, cable, or bolt. For this elementary case, the stress in the bar is uniformly distributed over the cross section of the tie bar and is given by:

$$\sigma = F/A \qquad \text{(Eq 1)}$$

where F is the applied force and A is the cross-sectional area of the bar. To avoid failure of the bar, this stress must be less than the failure stress, or strength, of the material:

$$\sigma = F/A < \sigma_f \qquad \text{(Eq 2)}$$

where σ_f is the stress at failure. The failure stress, σ_f, can be the yield strength, σ_o, if permanent deformation is the criterion for failure, or the ultimate tensile strength, σ_u, if fracture is the criterion for failure. In a ductile metal or polymer, the ultimate tensile strength is defined as the stress at which necking begins, leading to fracture. In a brittle material, the ultimate strength is simply the stress at fracture. Typical values of yield and ultimate tensile strength for various materials are summarized in Tables 1, 2, and 3. These typical values are intended only for general comparisons; design values should be based on statistically based minimum values or on minimum values published in the purchase specifications of materials (such as ASTM standards).

Equation 2 combines the performance of the part (load F) with the part geometry (cross-sectional area A) and the material characteristics (strength σ_f). The equation can be used several ways for design and material selection. If the material and its strength are specified, then, for a given load, the minimum cross-sectional area can be calculated; or, for a given cross-sectional area, the maximum load can be calculated. Conversely, if the force and area are specified, then materials with strengths satisfying Eq 2 can be selected.

Factor of Safety. Normally, designs involve the use of some type of a factor of safety. This factor, which is always greater than unity, is used in the design of components to ensure that the component can satisfactorily perform its intended purpose. The factor of safety is used to account for the uncertainties that exist in the real-world use of any component. Two main classifications of factors affect the factor of safety in a design, and they are these:

- Uncertainties associated with the material properties of the component itself, including the expected properties of the materials used to fabricate the component, as well as any uncertainties introduced by manufacturing and fabrication processing
- Uncertainties associated with the level and type of loading the component will see, as well as the actual service conditions and any environmental condition the component may experience

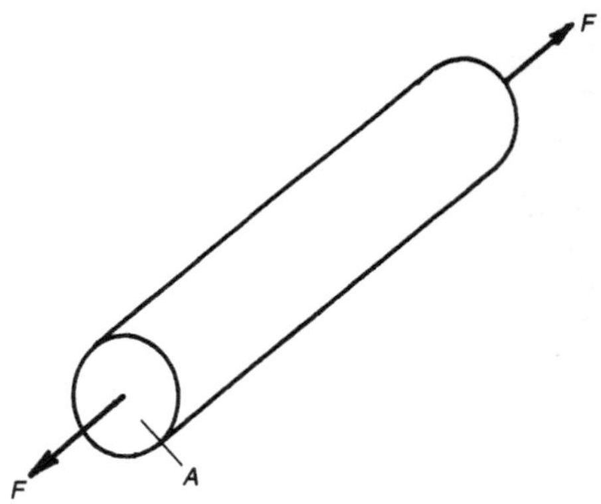

Fig. 1 Bar under axial tension

The factor of safety is used to establish a target stress level for the design. This is sometimes referred to as the allowable stress, the maximum allowable stress, or simply, the design stress. In order to determine this allowable stress condition, the failure stress is simply divided by the safety factor. Safety factors ranging from 1.5 to 10 are typical. The lower the uncertainty is, the lower the safety factor.

Design for Strength, Weight, and Cost

If minimum weight or minimum cost criteria must also be satisfied, Eq 2 can be modified by introducing other material parameters. To illustrate, the area A in Eq 2 is related to density and mass by $A = M/\rho L$, where M is the mass of the bar, L is the length of the bar, and ρ is the material density. Solving Eq 2 for F and substituting for A:

$$F < \sigma_f A = (\sigma_f/\rho)(M/L) \quad \text{(Eq 3)}$$

From Eq 3 it is clear that, to transmit a given load, F, the material mass will be minimized if the property ratio (σ_f/ρ) is maximized. The strength-to-weight ratio of a material is an important design and performance index; Fig. 2 is a plot developed by Ashby for comparison of materials by this design criterion. Similarly, material selection for minimum material cost can be obtained by maximizing the parameter $(\sigma_f/\rho c)$, or strength-to-cost ratio, where c represents the material cost per unit weight. These types of performance indexes for design and the use of materials property charts like Fig. 2 are described in more detail in Ref 7 and in the articles "Material Property Charts" and "Performance

Table 1 Typical room-temperature tensile properties of ferrous alloys and superalloys

Material	Strength in tension, MPa (ksi): 0.2% offset yield strength	Strength in tension, MPa (ksi): Ultimate	Modulus of elasticity, GPa (10^6 psi): Tension	Modulus of elasticity, GPa (10^6 psi): Shear	Elongation in 50 mm (2 in.), %
Cast irons					
Gray cast iron	...	140 (20)	105 (15)	40 (6)	1
White cast iron	...	415 (60)	140 (20)	55 (8)	...
Nickel cast iron, 1.5% nickel	...	310 (45)	140 (20)	55 (8)	1
Malleable iron	230 (33)	345 (50)	170 (25)	70 (10)	14
Ingot iron, annealed 0.02% carbon	165 (24)	290 (42)	205 (30)	85 (12)	45
Steels					
Wrought iron, 0.10% carbon	205 (30)	345 (50)	185 (27)	70 (10)	30
Steel, 0.20% carbon					
Hot-rolled	275 (40)	415 (60)	200 (29)	85 (12)	35
Cold-rolled	415 (60)	550 (80)	200 (29)	85 (12)	15
Annealed castings	240 (35)	415 (60)	200 (29)	85 (12)	25
Steel, 0.40% carbon					
Hot-rolled	290 (42)	485 (70)	200 (29)	85 (12)	25
Heat-treated for fine grain	415 (60)	620 (90)	200 (29)	85 (12)	25
Annealed castings	240 (35)	450 (65)	200 (29)	85 (12)	15
Steel, 0.60% carbon					
Hot-rolled	435 (63)	690 (100)	200 (29)	85 (12)	15
Heat-treated for fine grain	540 (78)	825 (120)	200 (29)	85 (12)	15
Steel, 0.80% carbon					
Hot-rolled	505 (73)	825 (120)	200 (29)	85 (12)	10
Oil-quenched, not drawn	860 (125)	1240 (180)	200 (29)	85 (12)	2
Steel, 1.00% carbon					
Hot-rolled	570 (83)	930 (135)	200 (29)	85 (12)	10
Oil-quenched, not drawn	965 (140)	1515 (220)	200 (29)	85 (12)	1
Nickel steel, 3.5% nickel, 0.40% carbon, max. hardness for machinability	1035 (150)	1170 (170)	200 (29)	85 (12)	12
Silicomanganese steel, 1.95% Si, 0.70% Mn, spring tempered	895 (130)	1200 (174)	200 (29)	85 (12)	1
Superalloys (wrought)					
A286 (bar)	760 (110)	1080 (157)	180 (26)	...	28
Inconel 600 (bar)	250 (36)	620 (90)	...	...	47
IN-100 (60 Ni-10Cr-15Co, 3Mo, 5.5Al, 4.7Ti)	850 (123)	1010 (147)	215 (31)	...	9
IN-738	915 (133)	1100 (159)	200 (29)	...	5

Source: Ref 3–5

Table 2 Typical room-temperature tensile properties of nonferrous alloys

Metal or alloy	Approximate composition, %	Condition	0.2% offset tensile yield strength, MPa (ksi)	Tensile strength, MPa (ksi)	Tensile modulus of elasticity, GPa (10^6 psi)	Elongation in 50 mm (2 in.), %
Heavy nonferrous alloys (~8–9 g/cm³)						
Copper	Cu	Annealed	33 (4.8)	209 (30)	125 (18)	60
		Cold drawn	333 (48)	344 (50)	112 (16)	14
Free-cutting brass	61.5 Cu, 35.5 Zn, 3 Pb	Annealed	125 (18)	340 (49)	85 (12)	53
		Quarter hard, 15% reduction	310 (45)	385 (56)	85 (12)	20
		Half hard, 25% reduction	360 (52)	470 (68)	95 (14)	18
High-leaded brass (1 mm thick)	65 Cu, 33 Zn, 2 Pb	Annealed, 0.050 mm grain	105 (15)	325 (47)	85 (12)	55
		Extra hard	425 (62)	585 (85)	105 (15)	5
Red brass (1 mm thick)	85 Cu, 15 Zn	Annealed, 0.070 mm grain	70 (10)	270 (39)	85 (12)	48
		Extra hard	420 (61)	540 (78)	105 (15)	4
Aluminum bronze	89 Cu, 8 Al, 3 Fe	Sand cast	195 (28)	515 (75)	...	40
		Extruded	260 (38)	565 (82)	125 (18)	25
Beryllium copper	97.9 Cu, 1.9 Be, 0.2 Ni	A (solution annealed)	...	500 (73)	125 (18)	35
		HT (hardened)	1035 (150)	1380 (200)	125 (18)	2
Manganese bronze (A)	58.5 Cu, 39 Zn, 1.4 Fe, 1 Sn, 0.1 Mn	Soft annealed	205 (30)	450 (65)	90 (13)	35
		Hard, 15% reduction	415 (60)	565 (82)	105 (15)	25
Phosphor bronze, 5% (A)	95 Cu, 5 Sn	Annealed, 0.035 mm grain	150 (22)	340 (49)	90 (13)	57
		Extra hard, 0.015 mm grain	635 (92)	650 (94)	115 (17)	5
Cupronickel, 30%	70 Cu, 30 Ni	Annealed at 760 °C	140 (20)	380 (55)	150 (22)	45
		Cold drawn, 50% reduction	540 (78)	585 (85)	150 (22)	15
Light nonferrous alloys (~2.7 g/cm³ for Al alloys; ~1.8 g/cm³ for Mg alloys)						
Aluminum	Al	Sand cast, 1100-F	40 (5.8 or 6)	75 (11)	60 (9)	22
		Annealed sheet, 1100-O	35 (5.075)	90 (13)	70 (10)	35
		Hard sheet, 1100-H18	145 (21)	165 (24)	70 (10)	5
Aluminum alloy 2024	93 Al, 4.5 Cu, 1.5 Mg, 0.6 Mn	Temper O	75 (11)	185 (27)	73 (11)	20
		Temper T36	395 (57)	495 (72)	73 (11)	13
Aluminum alloy 2014	93 Al, 4.4 Cu, 0.8 Si, 0.8 Mn, 0.4 Mg	Temper O	95 (14)	185 (27)	73 (11)	18
		Temper T6	415 (60)	485 (70)	73 (11)	13
Aluminum alloy 5052	97 Al, 2.5 Mg, 0.25 Cr	Temper O	90 (13)	195 (28)	69 (10)	30
		Temper H38	255 (37)	290 (42)	69 (10)	8
Aluminum alloy 5456	94 Al, 5.0 Mg, 0.7 Mn, 0.15 Cu, 0.15 Cr	Temper O	160 (23)	310 (45)	...	24
		Temper H321	255 (37)	350 (51)	...	16
Aluminum alloy 7075	90 Al, 5.5 Zn, 1.5 Cu, 2.5 Mg, 0.3 Cr	Temper O	105 (15)	230 (33)	...	17
		Temper T6	505 (73)	570 (83)	...	11
Magnesium	Mg	Cast	21 (3)	90 (13)	40 (6)	2–6
		Extruded	69–105 (10–15)	195 (28)	40 (6)	5–8
		Rolled	115–140 (17–20)	200 (29)	40 (6)	2–10
Magnesium alloy AM100A	90 Mg, 10 Al, 0.1 Mn	Cast, condition F	85 (12)	150 (22)	45 (7)	2
		Cast, condition T61	150 (22)	275 (40)	45 (7)	1
Magnesium alloy AZ63A	91 Mg, 6 Al, 3 Zn, 0.2 Mn	Cast, condition F	95 (14)	200 (29)	45 (7)	6
		Cast, condition T6	130 (19)	275 (40)	45 (7)	5

(continued)

Table 2 (continued)

Metal or alloy	Approximate composition, %	Condition	0.2% offset tensile yield strength, MPa (ksi)	Tensile strength, MPa (ksi)	Tensile modulus of elasticity, GPa (10^6 psi)	Elongation in 50 mm (2 in.), %
Titanium alloys (~4.5 g/cm³)						
Commercial ASTM grade 2 Ti	98 Ti	. . .	275 (40)	345 (50)	103 (15)	20
Ti-5Al-2.5Sn	92 Ti, 5 Al, 2.5 Sn	. . .	825 (120)	860 (125)	110 (16)	8–10
Ti-3Al-2.5V	94 Ti, 3 Al, 2.5 V	Annealed	560 (81)	655 (95)	103 (15)	29
		Cold worked and stress relieved	760 (110)	895 (130)	103 (15)	19
Ti-6Al-4V	90 Ti, 6 Al, 4 V	Solution treated and aged bar (1–2 in.)	965 (140)	1035 (150)	110 (16)	8
		Annealed bar	825 (120)	895 (130)	110 (16)	10
		Mill annealed	. . .	925 (134)	. . .	. . .

Indices" in *Materials Selection and Design,* Volume 20 of *ASM Handbook.*

Design for Stiffness in Tension

In addition to designing for strength, another important design criterion is often the stiffness or rigidity of a material. The elastic deflection of a component under load is governed by the stiffness of the material. For example, if a bridge or building is designed to avoid failure, it may still undergo motion under applied loads if it is not sufficiently rigid. As another example, if the tie bar in Fig. 1 were a bolt clamping a cap to a pressure vessel, excessive elastic change in length of the bolt under load might allow leakage through a gasket between the cap and vessel.

Elastic change in length occurs when an axial load is applied to the bar and is given by:

$$\Delta L = \varepsilon L \quad \text{(Eq 4)}$$

where ΔL is the change in length and ε is the strain in the bar. In the elastic range of deformation, axial stress is proportional to the strain:

$$\sigma = E\varepsilon \quad \text{(Eq 5)}$$

where the proportionality factor is E, the elastic modulus of the bar material.

The elastic modulus can be considered a physical property, because it is fundamentally related to the bond strength between the atoms or molecules in the material; that is, the stronger the bond, the higher the elastic modulus. Thus,

Table 3 Typical room-temperature tensile properties of plastics

Material	Tensile strength, MPa (ksi)	Elongation, %	Modulus of elasticity, GPa (10^6 psi)
Thermosets			
EP, reinforced with glass cloth	350 (51)	. . .	175 (25)
MF, alpha-cellulose filler	50–90 (7–13)	0.6–0.9	9 (1)
PF, no filler	50–55 (7–9)	1.0–1.5	5–7 (0.7–1)
PF, wood flour filler	45–60 (7–9)	0.4–0.8	6–8 (0.87–1.16)
PF, macerated fabric filler	25–65 (4–9)	0.4–0.6	6–9 (0.87–1)
PF, cast, no filler	40–65 (6–9)	1.5–2.0	3 (0.43)
Polyester, glass-fiber filler	35–65 (5–9)	. . .	11–14 (1.6–2.0)
UF, alpha-cellulose filler	55–90 (8–13)	0.5–1.0	10 (1.5)
Thermoplastics			
ABS	35–45 (5–7)	15–60	1.7–2.2 (0.25–0.32)
CA	15–60 (2–9)	6–50	0.6–3.0 (0.1–0.4)
CN	50–55 (7–9)	40–45	1.3–15.0 (0.18–2)
PA	80 (12)	90	3.0 (0.43)
PMMA	50–70 (7–10)	2–10	. . .
PS	35–60 (5–9)	1–4	3.0–4.0 (0.4–0.6)
PVC, rigid	40–60 (6–9)	5	2.4–2.7 (0.3–0.4)
PVCAc, rigid	50–60 (7–9)	. . .	2.0–3.0 (0.3–0.4)

ABS, acrylonitrile-butadiene-styrene; CA, cellulose acetate; CN, cellulose nitrate; EP, epoxy; MF, melamine formaldehyde; PA, polyamide (nylon); PF, phenol formaldehyde; PMMA, polymethyl methacrylate; PS, polystyrene; PVC, polyvinyl chloride; PVCAc, polyvinyl chloride acetate; UF, urea formaldehyde. Source: Ref 6 the elastic modulus does not vary much in material with a given type of crystal structure or microstructure. For example, the elastic modulus of most steels is typically about 200 GPa (29 $\times$ 10^6 psi) for steels of various composition and strength levels (Fig. 3). However, the modulus can vary with direction if the material has an anisotropic structure. For example, Fig. 4 is a plot of the tensile and compressive modulus for type 301 austenitic stainless steel. Transverse and longitudinal values vary, as do values for tensile and compressive loads. At low stresses, the tension and compressive moduli are, by theory and experiment, identical. At higher stresses, however, differences in the compressive and tensile moduli can be observed due to the effects of deformation (e.g., elongation in tension). Typical values of elastic moduli are given in Table 4 for various alloys and metals.

Equations 1 and 5 can be combined with Eq 4 to give the design equation:

$$\Delta L = FL/AE < \delta \qquad \text{(Eq 6)}$$

where δ is the design limit on change in length of the bar. Just as the strength, or load-carrying capacity, of the tie bar is related to geometry and material strength (Eq 2), the stiffness of the bar is related to geometry and the elastic modulus of the material. Again, part performance (force,

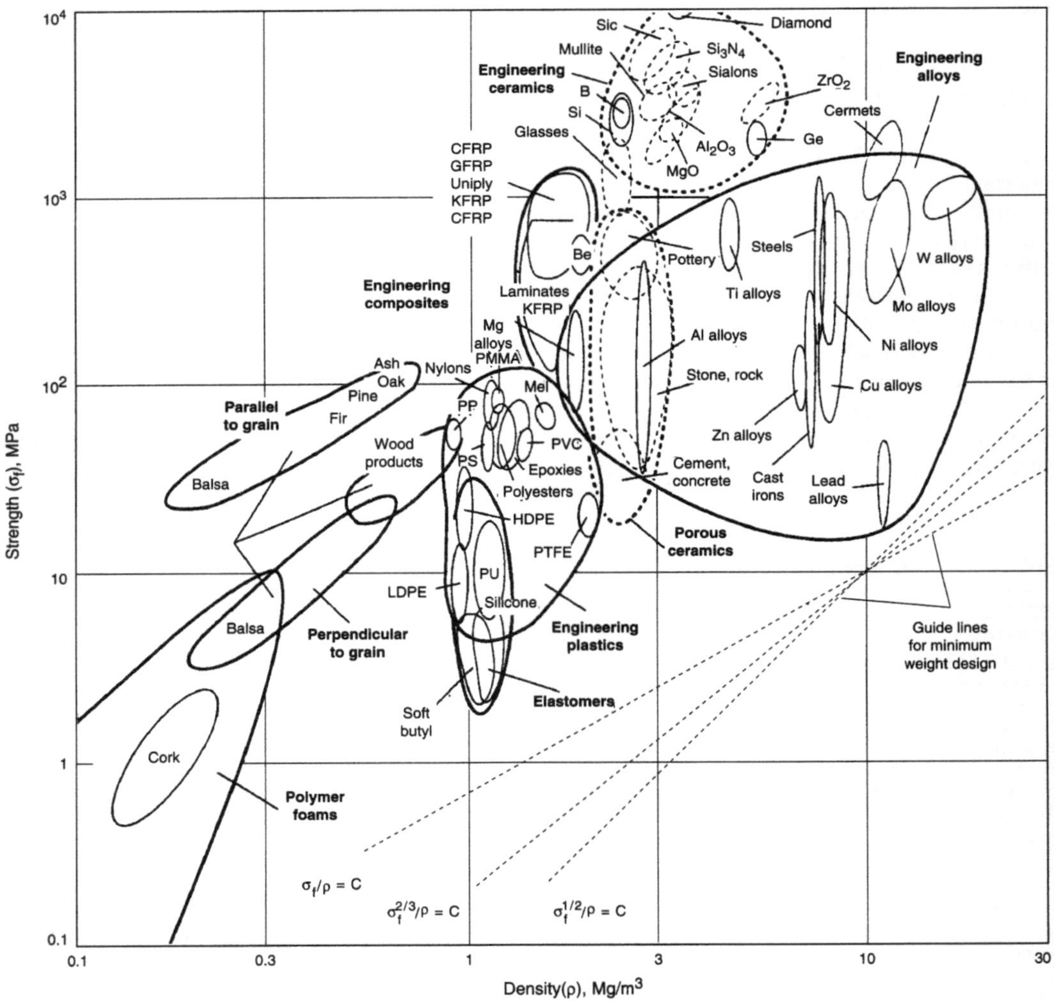

Fig. 2 Strength, σ_f, plotted against density, ρ, for various engineered materials. Strength is yield strength for metals and polymers, compressive strength for ceramics, tear strength for elastomers, and tensile strength for composites. Superimposing a line of constant σ_f/ρ enables identification of the optimum class of materials for strength at minimum weight.

F, and deflection, δ) is combined with part geometry (length, *L*, and cross-sectional area, *A*) and material characteristics (elastic modulus, *E*) in this design equation. To assure that the change in length is less than the allowable limit for a given force and material, the geometry parameters *L* and *A* can be calculated; or, for given dimensions, the maximum load can be calculated. Alternatively, for a given force and geometric parameters, materials can be selected whose elastic modulus, *E*, meets the design criterion given in Eq 6.

Similar to design for strength, additional criteria involving minimum weight or cost can be incorporated into design for stiffness. These criteria lead to the material selection parameters modulus-to-weight ratio (E/ρ) and modulus-to-cost ratio ($E/\rho c$), values that can be found in Ref 7 and *ASM Handbook,* Volume 20.

Mechanical Testing for Stress at Failure and Elastic Modulus

In Eq 2 and 6, the material properties σ_f and *E* play critical roles in design of the tie bar. These properties are determined from a simple tensile test described in detail in Chapter 3, "Uniaxial Tensile Testing." The elastic modulus *E* is determined from the slope of the elastic part of the tensile stress strain curve, and the failure stress, σ_f, is determined from the tensile yield strength, σ_o, or the ultimate tensile strength, σ_u.

Tensile-test specimens are cut from representative samples, as described in more detail in Chapter 3. In the example of the tie bar, test pieces would be cut from bar stock that has been processed similarly to the tie bar to be used in the product. In addition, the test piece should be

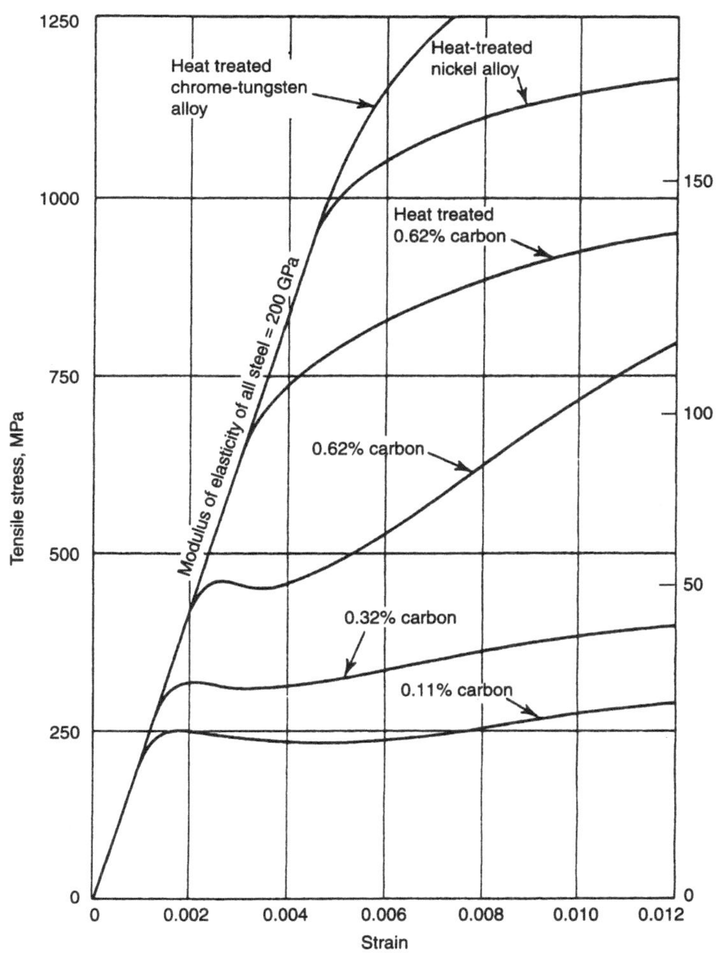

Fig. 3 Stress-strain diagram for various steels. Source: Ref 8

machined such that its gage length is parallel to the axis of the bar. This ensures that any anisotropy of the microstructural features will affect performance of the tie bar in the same way that they influence the measurements in the tensile test. For example, test pieces cut longitudinally and transverse to the rolling direction of hot rolled steel plates will exhibit the same elastic modulus and yield strength, but the tensile strength and ductility will be lower in the transverse direction because the stresses will be perpendicular to the alignment of inclusions caused by hot rolling (Ref 10).

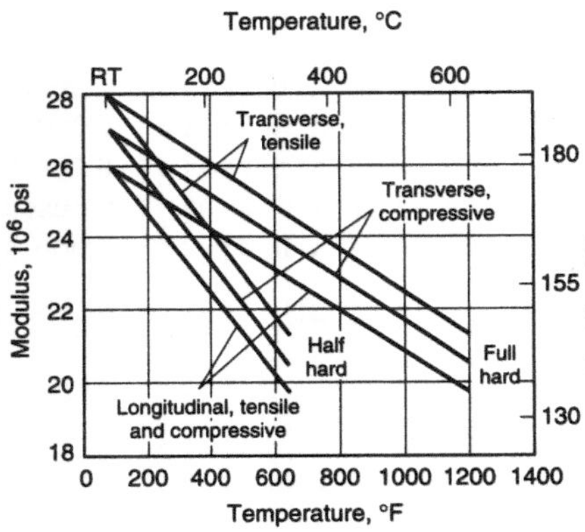

Fig. 4 Tensile and compressive modulus at half-hard and full-hard type 301 stainless steel in the transverse and longitudinal directions. Source: Ref 5

During tension testing of a material to measure E and σ_f, in addition to the change in length due to the applied axial tensile loads, the material will undergo a decrease in diameter. This reflects another elastic property of materials, the Poisson ratio, given by:

$$\nu = -\varepsilon_t/\varepsilon_l \quad \text{(Eq 7)}$$

where ε_t is the transverse strain and ε_l is the longitudinal strain measured during the elastic part of the tension test. Typical values of ν range from 0.25 to 0.40 for most structural materials, but ν approaches zero for structural foams and approaches 0.5 for materials undergoing plastic deformation. While the Poisson effect is of no consequence in the overall behavior of the tie bar (since the decrease in diameter has a negligible effect on the stress in the bar), the Poisson ratio is a very important material parameter in parts subjected to multiple stresses. The stress in one direction affects the stress in another direction via ν. Therefore, accurate measurements of the Poisson ratio are essential for reliable design analyses of the complex stresses in actual part geometries, as described later. Typical values of Poisson's ratio are given in Table 4.

Table 4 Elastic constants for polycrystalline metals at 20 °C (68 °F)

	Elastic modulus (E)		Bulk modulus (K)		Shear modulus (G)		
Metal	GPa	10^6 psi	GPa	10^6 psi	GPa	10^6 psi	Poisson's ratio, ν
Aluminum	70	10.2	75	10.9	26	3.80	0.345
Brass, 30 Zn	101	14.6	112	16.2	37	5.41	0.350
Chromium	279	40.5	160	23.2	115	16.7	0.210
Copper	130	18.8	138	20.0	48	7.01	0.343
Iron, soft	211	30.7	170	24.6	81	11.8	0.293
Iron, cast	152	22.1	110	15.9	60	8.7	0.27
Lead	16	2.34	46	6.64	6	0.811	0.44
Magnesium	45	6.48	36	5.16	17	2.51	0.291
Molybdenum	324	47.1	261	37.9	125	18.2	0.293
Nickel, soft	199	28.9	177	25.7	76	11.0	0.312
Nickel, hard	219	31.8	188	27.2	84	12.2	0.306
Nickel-silver, 55Cu-18Ni-27Zn	132	19.2	132	19.1	34	4.97	0.333
Niobium	104	15.2	170	24.7	38	5.44	0.397
Silver	83	12.0	103	15.0	30	4.39	0.367
Steel, mild	211	30.7	169	24.5	82	11.9	0.291
Steel, 0.75 C	210	30.5	169	24.5	81	11.8	0.293
Steel, 0.75 C, hardened	201	29.2	165	23.9	78	11.3	0.296
Steel, tool steel	211	30.7	165	24.0	82	11.9	0.287
Steel, tool steel, hardened	203	29.5	165	24.0	79	11.4	0.295
Steel, stainless, 2Ni-18Cr	215	31.2	166	24.1	84	12.2	0.283
Tantalum	185	26.9	197	28.5	69	10.0	0.342
Tin	50	7.24	58	8.44	18	2.67	0.357
Titanium	120	17.4	108	15.7	46	6.61	0.361
Tungsten	411	59.6	311	45.1	161	23.3	0.280
Vanadium	128	18.5	158	22.9	46.7	6.77	0.365
Zinc	105	15.2	70	10.1	42	6.08	0.249

Source: Ref 9

Sonic methods also offer an alternative and more accurate measurement of elastic properties, because the velocity of an extensional sound wave (i.e., longitudinal wave speed, V_L) is directly related to the square root of the ratio of elastic modulus and density as follows:

$$V_L = (E/\rho)^{1/2} \qquad \text{(Eq 8)}$$

By striking a sample of material on one end and measuring the time for the pulse to travel to the other end, the velocity can be calculated. Combining this with independent measurement of the density, Eq 8 can be used to calculate the elastic modulus (Ref 8).

Hardness-Strength Correlation

Correlation of hardness and strength has been examined for several materials as summarized in Ref 11. In hardness testing, a simple flat, spherical, or diamond-shaped indenter is forced under load into the surface of the material to be tested, causing plastic flow of material beneath the indenter as illustrated in Fig. 5. It would be expected, then, that the resistance to indentation or hardness is proportional to the yield strength of the material. Plasticity analysis (Ref 12) and empirical evidence (summarized in Ref 11) show that the pressure on the indenter is approximately three times the tensile yield strength of the material. However, correlation of hardness and yield strength is only straightforward when the strain-hardening coefficient varies directly with hardness. For carbon steels, the following relation has been developed to relate yield strength (YS) to Vickers hardness (HV) data (Ref 11):

$$\text{YS (in kgf/mm}^2) = \tfrac{1}{3}\ \text{HV}\ (0.1)^{m-2}$$

where m is Meyer's strain-hardening coefficient (Ref 13). To convert kgf/mm^2 values to units of lbf/in.^{2}, multiply the former by 1422. This relation applies only to carbon steels. Correlation of yield strength and hardness depends on the strengthening mechanism of the material. With aluminum alloys, for example, aged alloys exhibit higher strain-hardening coefficients and lower yield strengths than cold worked alloys (Ref 11).

For many metals and alloys, there has been found to be a reasonably accurate correlation between hardness and tensile strength, σ_u (Ref 11). Several studies are cited and described in Ref 11 and 14, and Tables 5 and 6 summarize hardness-tensile strength multiplying factors for various materials. It must be emphasized, however, that these are empirically based relationships, and so testing may still be warranted to confirm a correlation of tensile strength and hardness for a particular material (and/or material condition).

Table 5 Hardness-tensile strength conversions for steel

Material	Multiplying factor(a)
Heat-treated alloy steel (250–400 HB)	470 HB
Heat-treated carbon and alloy steel (<250 HB)	482 HB
Medium carbon steel (as-rolled, normalized, or annealed)	493 HB

(a) Tensile strength (in psi) = multiplying factor × HB. Source: Ref 11

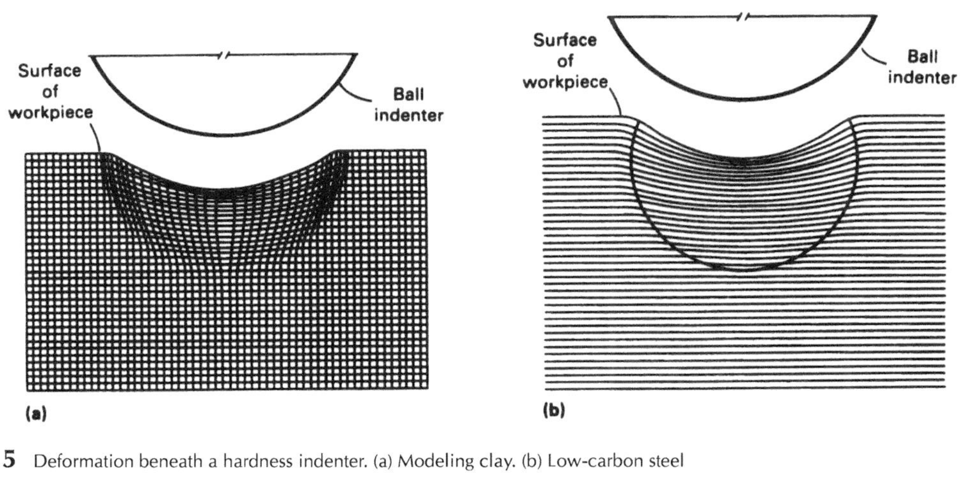

Fig. 5 Deformation beneath a hardness indenter. (a) Modeling clay. (b) Low-carbon steel

Table 6 Multiplying factors for obtaining tensile strength from hardness

Material	Multiplying factor range(a)
Heat treated carbon and alloy steel	470–515 HB
Annealed carbon steel	515–560 HB
All steels	448–515 HV
Ni-Cr austenitic steels	448–482 HV
Steel; sheet, strip, and tube	414–538 HV
Aluminum alloys; bar and extrusions	426–650 HB
Aluminum alloys; bar and extrusions	414–605 HV
Aluminum alloys; sheet, strip, and tube	470–582 HV
Al-Cu castings	246–426 HB
Al-Si-Ni castings	336–426 HB
Al-Si castings	381–538 HB
Phosphor bronze castings	336–470 HB
Brass castings	470–672 HB

(a) Tensile strength (in psi) = multiplying factor × hardness. Source: Ref 11, 15

A correlation with hardness may not be evident. For example, magnesium alloy castings did not exhibit a hardness-strength correlation in a study by Taylor (Ref 15).

More detailed information on hardness tests and the estimation of mechanical properties can be found in Ref 11, 13, and 14. The various types of hardness tests and their selection and application are described in *Mechanical Testing and Evaluation,* Volume 8, of the *ASM Handbook* series.

ACKNOWLEDGMENT

This chapter was adapted from H.A. Kuhn, Overview of Mechanical Properties and Testing for Design, *Mechanical Testing and Evaluation,* Vol 8, *ASM Handbook,* ASM International, 2000, p 49–69.

REFERENCES

1. Cambridge Engineering Selector, Granta Design Ltd., Cambridge, UK, 1998
2. G.E. Dieter, *Engineering Design: A Materials and Processing Approach,* McGraw Hill, 1991, p 1–51, 231–271
3. *Metals Handbook,* American Society for Metals, 1948
4. F.B. Seely, *Resistance of Materials,* John Wiley & Sons, 1947
5. *Properties and Selection of Metals,* Vol 1, *Metals Handbook,* 8th ed., American Society for Metals, 1961, p 503
6. *Modern Plastics Encyclopedia,* McGraw Hill, 2000
7. M.F. Ashby, *Materials Selection for Mechanical Design,* 2nd ed., Butterworth-Heinemann, 1999
8. H. Davis, G. Troxell, and G. Hauck, *The Testing of Engineering Materials,* 4th ed., McGraw Hill, 1982, p 314
9. G. Carter, *Principles of Physical and Chemical Metallurgy,* American Society for Metals, 1979, p 87
10. M.A. Meyers and K.K. Chawla, *Mechanical Metallurgy,* Prentice-Hall, Edgewood Cliffs, NJ, 1984, p 626–627
11. George Vander Voort, *Metallography: Principles and Practices,* ASM International, 1999, p 383–385, 391–393
12. R.T. Shield, On the Plastic Flow of Metals under Conditions of Axial Symmetry, *Proc. R. Soc.,* Vol A233, 1955, p 267
13. A. Fee, Selection and Industrial Applications of Hardness Tests, *Mechanical Testing and Evaluation,* Vol 8, *ASM Handbook,* ASM International, 2000, p 260–277
14. J. Datsko, L. Hartwig, and B. McClory, On the Tensile Strength and Hardness Relation for Metals, *Journal of Materials Engineering and Performance,* Vol 10 (6), Dec 2001, p 718–722
15. W.J. Taylor, The Hardness Test as a Means of Estimating the Tensile Strength of Metals, *J.R. Aeronaut. Soc.,* Vol 46 (No. 380), 1942, p 198–202

SELECTED REFERENCES

- M. Ashby, *Materials Selection for Mechanical Design,* 2nd ed., Butterworth-Heinemann, 1999
- N. Dowling, *Mechanical Behavior of Materials: Engineering Methods for Deformation, Fracture, and Fatigue,* Prentice Hall, 1999
- R.C. Juvinall and K.M. Marshek, *Fundamentals of Machine Component Design,* 2nd ed., John Wiley & Sons, 1991
- J.E. Shigley and L.D. Mitchell, *Mechanical Engineering Design,* 4th ed., McGraw-Hill, 1983

CHAPTER 6

Tensile Testing for Determining Sheet Formability

THE TERM *FORMABILITY* refers to the ease with which a metal can be shaped through plastic deformation. Evaluation of the formability of a metal involves measurement of strength, ductility, and the amount of deformation required to cause fracture. The term "workability" is used interchangeably with formability; however, formability refers to the shaping of sheet metal, while workability refers to shaping materials by bulk forming processes such as forging and extrusion.

Sheet metal forming operations consist of a large family of processes, ranging from simple bending to stamping and deep drawing of complex shapes. Because sheet forming operations are so diverse in type, extent, and rate, no single test provides an accurate indication of the formability of a material in all situations. However, as will be discussed in this chapter, the uniaxial tensile test is one of the most widely used tests for determining sheet metal formability. It should also be noted that tensile testing at elevated temperatures is also widely used to determine the workability of materials. See Chapter 13, "Hot Tensile Testing," for details.

Effect of Material Properties on Formability

The properties of sheet metals vary considerably, depending on the base metal (steel, aluminum, copper, and so on), alloying elements present, processing, heat treatment, gage, and level of cold work. In selecting material for a particular application, a compromise usually must be made between the functional properties required in the part and the forming properties of the available materials. For optimal formability in a wide range of applications, the work material should:

- Distribute strain uniformly
- Reach high strain levels without necking or fracturing
- Withstand in-plane compressive stresses without wrinkling
- Withstand in-plane shear stresses without fracturing
- Retain part shape upon removal from the die
- Retain a smooth surface and resist surface damage

Some production processes can be successfully operated only when the forming properties of the work material are within a narrow range. More frequently, the process can be adjusted to accommodate shifts in work material properties from one range to another, although sometimes at the cost of lower production and higher material waste. Some processes can be successfully operated using work material that has a wide range of properties. In general, consistency in the forming properties of the work material is an important factor in producing a high output of dimensionally accurate parts.

Strain Distribution

Three material properties determine the strain distribution in a forming operation:

- The strain-hardening coefficient (also known as the work-hardening coefficient or exponent) or n value
- The strain rate sensitivity or m value
- The plastic strain ratio (anisotropy factor) or r value

The ability to distribute strain evenly depends on the n value and the m value. The ability to reach high overall strain levels depends on many factors, such as the base material, alloying elements, temper, n value, m value, r value, thickness, uniformity, and freedom from defects and inclusions.

The *n* value, or strain-hardening coefficient, is determined by the dependence of the flow (yield) stress on the level of strain. In materials with a high n value, the flow stress increases rapidly with strain. This tends to distribute further strain to regions of lower strain and flow stress. A high n value is also an indication of good formability in a stretching operation.

In the region of uniform elongation, the n value is defined as:

$$n = \frac{d \ln \sigma_T}{d \ln \varepsilon} \quad \text{(Eq 1)}$$

where σ_T is the true stress (load/instantaneous area). This relationship implies that the true stress-strain curve of the material can be approximated by a power law constitutive equation proposed in Ref 1:

$$\sigma_T = k\varepsilon^n \quad \text{(Eq 2)}$$

where k is a constant known as the strength coefficient.

Equation 2 provides a good approximation for most steels, but is not very accurate for dual-phase steels and some aluminum alloys. For these materials, two or three n values may need to be calculated for the low, intermediate, and high strain regions.

When Eq 2 is an accurate representation of material behavior, $n = \ln(1 + e_u)$, where e_u is the uniform elongation, or elongation at maximum load in a tensile test. By definition, $\ln(1 + e_u)$ is identical to ε_u, which is the true strain at uniform elongation.

Most steels with yield strengths below 345 MPa (50 ksi) and many aluminum alloys have n values ranging from 0.2 to 0.3. For many higher yield strength steels, n is given by the relationship (Ref 2):

$$n \simeq \frac{70}{\text{(yield strength in MPa)}} \quad \text{(Eq 3)}$$

A high n value leads to a large difference between yield strength and ultimate tensile strength (engineering stress at maximum load in a tensile test). The ratio of these properties therefore provides another measure of formability.

The *m* value, or strain rate sensitivity, is defined by:

$$m = \frac{d \ln \sigma_T}{d \ln \dot{\varepsilon}} \quad \text{(Eq 4)}$$

where $\dot{\varepsilon}$ is the strain rate, $d\varepsilon/dt$. This implies a relationship of the form:

$$\sigma_T = f(\varepsilon) \cdot \dot{\varepsilon}^m$$

or

$$\sigma_T = k\varepsilon^n \cdot \dot{\varepsilon}^m \quad \text{(Eq 5)}$$

where Eq 5 incorporates Eq 2 between stress and strain.

A positive strain rate sensitivity indicates that the flow stress increases with the rate of deformation. This has two consequences. First, higher stresses are required to form parts at higher rates. Second, at a given forming rate, the material resists further deformation in regions that are being strained more rapidly than adjacent regions by increasing the flow stress in these regions. This helps to distribute the strain more uniformly.

The need for higher stresses in a forming operation is usually not a major consideration, but the ability to distribute strains can be crucial. This becomes particularly important in the post-uniform elongation region, where necking and high strain concentrations occur. An approximately linear relationship has been reported between the m value and the post-uniform elongation for a variety of steels and nonferrous alloys (Ref 3). As m increases from -0.01 to $+0.06$, the post-uniform elongation increases from 2 to 40%.

Metals in the superplastic range have high m values of 0.2 to 0.7, which are one to two orders of magnitude higher than typical values for steel. At ambient temperatures, some metals, such as aluminum alloys and brass, have low or slightly negative m values, which explains their low post-uniform elongation.

High n and m values lead to good formability in stretching operations, but have little effect on drawability. In a drawing operation, metal in the flange must be drawn in without causing fracture in the wall. In this case, high n and m values strengthen the wall, which is beneficial, but they also strengthen the flange and make it harder to draw in, which is detrimental.

The *r* value, or plastic strain ratio, relates to drawability and is known as the anisotropy factor. This is defined as the ratio of the true width strain to the true thickness strain in the uniform elongation region of a tensile test:

$$r = \frac{\varepsilon_w}{\varepsilon_t} = \frac{\ln\left(\frac{w}{w_o}\right)}{\ln\left(\frac{t}{t_o}\right)} \quad \text{(Eq 6)}$$

The r value is a measure of the ability of a material to resist thinning. In drawing, material in the flange is stretched in one direction (radially) and compressed in the perpendicular direction (circumferentially). A high r value indicates a material with good drawing properties.

The r value frequently changes with direction in the sheet. In a cylindrical cup drawing operation, this variation leads to a cup with a wall that varies in height, a phenomenon known as earing (Fig. 1). It is therefore common to measure the average r value, or average normal anisotropy, r_m, and the planar anisotropy, Δr.

The property r_m is defined as $(r_0 + 2r_{45} + r_{90})/4$, where the subscripts refer to the angle between the tensile specimen axis and the rolling direction. The value Δr is defined as $(r_0 - 2r_{45} + r_{90})/2$. It is a measure of the variation of r with direction in the plane of a sheet. The value r_m determines the average depth (that is, the wall height) of the deepest draw possible. The value Δr determines the extent of earing. A combination of a high r_m value and a low Δr value provides optimal drawability.

Hot-rolled low-carbon steels have r_m values ranging from 0.8 to 1.0, cold-rolled rimmed steels range from 1.0 to 1.4, and cold-rolled aluminum-killed (deoxidized) steels range from 1.4 to 2.0. Interstitial-free steels have values ranging from 1.8 to 2.5, and aluminum alloys range from 0.6 to 0.8. The theoretical maximum r_m value for a ferritic steel is 3.0; a measured value of 2.8 has been reported (Ref 4).

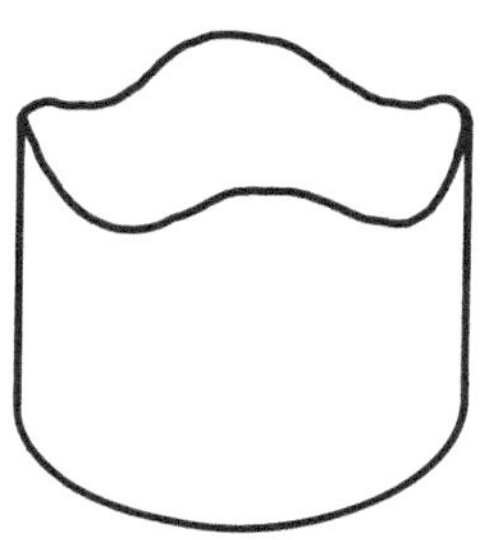

Fig. 1 Drawn cup with ears in the directions of high r value

Maximum Strain Levels: The Forming Limit Diagram

Each type of steel, aluminum, brass, or other sheet metal can be deformed only to a certain level before local thinning (necking) and fracture occur. This level depends principally on the combination of strains imposed, that is, the ratio of major and minor strains. The lowest level occurs at or near plane strain, that is, when the minor strain is zero.

This information was first represented graphically as the forming limit diagram, which is a graph of the major strain at the onset of necking for all values of the minor strain that can be realized (Ref 5, 6). Figure 2 shows a typical forming limit diagram for steel. The diagram is used in combination with strain measurements, usually obtained from circle grids, to determine how close to failure (necking) a forming operation is or whether a particular failure is due to inferior work material or to a poor die condition (Ref 7).

For most low-carbon steels, the forming limit diagram has the same shape as the one shown in Fig. 2, but the vertical position of the curve depends on the sheet thickness and the n value. The intercept of the curve with the vertical axis, which represents plane strain and is also the minimum point on the curve, has a value equal to n in the (extrapolated) zero thickness limit.

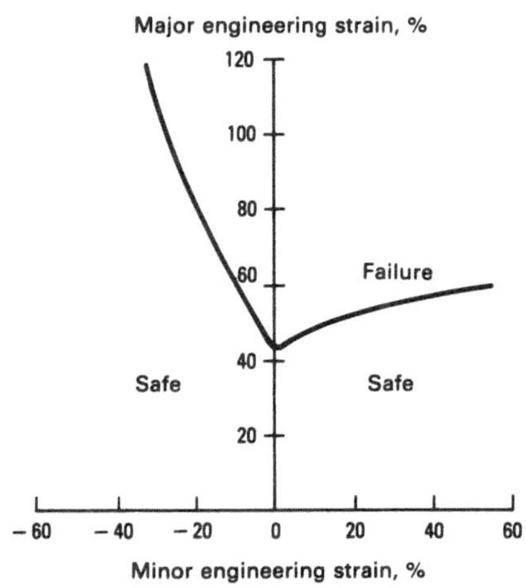

Fig. 2 Typical forming limit diagram for steel

The intercept increases linearly with thickness to a thickness of about 3 mm (0.12 in.).

The rate of increase is proportional to the n value up to $n = 0.2$, as shown in Fig. 3. Beyond these limits, further increases in thickness and n value have little effect on the position of the curve. The level of the forming limits also increases with the m value (Ref 3).

The shape of the curve for aluminum alloys, brass, and other materials differs from that in Fig. 2 and varies from alloy to alloy within a system. The position of the curve also varies and rises with an increase in the thickness, n value, or m value, but at rates that are generally not the same as those for low-carbon steel.

The forming limit diagram is also dependent on the strain path. The standard diagram is based on an approximately uniform strain path. Diagrams generated by uniaxial straining followed by biaxial straining, or the reverse, differ considerably from the standard diagram. Therefore, the effect of the strain path must be taken into account when using the diagram to analyze a forming problem.

Material Properties and Wrinkling

The effect of material properties on the formation of buckles or wrinkles is the subject of extensive research. In drawing operations, there is general agreement, based primarily on experiments with conical and cylindrical cups, that a high r_m value and a low Δr value reduce buckling in both flanges and walls (Ref 9–11). In addition to the above correlations, a low flow-stress-to-elastic-modulus ratio (σ_F/E) decreases wall wrinkling (Ref 12). The n value has an indirect effect. When the binder force is kept constant, the n value has no effect. However, high n values enable higher binder forces to be used, which reduces buckling.

In stretching operations, the situation appears to be different. A close correlation between the formation of buckles at low strain levels and the yield-strength-to-tensile-strength ratio (YS/TS) has been reported, as well as an inverse correlation with the low strain n value and an absence of correlation with the r_m value and uniform elongation (Ref 13). Some of the differences between these results may be attributed to the fact that the experiments with cups involved high strains and high compressive stresses, while the stretching experiments were conducted at low strain and low compressive stress levels. In both situations, the problem becomes significantly more severe as the sheet thickness decreases.

Material Properties and Shear Fracture

Shear fractures due to in-plane shear stresses are more prevalent in high-strength cold-worked materials, particularly when internal defects such as inclusions are present. Typical strain combinations that cause shear fracture are shown on the forming limit diagram in Fig. 4. For this material, Fig. 4 shows that, at high strain levels in the regions close to $\varepsilon_2 = \pm\varepsilon_1$, failure occurs by shearing before the initiation of necking.

The position and shape of the shear fracture curve depends on the material, its temper, and the type and degree of prestrain or cold work (Ref 14–16). Limited data are available on shear fracture.

Material Properties and Springback

Material properties that control the amount of springback that occurs after a forming operation are:

- Elastic modulus, E
- Yield stress, σ_y
- Slope of the true stress/strain curve, or tangent modulus, $d\sigma_T/d\varepsilon$

Springback is best described by means of three examples involving a rectangular beam: elastic bending below the yield stress, simple bending with the yield stress exceeded in the outer layers of the beam, and combined stretching and bending. In an actual part, springback is determined by the complex interaction of the residual internal elastic stresses, subject to the constraints of the part geometry.

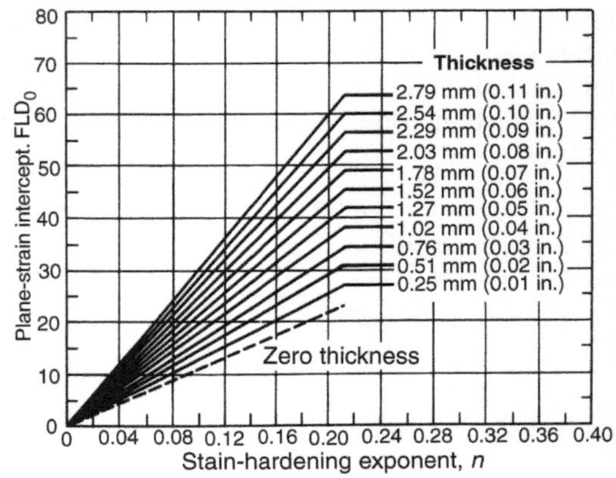

Fig. 3 Effect of thickness and n value on the plane-strain intercept of a forming limit diagram. Source: Ref 8

Elastic Bending Below the Yield Stress. Tensile elastic stresses are generated on the outside of the bend. These stresses decrease linearly from a maximum at the surface to zero at the center (neutral axis). They then become compressive and increase linearly to a maximum at the inner surface. Upon removal of the externally applied bending forces, the internal elastic forces cause the beam to unbend as they decrease to zero throughout the cross section (Fig. 5a).

The maximum amount of elastic deflection that can be produced without entering the plastic range is proportional to the yield stress divided by the elastic modulus. The strain at the yield point is equal to σ_y/E ($E = \sigma/\varepsilon$). The springback moment for a given deflection is therefore proportional to the elastic modulus ($\sigma = E\varepsilon$).

Simple Bending. In this example, the yield stress is exceeded in the outer layers of the beam. The outer layers deform plastically, and their stored elastic stresses continue to increase, but at a much lower rate that is proportional to the slope of the true stress-strain curve, or tangent modulus, $d\sigma_T/d\varepsilon$, instead of the elastic modulus. Figure 5(b) illustrates this condition

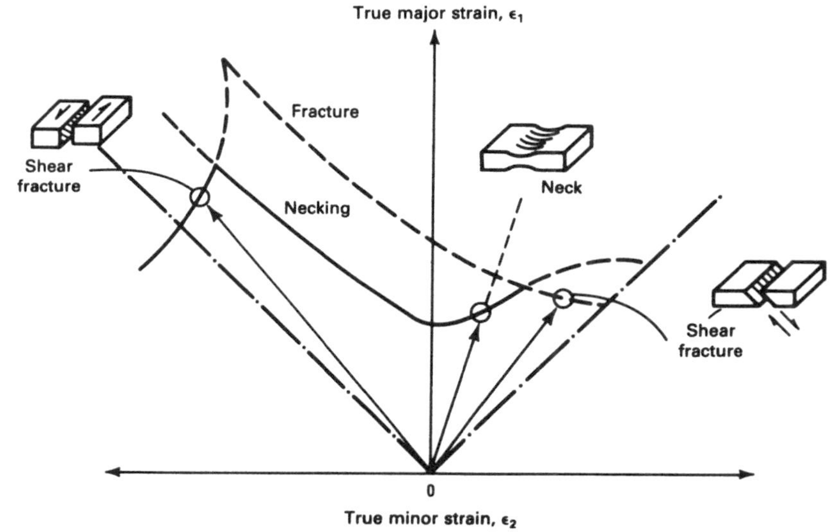

Fig. 4 Forming limit diagram including shear fracture. Source: Ref 14

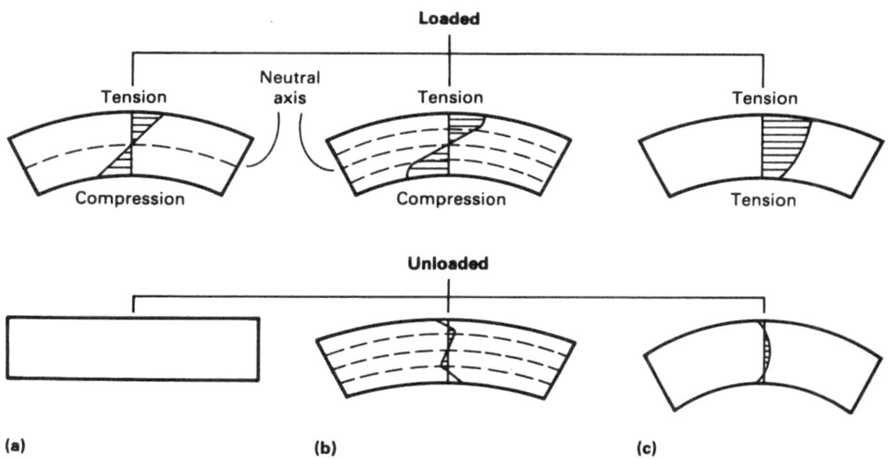

Fig. 5 Springback of a beam in simple bending. (a) Elastic bending. (b) Elastic and plastic bending. (c) Bending and stretching

for a beam bent so that 50% of its volume is in the plastic range.

Upon removal of the externally applied bending forces, the stored elastic stresses cause the beam to unbend until their combined bending moment is zero. This produces compressive stresses at the outer surface and tensile stresses at the inner surface.

The springback in this case is less than for a material whose yield strength is not exceeded at the same strain level. This can result from either a higher yield stress or a lower elastic modulus. It is also apparent that higher values of the tangent modulus cause greater springback when the yield strength is exceeded.

In actual conditions, the neutral axis moves inward upon bending because the outer part of the beam is stretched and becomes thinner and because the inner part is compressed and becomes thicker. This effect is analyzed in detail in Ref 17.

Combined Stretching and Bending. In this case, the entire beam can be plastically deformed in tension by as little as 0.5% stretching. However, a stress gradient still exists from the outer to the inner surface (Fig. 5c). Upon removing the external forces, the internal elastic stresses recover.

This causes unbending, but to a lesser extent than in the previous cases. As the level of stretching is increased, the amount of springback decreases because the tangent modulus and therefore the stress gradient through the beam decrease at higher strains. The yield strength ceases to be a factor in springback once all regions are plastically deformed in tension.

In the bending of wide sheets, the metal is deformed in plane strain, and the plane-strain properties (elastic modulus, yield stress, and tangent modulus) should be used. The effects of a low elastic modulus and a high yield stress and tangent modulus in increasing springback have been experienced in forming operations. Springback is more severe with aluminum alloys than with low-carbon steel (1 to 3 modulus ratio). High-strength steels exhibit more springback than low-carbon steels (~2 to 1 yield strength ratio), and dual-phase steels spring back more than high-strength steels of the same yield strength (higher tangent modulus).

The effect of stretching in reducing springback to very low levels has also been reported (Ref 18). Springback is also greatly influenced by geometrical factors, and it increases as the bend angle and ratio of bend radius to sheet thickness increase.

Surface Quality

The previously mentioned conditions that lead to undesirable surface textures can be minimized or prevented. The formation of orange peel in heavily deformed regions can be minimized by using a fine-grain material. The development of Lüders lines in rimmed steels can be prevented by temper rolling to 0.25 to 1.25% extension or by flex rolling, which produces mobile dislocations for a limited period of time, until they are trapped by nitrogen atoms. This also reduces elongation slightly. This problem is becoming less common with the increased use of continuous casting, which requires killed steels. These steels have less free nitrogen to interact with the dislocations and do not develop Lüders lines. Similar treatments can be applied to aluminum-magnesium alloys to prevent this defect.

Effect of Temperature on Formability

A change in the overall temperature alters the properties of the material, which thus affects formability. In addition, local temperature differences within a deforming blank lead to local differences in properties that affect formability.

At high temperatures, above one-half of the melting point on the absolute temperature scale, extremely fine-grain aluminum, copper, magnesium, nickel, stainless steel, steel, titanium, zinc, and other alloys become superplastic. Superplasticity is characterized by extremely high elongation, ranging from several hundred to more than 1000%, but only at low strain rates (usually below about $10^{-2}/s^{-1}$) at high temperatures.

The requirements of high temperatures and low forming rates have limited superplastic forming to low-volume production. In the aerospace industry, titanium is formed in this manner. The process is particularly attractive for zinc alloys because they require comparatively low temperatures (~270 °C, or ~520 °F).

At intermediate elevated temperatures, steels and many other alloys have less ductility than at room temperature (Ref 19, 20). Aluminum and magnesium alloys are exceptions and have minimum ductility near room temperature. Alloys of these metals have been formed commercially at slightly elevated temperatures (~250 °C, or ~480 °F). The strain rate sensitivity (*m* value) and post-uniform elongation for aluminum-magnesium alloys have been found to increase significantly in this temperature range (Ref 21).

Low-temperature forming has potential advantages for some materials, based on their tensile properties, but practical problems have limited application. Local increases in temperature occur during forming because of the surface friction and internal heating produced by the deformation. Generally, this is detrimental because it lowers the flow stress in the area of greatest strain and tends to localize deformation.

A method of improving drawability by creating local temperature differences has been developed and is being used commercially (Ref 22). It involves water cooling the punch in a deep-drawing operation. This lowers the temperature of the blank where it contacts the punch, which is the principal failure zone, and increases the local flow stress. Heating the die in order to lower the flow stress in the deformation zone at the top of the draw wall has also been found to be beneficial. The combination of these procedures has produced an increase of over 20% in the drawability of an austenitic stainless steel.

Types of Formability Tests

Formability tests are of two basic types: intrinsic and simulative (Ref 23). Intrinsic tests measure the basic characteristic properties of materials that can be related to their formability. Examples include the uniaxial tensile test and the plane-strain tensile test which will be subsequently described in this chapter. Other intrinsic tests reviewed in Ref 23 are the Marciniak stretching and sheet torsion tests, the hydraulic bulge test, the Miyauchi shear test, and hardness tests. Simulative tests subject the material to deformation that closely resembles the deformation that occurs in a particular forming operation. Many simulative tests, such as the Olsen and Swift cup tests, have been used extensively for many years with good correlation to production in specific cases. A number of simulative tests are described in Ref 23.

Uniaxial Tensile Testing

The most widely used intrinsic test of sheet metal formability is the uniaxial tensile test. A specimen such as that illustrated in Fig. 6 is used; its sides are accurately parallel over the gage length, which is usually 50.8 mm (2.00 in.) long and 12.7 mm (0.50 in.) wide. The specimen is gripped at each end and stretched at a constant rate in a tensile machine until it fractures, as described in ASTM E 8, "Standard Test Methods for Tension Testing of Metallic Materials." The applied load and extension are measured by means of a load cell and strain gage extensometer.

The load extension data can be plotted directly. However, data are usually converted into engineering (conventional) stress, σ_E (load/original cross section), and engineering strain, e (elongation/original length), or to true stress, σ_T (load/instantaneous cross section), and true strain, ε (natural logarithm of strained length/original length).

In addition, for formability testing, it is common practice to measure the width of the specimen during the test. This is done either intermittently by interrupting the test at preselected elongations to make measurements manually or continuously by means of width extensometers. From these measurements, the plastic strain ratio (anisotropy factor), or r value, can be determined.

During the rolling process used to produce metals in sheet form and the subsequent annealing, the grains and any inclusions present become elongated in the rolling direction, and a preferred crystallographic orientation develops. This causes a variation of properties with direction. Therefore, it is common practice to test specimens cut parallel to the rolling direction and at 45 and 90° to this direction. These are known as longitudinal, diagonal, and transverse specimens, respectively. This also enables the values of r_m and Δr to be calculated. Because the mechanical properties and elongation tend to be lower in the transverse direction, tests in this direction are often used as the basis for specifications.

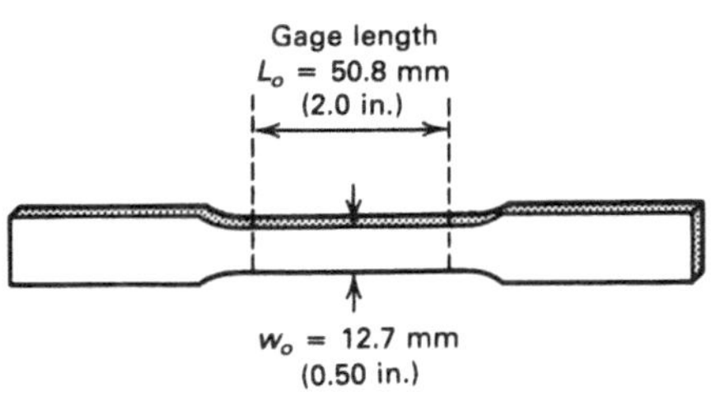

Fig. 6 Sheet tensile test specimen

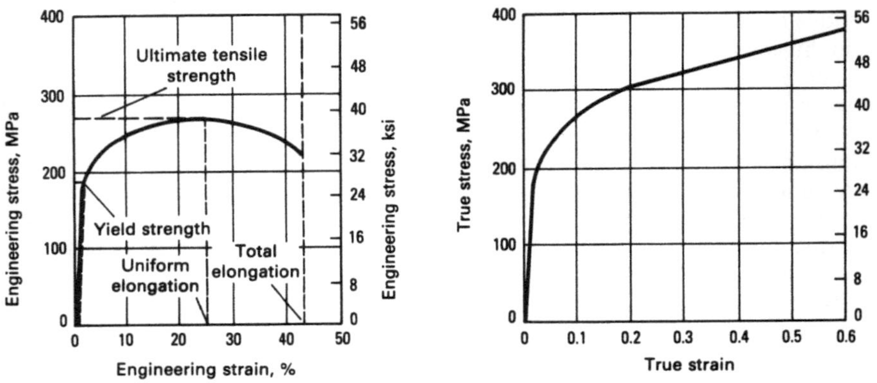

Fig. 7 Typical engineering and true stress-strain curves

The rate at which the test is performed can have a significant effect on the end results. Two methods are commonly used to determine this effect. In the first method, replicate samples are tested at different rates, and the results are influenced by variations between the samples. In the second method, the test rate is alternated between two levels. This approach avoids the problem of variation between samples, but it cannot be used at very high rates and is complicated by transients, which occur each time the rate is changed. The strain rate sensitivity, or *m* value, can be calculated from these tests.

Figure 7 shows a typical engineering stress-strain curve and the corresponding true stress-strain curve for a material that has a smooth transition between the very low strain (elastic) and the higher strain (plastic) regions of the curve. When the load is removed in the elastic region, the sample returns to its original dimensions. When this is done in the plastic region, the sample retains permanent deformation.

In the tensile test, the load increases to a maximum value and then decreases prior to fracture. The decrease is due to the localization of the deformation, which causes a reduction in cross section. This reduction has a greater effect than the opposing increase in flow stress due to strain hardening.

Some materials such as aged rimmed steels do not have a smooth transition between the elastic and plastic regions of the stress-strain curve. The load they can support decreases at the beginning of the plastic region and remains approximately constant for up to about 7% elongation. Subsequently, the load increases to a maximum and then decreases again at high elongations. This type of stress-strain curve is shown in Fig. 8. With the increasing use of continuous casting, which requires killed steels (steels deoxidized by small additions of aluminum, for example), rimmed steels are becoming less common.

Test Procedure

For accurate and reproducible results, uniaxial tensile testing must be performed in a carefully controlled manner. The main steps in the procedure are discussed in detail in Chapter 3, "Uniaxial Tensile Testing." These procedures are summarized below.

Specimen Preparation. The surfaces of the specimen should be free from scratches or other damage that can act as stress raisers and cause early failure. The edges should be smooth and free from irregularities. Care should be taken not to cold work the edges, or to ensure that any cold work introduced is removed in a subsequent operation, because this changes mechanical properties and lowers ductility.

It is common practice to mill and grind the edges, but other procedures such as fine milling, nibbling, and laser cutting are also used. When a new method is used, initial tests should be per-

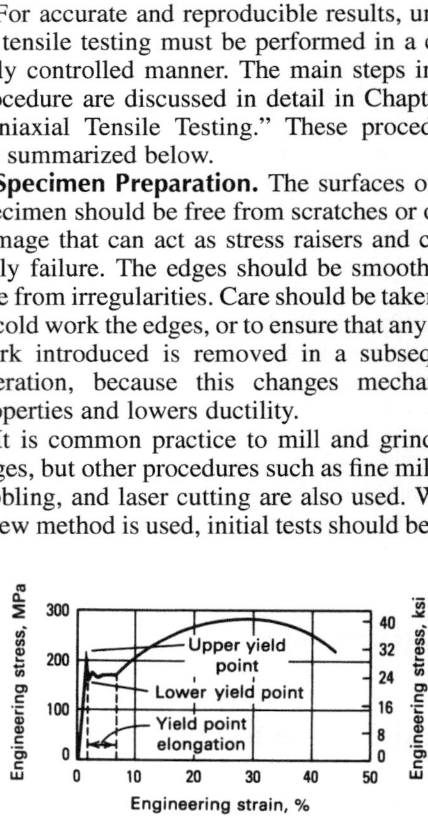

Fig. 8 Engineering stress-strain curve for rimmed steel

formed to compare the results with those obtained by conventional methods.

The width of a nominally 12.7 mm (0.50 in.) wide specimen should be measured to the nearest 0.025 mm (0.001 in.), and the thickness for specimens in the range of 0.5 to 2.5 mm (0.02 to 0.1 in.) should be measured to the nearest 0.0025 mm (0.0001 in.). If this is impractical because of surface roughness, the thickness should be measured to the nearest 0.025 mm (0.001 in.).

The tensile test is sensitive to variations in the width of the specimen, which should be accurately controlled. For a specimen 12.7 mm (0.50 in.) wide, the width of the reduced section should not deviate by more than ±0.25 mm (±0.01 in.) from the nominal value and should not differ by more than ±0.05 mm (±0.002 in.) from end to end.

Some investigators intentionally taper the reduced section slightly toward the center to increase the probability that fracture will occur within the gage length. In this case, the center should not be narrower than the ends by more than 0.10 mm (0.004 in.).

Alignment of Specimens. The specimen should be accurately aligned with the centerline of the grips. The effect of small displacements (10% of the specimen width) of one or both ends from the centerline has been calculated (Ref 24). It has been determined that the latter case is the more serious, but both strongly affect the strain in the outermost fibers. It has also been concluded that the calculated stress-strain curve is not significantly affected at strains above 0.3%.

Measurement of Load and Elongation. The applied load is measured by means of a load cell in the test machine, for which the usual calibration procedures must be followed (ASTM E 4, "Standard Practices for Force Verification of Testing Machines"). Elongation is usually determined by using a clip-on strain gage extensometer (ASTM E 83, "Standard Practice for Verification and Classification of Extensometer System"). In addition, small scratches are often scribed across the specimen at the ends of the gage length so that the total elongation can be determined from the broken specimen.

Circle grids are sometimes etched or printed on the specimen. These can be used to measure the strain distribution and width strain as well as the overall strain. This can be done continuously by means of a video camera and data processing system if required. Optical extensometers are used for some applications, particularly high-speed testing. These units require well-illuminated boundaries that are clearly delineated by means of high-contrast coatings, such as black-and-white paint.

An approximate measure of elongation can be obtained from the crosshead travel. This involves errors due to elongation of the specimen outside the gage length and elastic strain in the grips, which can be compensated for to some extent. This method is used when the specimen is inaccessible, such as in nonambient testing.

The signals from the load cell and extensometer can be plotted on a chart recorder or processed by a data processing system to the required form, such as plots of stress versus strain or tables of mechanical and forming properties.

Measurement of Width and Thickness. In addition to the initial measurements of specimen width and thickness, which are required to calculate the stress, measurements can be made at intervals during the test to determine the r value (ASTM E 517, "Standard Test Method for Plastic Strain Ratio r for Sheet Metal") and to determine the reduction in area and true strain. The r value is measured at a specified strain level between the yield point and the uniform elongation (for example, at 15% elongation). It can be measured by stopping the test at this strain level and then measuring the width accurately (±0.013 mm, or ±0.0005 in.) at a minimum of three equally spaced points in the gage length (for a 50.8 mm, or 2.0 in., gage length). In practice, the thickness is calculated from the specimen width and length, assuming no change in volume.

Alternatively, width measurements can be made during the test using width extensometers, although this is a more complicated procedure. Attempts are underway to develop combined width and length extensometers to simplify this method.

Reduction in area is the ratio $(A_o - A)/A_o$, where A is the instantaneous cross-sectional area and A_o is the original cross-sectional area. It is used to calculate the true strain in the region of post-uniform elongation. A large reduction in area at fracture correlates with a small minimum bend radius, a high m value, and high energy absorption. To calculate the reduction in area, the width and thickness must be measured in the narrowest part of the necked region.

Effect of Gage Length on Elongation. In post-uniform elongation, part of the specimen is elongated uniformly, and the remainder is narrowed into a necked region of higher strain level. A change in the gage length alters the ratio of these two regions and has a significant effect on the total elongation measurement. This phenomenon is discussed in detail in Ref 25.

To obtain results that are comparable for different gage lengths, the ratio of the square root of the cross-sectional area to the length, $\sqrt{A}/L$, should be the same. When comparing samples of different thickness, this implies that the gage length or the width should be adjusted to maintain this ratio.

Rate of Testing. Most tensile tests are performed on screw-driven or hydraulic testing machines at strain rates of 10^{-5} to 10^{-2} s^{-1}. The strain rate is defined as the increase in length per unit length per second. These tests are known as low strain rate or static tests.

Most high-volume production forming operations are performed at considerably higher strain rates—in the range of 1 to 10^2 s^{-1}. To determine the tensile properties in this range, dynamic test machines, which operate at rates of 10^{-1} to 10^2 s^{-1}, are used (Ref 25). As mentioned previously, steels have higher tensile properties and lower elongations at high strain rates. The properties of aluminum alloys have little sensitivity to the strain rate.

Determination of Material Properties

The stress-strain curve determined by uniaxial tensile testing provides values of many formability-related material properties. These properties include the modulus of elasticity (Young's modulus), yield strength, tensile strength, uniform elongation, total elongation, and reduction in area. Determination of these properties is described in Chapter 3, "Uniaxial Tensile Testing." As discussed earlier in this chapter, other key properties are the strain-hardening exponent, the plastic strain ratio, and the strain rate sensitivity. Table 1 lists typical values of properties measured in tensile tests on thin (0.5 to 1.0 mm, or 0.02 to 0.04 in.) sheet materials.

Strain-Hardening Exponent. The n value, $d \ln \sigma_T / d \ln \varepsilon$, is given by the slope of a graph of the logarithm of the true stress versus the logarithm of the true strain in the region of uniform elongation. For materials that closely follow the Holloman constitutive equation (Eq 2), an approximate n value can be obtained from two points on the stress-strain curve by the Nelson-Winlock procedure (Ref 26). The two points commonly used are at 10% strain and at the maximum load. The ratio of the loads or stresses at these two points is calculated, and the n value and uniform elongation can then be determined from a table or graph. The accuracy of the n value determined in this way is ± 0.02.

The n value can be determined more accurately by linear regression analysis, as in ASTM E 646, "Standard Test Method for Tensile Strain-Hardening Exponents (n-Values) of Metallic Sheet Materials." For some materials, n is not constant, and initial (low strain), terminal (high strain), and sometimes intermediate n values are determined. The initial n value relates to the low deformation region, in which springback is often

Table 1 Typical tensile properties of selected sheet metals

	Young's modulus, E		Yield strength		Tensile strength		Uniform elongation,	Total elongation,	Strain-hardening exponent,	Average normal anisotropy,	Planar anisotropy,	Strain rate sensitivity,
Material	GPa	10^6 psi	MPa	ksi	MPa	ksi	%	%	n	r_m	Δr	m
Aluminum-killed drawing quality steel	207	30	193	28	296	43	24	43	0.22	1.8	0.7	0.013
Interstitial-free steel	207	30	165	24	317	46	25	45	0.23	1.9	0.5	0.015
Rimmed steel	207	30	214	31	303	44	22	42	0.20	1.1	0.4	0.012
High-strength low-alloy steel	207	30	345	50	448	65	20	31	0.18	1.2	0.2	0.007
Dual-phase steel	207	30	414	60	621	90	14	20	0.16	1.0	0.1	0.008
301 stainless steel	193	28	276	40	690	100	58	60	0.48	1.0	0.0	0.012
409 stainless steel	207	30	262	38	469	68	23	30	0.20	1.2	0.1	0.012
3003-O aluminum	69	10	48	7	110	16	23	33	0.24	0.6	0.2	0.005
6009-T4 aluminum	69	10	131	19	234	34	21	26	0.23	0.6	0.1	−0.002
70-30 brass	110	16	110	16	331	48	54	61	0.56	0.9	0.2	0.001

a problem. The terminal n value relates to the high deformation region, in which fracture may occur.

Plastic Strain Ratio. The r value, or anisotropy factor, is defined as the ratio of the true width strain to the true thickness strain in a tensile test. Generally, its value depends on the elongation at which it is measured. It is usually measured at 10, 15, or 20% elongation.

The r value is calculated from the measured width and length as:

$$\varepsilon_w = \ln\left(\frac{w}{w_o}\right)$$

$$\varepsilon_t = \ln\left(\frac{t}{t_o}\right) = \ln\left(\frac{L_o w_o}{Lw}\right) \quad \text{(Eq 7)}$$

where constancy of volume ($Lwt = L_o w_o t_o$) has been used and:

$$r = \frac{\varepsilon_w}{\varepsilon_t} = \frac{\ln\left(\frac{w}{w_o}\right)}{\ln\left(\frac{L_o w_o}{Lw}\right)} \quad \text{(Eq 8)}$$

The average r value, or normal anisotropy (r_m), and the planar anisotropy, or Δr value, can be calculated from the values of r in different directions using Eq 6 and 7.

Strain Rate Sensitivity. The m value, $d \ln \sigma_T / d \ln \dot{\varepsilon}$, is determined either from duplicate tensile tests performed at different strain rates or from a single test in which the rate is alternated between two levels during the test. These methods are shown schematically in Fig. 9. The m value can be determined at various strain levels in the region of uniform elongation:

$$m = \frac{\ln\left(\frac{\sigma_1}{\sigma_2}\right)}{\ln\left(\frac{\dot{\varepsilon}_1}{\dot{\varepsilon}_2}\right)} \quad \text{(Eq 9)}$$

In some materials, m is insensitive to strain (Ref 3, 27). In other materials, however, m is sensitive to strain and strain rate (Ref 28). In many materials, m increases and n decreases with an increase in temperature (Ref 29), sometimes to the extent that superplastic properties develop.

Plane-Strain Tensile Testing

In conventional uniaxial tensile testing, the sample is strained in the region of drawing; that is, the minor or width strain is negative. The test does not provide information on the response of sheet materials in the plane-strain state, in which the minor strain is zero. However, it can be modified to produce this strain state in part of the sample. This modification involves the use of a very wide, short sample or the use of knife-edges to prevent transverse (width) strain in part of the sample.

Wide Sample Methods. Increasing the width of the sample and decreasing the gage length changes the strain state from one with a large negative minor strain component toward the plane-strain state, in which the minor strain component is zero. In the rectangular sheet ten-

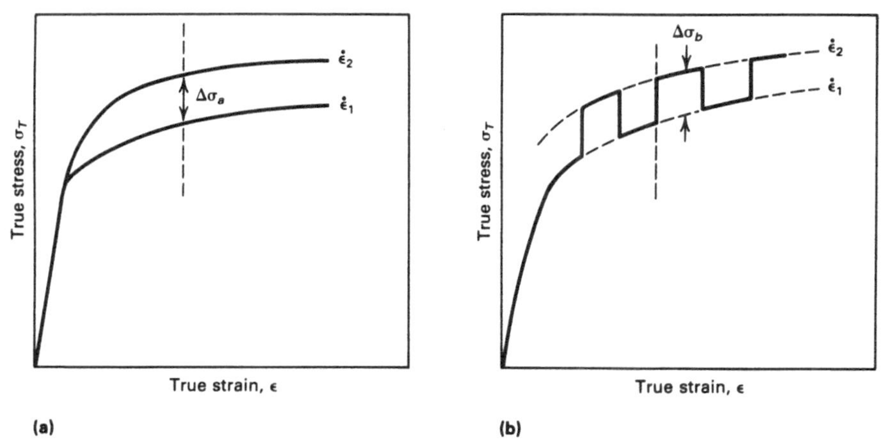

Fig. 9 Methods for determining strain-rate sensitivity (m value). (a) Duplicate test method. (b) Changing rate method

sion test, samples with length-to-width ratios of 1 to 1, 1 to 2, and 1 to 4 are used to approach the plane-strain conditions (Ref 30). Gage lengths are constrained further by reinforcements welded onto each side of the sample at both ends, thus making the samples three layers thick except in the gage length.

The minimum minor strain obtained with the 1 to 4 length-to-width ratio is −0.05 times the major strain, which is close to the plane-strain condition of zero minor strain. The in-plane strains are measured by means of grid markings on the samples, and through-thickness deformations can be observed by holographic interferometry.

A similar approach was used in testing many wide specimen designs to determine the effect of edge profile and length-to-width ratio on strain state (Ref 31–33). The specimen geometry that yielded the highest center strain at failure with a large region of plane strain is shown in Fig. 10. The plane-strain region, which is arbitrarily taken as the region where $|e_2/e_1|$ is less than 0.2, occupies about 80% of the specimen width. The outer part of the specimen deforms in a similar manner to a standard tensile test specimen.

Special grips were developed that exert a high clamping force at the inner contact lines. This minimizes distortion and slippage in these regions, giving the test well-defined boundary conditions. The results of both types of wide specimen tensile tests described above correlated well with stress-strain predictions obtained by finite-element modeling using material properties obtained in the standard tensile test (Ref 31, 34).

Width Constraint Method. In the width constraint method, a rectangular sample is used that has a central gage section reduced in width by circular notches (Ref 35). The gage section is clamped between two pairs of opposing parallel knife-edges (stingers) aligned with the sample axis. The knife-edges prevent transverse (width) strain in this region. The sample is pulled to fracture in a tension-testing machine, and the plane-strain limit (necking) and fracture strains are determined from thickness measurements made on the fractured sample. This procedure is described in detail in Ref 35. The use of a spring-loaded clamp around the knife-edges makes adjustment of the clamp during testing unnecessary.

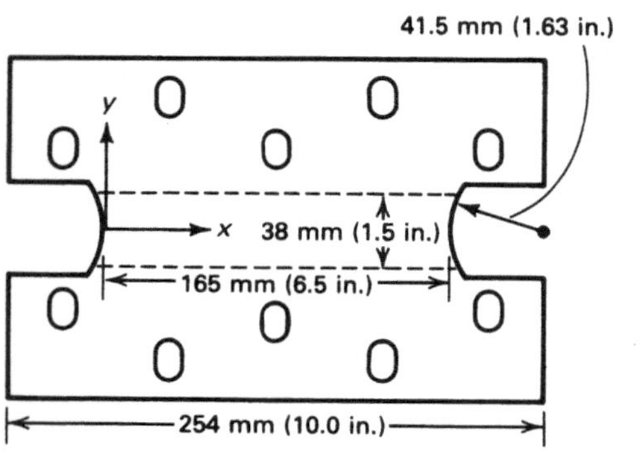

Fig. 10 Plane-strain tensile test specimen. Source: Ref 33

ACKNOWLEDGMENT

This chapter was adapted from B. Taylor, Formability Testing of Sheet Metals, *Forming and Forging,* Vol 14, *Metals Handbook,* 9th ed., ASM International, 1988, p 877–899.

REFERENCES

1. J.H. Holloman, Tensile Deformation, *Trans. AIME,* Vol 162, 1945, p 268–290
2. W.A. Backofen, Massachusetts Institute of Technology Industrial Liaison Symposium, Chicago, March 1974
3. A.K. Ghosh, The Influence of Strain Hardening and Strain-Rate Sensitivity on Sheet Metal Forming, *Trans. ASME,* Vol 99, July 1977, p 264–274
4. I.S. Brammar and D.A. Harris, Production and Properties of Sheet Steel and Aluminum Alloys for Forming Applications, *J. Austral. Inst. Met.,* Vol 20 (No. 2), 1975, p 85–100
5. S.P. Keeler and W.A. Backofen, Plastic Instability and Fracture in Sheets Stretched Over Rigid Punches, *Trans. ASM,* Vol 56 (No. 1), 1963, p 25–48
6. G.M. Goodwin, "Application of Strain Analysis to Sheet Metal Forming Problems in the Press Shop," Paper 680093, Society of Automotive Engineers, 1968
7. S.P. Keeler, Determination of Forming Limits in Automotive Stampings, *Sheet Met. Ind.,* Vol 42, Sept 1965, p 683–691
8. S.P. Keeler and W.G. Brazier, Relationship Between Laboratory Material Characterization and Press-Shop Formability, in *Microalloying 75 Proceedings,* Union Carbide Corporation, 1977, p 517–530

9. H. Naziri and R. Pearce, The Effect of Plastic Anisotropy on Flange-Wrinkling Behavior During Sheet Metal Forming, *Int. J. Mech. Sci.,* Vol 10, 1968, p 681–694
10. K. Yoshida and K. Miyauchi, Experimental Studies of Material Behavior as Related to Sheet Metal Forming, in *Mechanics of Sheet Metal Forming,* Plenum Press, 1978, p 19–49
11. W.F. Hosford and R.M. Caddell, in *Metal Forming, Mechanics and Metallurgy,* Prentice-Hall, 1983, p 273, 309
12. J. Havranek, The Effect of Mechanical Properties of Sheet Steels on the Wrinkling Behavior During Deep Drawing of Conical Shells, in *Sheet Metal Forming and Energy Conservation,* Proceedings of the 9th Biennial Congress of the International Deep Drawing Research Group, Ann Arbor, MI, American Society for Metals, 1976, p 245–263
13. J.S.H. Lake, The Yoshida Test—A Critical Evaluation and Correlation with Low-Strain Tensile Parameters, in *Efficiency in Sheet Metal Forming,* Proceedings of the 13th Biennial Congress, Melbourne, Australia, International Deep Drawing Research Group, Feb 1984, p 555–564
14. J.L. Duncan, R. Sowerby, and M.P. Sklad, "Failure Modes in Aluminum Sheet in Deep Drawing Square Cups," Paper presented at the Conference on Sheet Forming, University of Aston, Birmingham, England, Sept 1981
15. G. Glover, J.L. Duncan, and J.D. Embury, Failure Maps for Sheet Metal, *Met. Technol.,* March 1977, p 153–159
16. J.D. Embury and J.L. Duncan, Formability Maps, *Ann. Rev. Mat. Sci.,* Vol 11, 1981, p 505–521
17. J. Datsko, *Materials in Design and Manufacturing,* J. Datsko Consultants, 1977, p 7–16
18. N. Kuhn, On the Springback Behavior of Low-Carbon Steel Sheet After Stretch Bending, *J. Austral. Inst. Met.,* Vol 12 (No. 1), Feb 1967, p 71–76
19. G.V. Smith, Elevated Temperature Static Properties of Wrought Carbon Steel, in *Special Technical Publication on Temperature Effects,* STP 503, American Society for Testing and Materials, 1972
20. F.N. Rhines and P.J. Wray, Investigation of the Intermediate Temperature Ductility Minimum in Metals, *Trans. ASM,* Vol 54, 1961, p 117–128
21. B. Taylor, R.A. Heimbuch, and S.G. Babcock, Warm Forming of Aluminum, in *Proceedings of the Second International Conference on Mechanical Behavior of Materials,* American Society for Metals, 1976, p 2004–2008
22. W.G. Granzow, The Influence of Tooling Temperature on the Formability of Stainless Sheet Steel, in *Formability of Metallic Materials—2000 A.D.,* STP 753, J.R. Newby and B.A. Niemeier, Ed., American Society for Testing and Materials, 1981, p 137–146
23. B. Taylor, Formability Testing of Sheet Metals, *Forming and Forging,* Vol 14, *Metals Handbook,* 9th ed., ASM International, 1988, p 877–899
24. H.C. Wu and D.R. Rummler, Analysis of Misalignment in the Tension Test, *Trans. ASME,* Vol 101, Jan 1979, p 68–74
25. G.E. Dieter, *Mechanical Metallurgy,* 2nd ed., McGraw-Hill, 1976, p 347, 349, 681
26. R.L. Whitely, *Correlation of Deep Drawing Press Performance With Tensile Properties,* STP 390, American Society for Testing and Materials, 1965
27. S.J. Green, J.J. Langan, J.D. Leasia, and W.H. Yang, Material Properties, Including Strain-Rate Effects, as Related to Sheet Metal Forming, *Met. Trans. A,* Vol 2A, 1971, p 1813–1820
28. G. Rai and N.J. Grant, On the Measurements of Superplasticity in an Al-Cu Alloy, *Met. Trans A.,* Vol 6A, 1975, p 385–390
29. W.J. McGregor Tegart, in *Elements of Mechanical Metallurgy,* Macmillan, 1966, p 29–38
30. M.L. Devenpeck and O. Richmond, Limiting Strain Tests for In-Plane Sheet Stretching, in *Novel Techniques in Metal Deformation Testing,* The Metallurgical Society, 1983, p 79–88
31. R.H. Wagoner and N.M. Wang, An Experimental and Analytical Investigation of In-Plane Deformation of 2036-T4 Aluminum, *Int. J. Mech. Sci.,* Vol 21, 1979, p 255–264
32. R.H. Wagoner, Measurement and Analysis of Plane-Strain Work Hardening, *Met. Trans. A,* Vol 11A, Jan 1980, p 165–175
33. R.H. Wagoner, Plane-Strain and Tensile Hardening Behavior of Three Automotive Sheet Alloys, in *Experimental Verification of Process Models,* Symposium proceedings, Cincinnati, OH, Sept 1981, American Society for Metals, 1983, p 236
34. E.J. Appleby, M.L. Devenpeck, L.M.

O'Hara, and O. Richmond, Finite Element Analysis and Experimental Examination of the Rectangular-Sheet Tension Test, in *Applications of Numerical Methods to Forming Processes,* Vol 28, Proceedings of the ASME Winter Annual Meeting, San Francisco, Applied Mechanics Division, American Society of Mechanical Engineers, Dec 1978, p 95–105

35. H. Sang and Y. Nishikawa, A Plane Strain Tensile Apparatus, *J. Met.,* Feb 1983, p 30–33

CHAPTER 7

Tensile Testing of Metals and Alloys

THE TENSILE TEST provides a relatively easy, inexpensive technique for developing mechanical property data for the selection, qualification, and utilization of metals and alloys in engineering service. This data may be used to establish the suitability of the alloy for a particular application, and/or to provide a basis for comparison with other candidate materials. Design guidelines generally require that the tensile properties of metals and alloys meet specific, well-defined criteria. ASME has established code requirements for the strengths and ductilities of many classes of metals and alloys. Step-by-step procedures for conducting the tensile test are defined in various ASTM standards (see, for example, ASTM E 8, "Standard Test Methods for Tension Testing of Metallic Materials"). Descriptions of the test methodology and discussions on the importance of both material and test variables on the measured tensile properties can be found in Chapter 3, "Uniaxial Tensile Testing." Because such variables have significant influences on the measured tensile properties, an understanding of the influences is necessary for accurate interpretation and use of most tensile data.

The elastic moduli of cast iron, carbon steel, and many other engineering materials are dependent on the rate at which the test specimen is stretched (strain rate). The yield strength or stress at which a specified amount of plastic strain takes place is also dependent on the test strain rate. Alloy composition, grain size, prior deformation, test temperature and heat treatment may also influence the measured yield strength. Generally, factors that increase the yield strength decrease the tensile ductility because these factors also inhibit plastic deformation. However, a notable exception to this trend is the increase in ductility that accompanies an increase in yield strength when the grain size is reduced.

Most structural metals and alloys, when strained to failure in a tensile test, fracture by ductile processes. The fracture surface is formed by the coalescence or combination of microvoids. These microvoids generally nucleate during plastic deformation processes, and coalescence begins after the plastic deformation processes become highly localized. Strain rate, test temperature, and microstructure influence the coalescence process and, under selected conditions (decreasing temperature, for example), the fracture may undergo a transition from ductile to brittle processes. Such transitions may limit the utility of the alloy and may not be apparent from strength measurements. The tensile test, therefore, may require interpretation, and interpretation requires a knowledge of the factors that influence the test results. This chapter provides a metallurgical perspective for such interpretation. Additional information can also be found in Chapter 2, "Mechanical Behavior of Materials Under Tensile Loads."

Elastic Behavior

Most structures are designed so that the materials of construction undergo elastic loadings under normal service conditions. These loads produce elastic or reversible strains in the structural materials. The upward movement of a wing as an airplane takes off and the sway of a tall building in a strong wind are examples in which the elastic strains are readily apparent. Bending of an automobile axle and stretching of a bridge with the passing of a car are less noticeable examples of elastic strains. The magnitude of the strain is dependent on the elastic moduli of the material supporting the load. Although elastic moduli are not generally determined by tensile testing, tensile behavior can be used to illustrate the importance of elastic properties in the selection and use of metals and alloys.

Young's modulus for iron (207 GPa, or 30 × 10^6 psi) is approximately three times that of aluminum (69 GPa, or 10 × 10^6 psi) and almost twice that of copper (117 GPa, or 17 × 10^6 psi). This variation in elastic behavior is illustrated in Fig. 1. Because of its higher value of Young's modulus, an iron component will deflect less than an "identical" copper or aluminum component that undergoes an equivalent load. In a tensile test, for example, the elastic tensile strains in 12.8 mm (0.505 in.) diam tensile bars of iron, copper, and aluminum loaded to 455 kg (1000 lb) will be 1.6 × 10^{-4} mm/mm (in./in.) for iron, 2.9 × 10^{-4} mm/mm (in./in.) for copper, and 5 × 10^{-4} mm/mm (in./in.) for aluminum.

The ability of a material to resist elastic deformation is termed "stiffness," and Young's modulus (E) is one measure of that ability. Engineering applications that require very rigid structures, such as microscopes, antennas, satellite dishes, and radio telescopes, must be constructed from either very massive components or selected materials that have high values of elastic moduli. The elastic modulus of iron is higher than those of many metals and alloys, and thus iron and iron alloys are frequently used for applications that require high stiffness.

The equation that defines Young's modulus, $\sigma = E\varepsilon$, is based on the observation that tensile strain (ε) is linearly proportional to the applied stress (σ). This linear relationship provides an adequate description of the behavior of metals and alloys under most practical situations. However, when materials are subjected to cyclic or vibratory loading, even slight departures from truly linear elastic behavior may become important. One measure of the departure from linear elasticity is the anelastic response of a material.

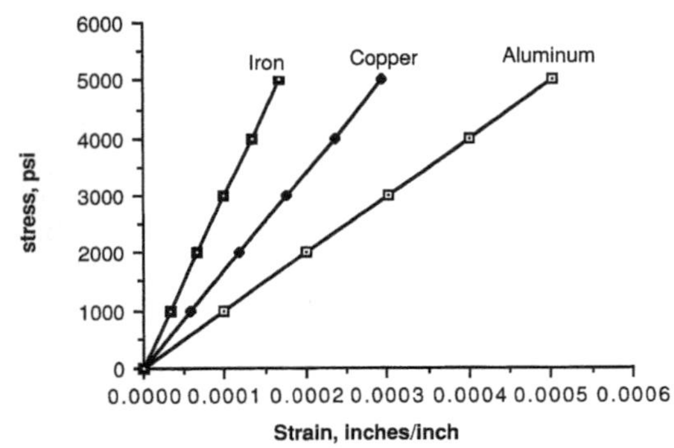

Fig. 1 Schematic representation of the elastic portions of the stress-strain curves for iron, copper, and aluminum

Anelasticity

Anelasticity is time-dependent, fully reversible deformation. The time dependence results from the lack of instantaneous atom movement during the application of a load. There are several mechanisms for time-dependent deformation processes, including the diffusive motion of alloy and/or impurity atoms. This diffusive motion may simply be atoms jumping to nearby lattice sites made favorable by the application of a load.

Tensile loading of an iron-carbon alloy will produce elastic strains in the alloy, and its body-centered-cubic structure will be distorted to become body-centered-tetragonal. Carbon, in solid solution, produces a similar distortion of the iron lattice. There is one basic difference between the distortions introduced by tensile loads and those introduced by dissolving carbon. The average distortion of a metallic lattice during a tensile test is anisotropic: each unit cell of the structure is elongated in the direction of the tensile load and, because of Poisson's ratio, the material also contracts in the lateral direction. In contrast, the average lattice distortion resulting from the solution of carbon is isotropic even though each individual carbon atom produces a localized anisotropic distortion.

Carbon atoms, in solid solution in iron, are located at the interstitial sites shown schematically in Fig. 2. Because the dissolved carbon atoms are too big for the interstitial sites, a carbon atom at site X would push the iron atoms A and

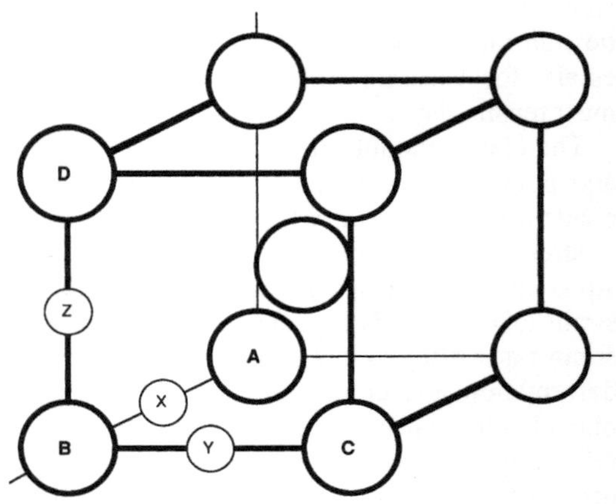

Fig. 2 Interstitial sites in an iron lattice. The large spheres at the corners and center of the cube represent iron atoms, and the small spheres (X, Y, and Z) represent interstitial sites for carbon. There are duplicate interstitial sites at the corners of the cube or unit cell.

B apart and cause the unit cell to elongate in the *x* direction. Similarly, a carbon atom at site Y would push iron atoms B and C apart and cause elongation in the *y* direction, and a carbon atom at site Z would cause elongation in the *z* direction. Within any given unstressed iron or alpha grain, carbon atoms are randomly distributed in X, Y, and Z sites. Thus, although each unit cell is distorted in one specific direction, the over-all distortion of the unstressed grain is basically isotropic, or equal in all directions.

The application of a tensile stress causes specific interstitial sites to be favored. If the tensile stress is parallel to the *x* direction, type X sites are expanded and become favored sites for the carbon atoms. Type Y sites become favored if the stress is in the *y* direction, and type Z sites are favored when stresses are in the *z* direction. During a tensile test, carbon atoms will migrate or diffuse to the sites made favorable by the application of the tensile load. This migration is time and temperature dependent and can be the cause of anelastic deformation.

The sudden application of the tensile load may elastically strain the iron lattice at such a high rate that carbon migration to favored sites cannot occur as the load is applied. However, if the material remains under load, the time-dependent migration to favored sites will produce additional lattice strain because of the tendency for the interstitial carbon to push iron atoms in the direction of the applied stress. These additional strains are the anelastic strains in the material. Similarly, if the load is suddenly released, the elastic strains will be immediately recovered whereas recovery of the anelastic strains will require time as the interstitial carbon atoms relocate from the previously favorable sites to form a uniform distribution in the iron lattice. The time dependence of the elastic and anelastic strains is shown schematically in Fig. 3.

The combination of the elastic and anelastic strains may cause Young's modulus, as determined in a tensile test, to be loading-rate (or strain-rate) dependent and may produce damping or internal friction in a metal or alloy subjected to cyclic or vibratory loads. Anelastic strains are one cause of stress relaxation in a tensile test when the test specimen is loaded and held at a fixed displacement. This stress relaxation is frequently called an "elastic aftereffect" and results in a time-dependent load drop because the load necessary to maintain the fixed displacement will decrease as atoms move to favored sites and anelastic deformation takes place. This elastic aftereffect, illustrated in Fig. 4, demonstrates the importance of time or loading rate on test results.

The total reversible strain that accompanies the application of a tensile load to a test specimen is the sum of the elastic and anelastic strains. Rapid application of the load will cause

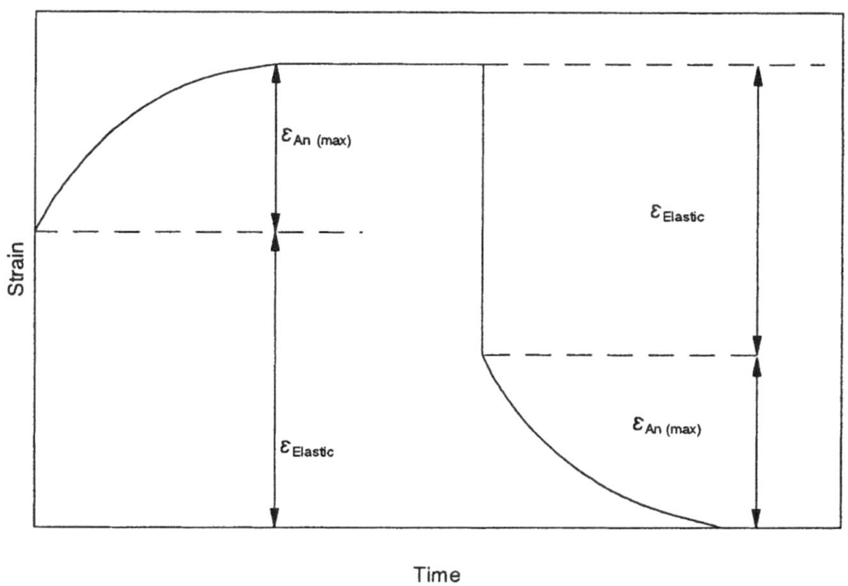

Fig. 3 A relationship between elastic and anelastic strains. The elastic strains develop as soon as the load is applied, whereas the anelastic strains are time dependent.

the anelastic strain to approach zero (the test time is not sufficient for anelastic strain), thus the total strain during loading will equal the true elastic strain. Very slow application of the same load will allow the anelastic strain to accompany the loading process, thus the total reversible strain in this test will exceed the reversible strain during rapid loading. The measured value of Young's modulus in the low-strain-rate test will be lower than that measured in the high-strain-rate test, and the measured modulus of elasticity will be strain-rate dependent. This dependency is illustrated in Fig. 5. The low value of Young's modulus is termed the "relaxed modulus," and the modulus measured at high strain rates is termed the "unrelaxed modulus."

Damping

Tensile tests and cyclic loadings frequently are made at strain or loading rates that are intermediate between those required for fully relaxed behavior and those required for fully unrelaxed behavior. Therefore, on either loading or unloading, the initial or short-time portion of the stress-strain curve will produce unrelaxed behavior whereas the later, longer-time portions of the curve will produce more relaxed behavior. The transition from unrelaxed to relaxed behavior produces a loading-unloading hysteresis in the stress-strain curve (Fig. 6).

This hysteresis represents an energy loss during the load-unload cycle. The amount of energy loss is proportional to the magnitude of the hysteresis. Such energy losses that may be attributed to anelastic effects within the metal lattice are termed "internal friction." Internal friction plays a major role in the ability of a material to absorb vibrational energy. Such absorption may cause the temperature of a material to rise during the loading-unloading cycle. One measurement of the susceptibility of a material to internal friction is the damping capacity. Because anelasticity and internal friction are dependent on time and temperature, the damping capacity of a metal or alloy is both temperature and strain-rate dependent.

Internal friction and damping play major roles in the response of a metal or alloy to vibrations. Materials tested under conditions that cause significant internal friction during loading-unloading cycles undergo large energy losses and are said to have high damping capacities. Such materials are useful for the absorption of vibrations. Gray cast iron, for example, has a very high damping capacity and is frequently used for the bases of instruments and equipment that must be isolated from room vibrations. Lathes, presses, and other pieces of heavy machinery also use

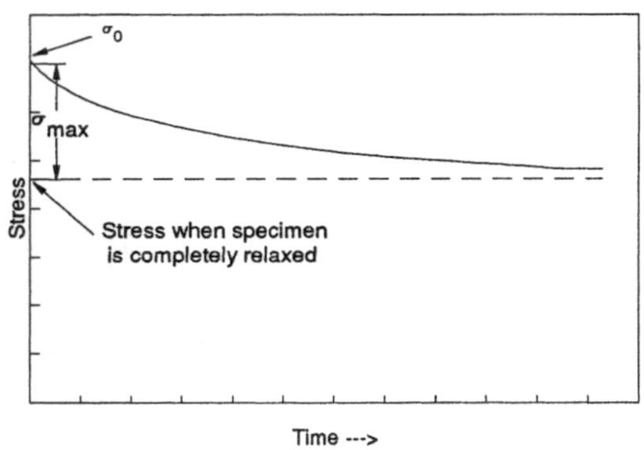

Fig. 4 The elastic aftereffect. The tensile specimen was loaded to a stress of σ_0 and then held. The time-dependent drop in stress results from a decrease in the load required to maintain a fixed displacement. This decrease results from anelastic strains that increase the length of the test specimen. When the anelastic straining process is complete, the stress has relaxed by a value of σ_{max}.

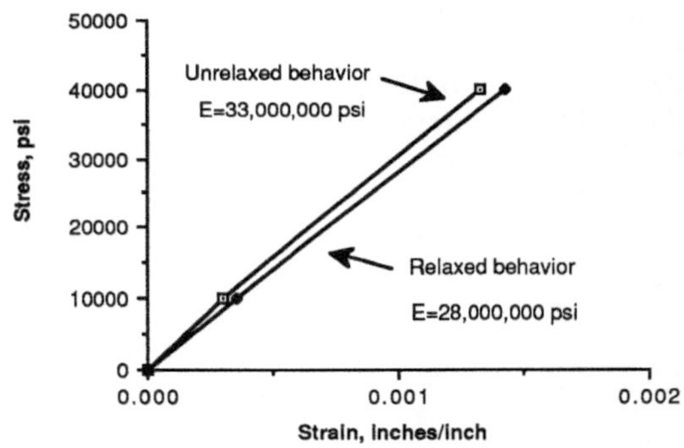

Fig. 5 Loading-rate effects on Young's modulus

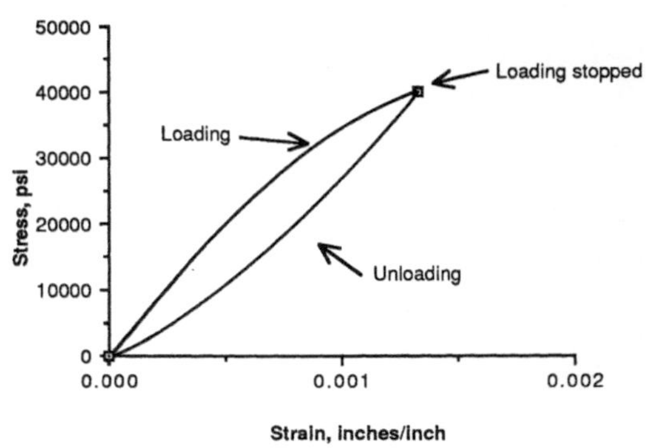

Fig. 6 Hysteresis in the loading-unloading curve

cast iron bases to minimize transmission of machine vibrations to the floor and surrounding area. However, a high damping capacity is not always a useful material quality. Bells, for example, are constructed from materials with low damping capacities because both the length of bell ring and the loudness of the tone will increase as the damping capacity decreases.

Anelasticity, damping, stress relaxation, and the elastic moduli of most metals and alloys are dependent on the microstructure of the material as well as on test conditions. These properties are not typically determined by tensile-testing techniques. However, these properties, as well as the machine parameters, influence the shape of the stress-strain curve. Therefore, an awareness of these phenomena may be useful in the interpretation of tensile-test data.

The Proportional Limit

The apparent stress necessary to produce the onset of curvature in the tensile stress-strain relationship is the proportional limit. The proportional limit is defined as the maximum stress at which strain remains directionally proportional to stress. Departures from proportionality may be attributed to anelasticity and/or the initiation of plastic deformation. The ability to detect the occurrence of these phenomena during a tensile test is dependent on the accuracy with which stress and strain are measured. The measured value of the proportional limit decreases as the accuracy of the measurement increases (Fig. 7). Because the measured value of proportional limit is dependent on test accuracy, the proportional limit is not generally reported as a tensile property of metals and alloys. Furthermore, values of proportional limit have little or no utility in the selection, qualification, and use of metals and alloys for engineering service. A far more reproducible and practical stress is the yield strength of the material.

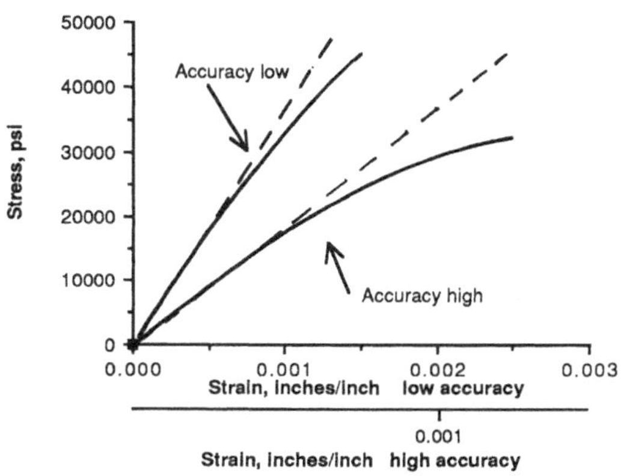

Fig. 7 Effect of accuracy of measurement on the determination of the proportional limit

Yielding and the Onset of Plasticity

The yield strength of a metal or alloy may be defined as the stress at which that metal or alloy exhibits a specified deviation from the proportionality between stress and strain. Very small deviations from proportionality may be caused by anelastic effects, but these departures from linear behavior are fully reversible and do not represent the onset of significant plastic (nonreversible) deformation or yielding. Theoretical values of yield strength, σ_{theor}, are calculated from equations such as

$$\sigma_{theor} = \frac{E}{2\pi}$$

Based on these calculations, yielding should not take place until the applied stress is a significant fraction of the modulus of elasticity. These estimates for yielding generally overpredict the measured yield strengths by factors of at least 100, as summarized in Table 1.

Table 1 Young's modulus and theoretical and measured yield strengths of selected metals at 20 °C (68 °F)

Metal	Young's modulus		Yield strength: Theoretical		Yield strength: Measured(a)	
	GPa	10^6 psi	GPa	10^6 psi	MPa	ksi
Aluminum	70.3	10.2	11	1.6	26	4
Nickel	199	28.9	32	4.6	234	34
Silver	82.7	12.0	13	1.9	131	19
Steel (mild)	212	30.7	34	4.9	207	30
Titanium	120	17.4	19	2.7	172	25

(a) Measured values of yield strength are dependent on the metallurgical condition of the material.

The discrepancy between the theoretical and actual yield strength results from the motion of dislocations. Dislocations are defects in the crystal lattice, and the motion of these defects is a primary mechanism of plastic deformation in most metals and alloys.

There are three very broad categories of crystal defects in metallic solids:

1. Point defects, including vacancies and alloy or impurity atoms
2. Line defects of dislocations
3. Area defects, including grain and twin boundaries, phase boundaries, inclusion-matrix interfaces, and even external surfaces

The characterization of these defects in any particular material may be accomplished through metallography. Optical metallography is used to characterize area defects or grain structure, as shown in Fig. 8.

Transmission electron microscopy is used to characterize line defects or dislocation substructure, as shown in Fig. 9. More specialized metallographic techniques, such as field ion microscopy, are used to characterize the point defects. Interaction among defects is common, and most techniques that alter the yield strengths of metals and alloys are dependent on defect interactions to alter the ease of dislocation motion.

Dislocation mobility is dependent on the alloy content, the extent of cold work, the size, shape, and distribution of inclusions and second phase particles, and the grain size of the alloy. The strength of most metals increases as alloy content increases, because the alloy (or impurity) atoms interact with dislocations and inhibit subsequent motion. Thus, this type of strengthening results from the interaction of point defects with line defects. Such strengthening was discovered by ancient metallurgists and was the basis for the Bronze Age. The strength, and therefore the utility, of copper was significantly increased by dissolving tin to form bronze. The yield strength of the copper-tin alloys (bronze) was sufficiently high for the manufacture of tools and spear points. This strengthening mechanism was not discovered by the native Americans, and on the American Continents, copper was used for jewelry but not for more practical purposes. Bronzes (Cu-Sn alloys), brasses (Cu-Zn alloys), Monels (Ni-Cu alloys), and many other alloy systems are dependent on solid-solution strengthening to control the yield strength of the material. The effects of nickel and zinc additions on the yield strength of copper are illustrated in Fig. 10.

Cold work is another effective technique for increasing the strength of metals and alloys. This strengthening mechanism is effective because the number of dislocations in the metal increases

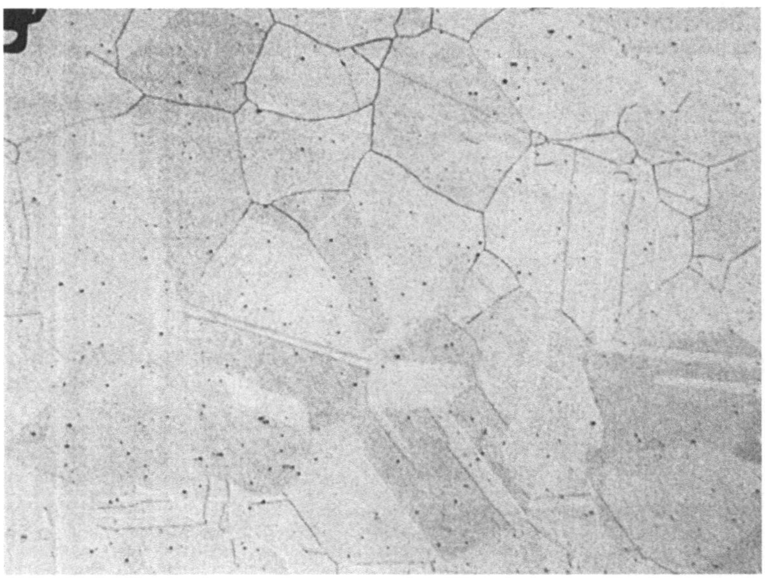

Fig. 8 Optical photomicrograph of type 304 stainless steel. The apparent defects include grain boundaries, twin boundaries, and inclusions. 100×

as the percentage of cold work increases. These additional dislocations inhibit the continued motion of other dislocations in much the same manner as increased traffic decreases the mobility of cars along a highway system. Cold work is an example of strengthening because of line defects interacting with other line defects in a crystal lattice. Many manufactured components depend on cold work to raise their strength to the required level. Rolling, stamping, forging, drawing, swaging, and even extrusion may be used to provide the necessary cold work. The effects of cold work on the hardness and strength of a 70%Cu-30%Zn alloy, iron and copper are illustrated in Fig. 11. The yield and tensile strengths follow nearly identical trends, with the yield strength increasing slightly faster than the tensile strength.

Grain and phase boundaries also block dislocation motion. Thus, the yield strength of most metals and alloys increases as the number of grain boundaries increases and/or as the percentage of second phase in the structure increases. A decrease in the grain size increases

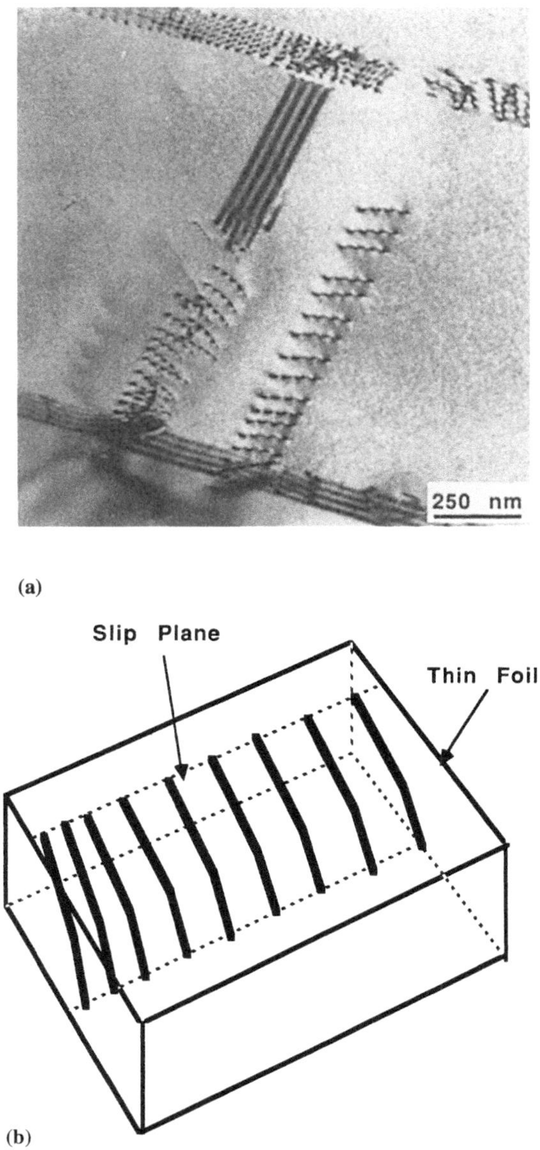

Fig. 9 Dislocations. (a) Transmission electron micrograph of type 304 stainless steel showing dislocation pileups at an annealing twin boundary. (b) Schematic representation of dislocations on a slip plane

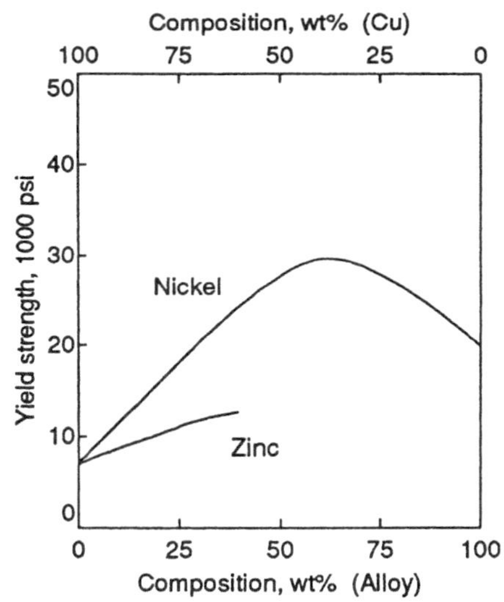

Fig. 10 Effects of nickel and zinc contents on the yield strengths of copper alloys

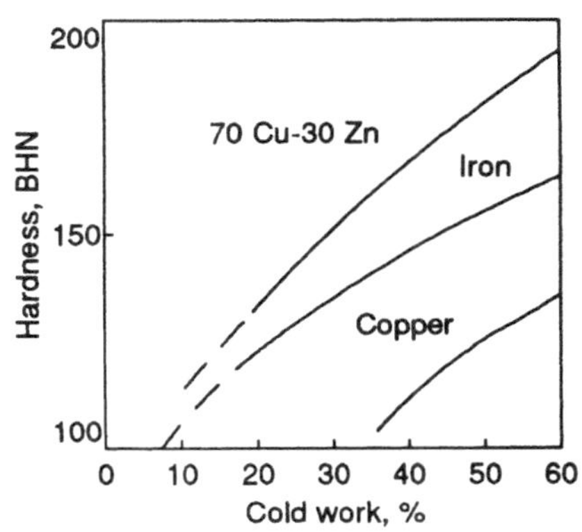

Fig. 11 Effects of cold work on the hardnesses and strengths of brass, iron, and copper

the number of grain boundaries per unit volume, thus increasing the density of area defects in the metal lattice. Because interactions between area defects and line defects inhibit dislocation mobility, the yield strengths of most metals and alloys increase as the grain size decreases and as the number of second-phase particles increases. The effects of grain size are illustrated in Fig. 12. Because of these and other strengthening mechanisms, any given alloy may show a wide range of yield strengths. The range will be dependent on the grain size, percentage of cold work, distribution of second-phase particles, and other relatively easily quantified, microstructural parameters. The values of these microstructural parameters depend on the thermomechanical history of the material; thus a knowledge of these very important metallurgical variables is almost a necessity for intelligent interpretation of yield-strength data and for the design and utilization of metallic structures and components.

The most common definition of yield strength is the stress necessary to cause a plastic strain of 0.002 mm/mm (in./in.). This strain represents a readily measurable deviation from proportionality, and the stress necessary to produce this deviation is the 0.2% offset yield strength (see Chapter 3, "Uniaxial Tensile Testing," for a detailed description of the 0.2% offset yield strength). A significant amount of dislocation motion is required before a 0.2% deviation from linear behavior is reached. Therefore, in a standard tensile test, the 0.2% offset yield strength is almost independent of test-machine variables, gripping effects and reversible nonlinear strains such as anelasticity. Because of this independence, the 0.2% offset yield strength is a reproducible material property that may be used in the characterization of the mechanical properties of metals and alloys. However, it is vital to realize that the magnitude of the yield strength, or any other tensile property, is dependent on the defect structure of the material tested. Therefore, the thermomechanical history of the metal or alloy must be known if yield strength is to be a meaningful design parameter.

The Yield Point

The onset of dislocation motion in some alloys, particularly low-carbon steels tested at room temperature, is sudden, rather than a relatively gradual process. This sudden occurrence of yielding makes the characterization of yielding by a 0.2% offset method impractical. Because of the sudden yielding, the stress-strain curve for many mild steels has a yield point, and the yield strength is characterized by lower yield stress. The yield point develops because of interactions between the solute (dissolved) atoms and dislocations in the solvent (host) lattice. The solute-dislocation interaction in mild steels involves carbon migration to and interaction with dislocations. Because the interaction causes the concentration of solute to be high in the vicinity of the dislocations, the yield point is said to develop because of segregation of carbon to the dislocations.

Many of the interstitial sites around dislocations are enlarged and are therefore low-energy or favored sites for occupancy by the solute atoms. When these enlarged sites are occupied, a high concentration or atmosphere of solute is associated with the dislocation. In mild steels, the solute segregation produces carbon-rich atmospheres at dislocations. Motion of the dislocations is inhibited because such motion requires

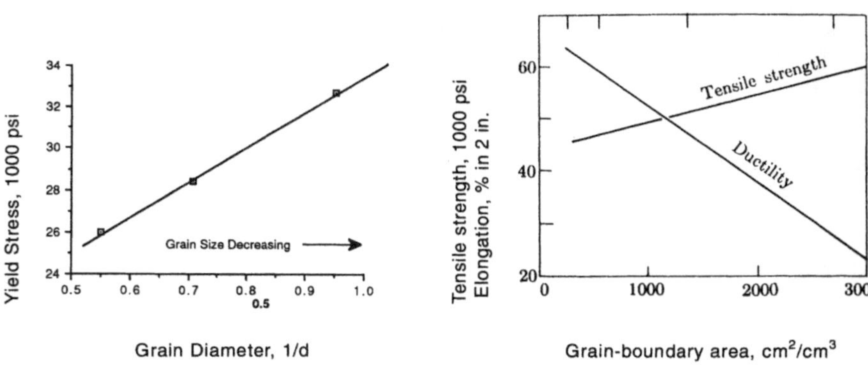

Fig. 12 The effects of grain size on the strengths and ductilities of metals and alloys

the separation of the dislocations from the carbon atmospheres. As soon as the separation takes place, the stress required for continued dislocation motion decreases and, in a tensile test, the lower yield strength is reached. This yielding process involves dislocation motion in localized regions of the test specimen. Because dislocation motion is plastic deformation, the regions in which dislocations moved represent deformed regions or bands in the metal. These localized, deformed bands are called Lüders bands (see Fig. 4 in Chapter 2). Once initiated, additional strain causes the Lüders bands to propagate throughout the gage length of the test specimen. This propagation takes place at a constant stress which, is the lower yield strength of the steel. When the entire gage section has yielded, the stress-strain curve begins to rise because of the interaction of dislocations with other dislocations, and strain-hardening initiates.

The existence of a yield point and Lüders band is particularly important because of the impact of the sudden softening and localized straining on processing techniques. For example, sudden localized yielding will cause jerky material flow. Jerky flow is undesirable in a drawing operation because the load on the drawing equipment would change rapidly, causing large energy releases that must be absorbed by the processing equipment. Furthermore, localized Lüders strains will produce stretch marks in stamped materials. These stretch marks are termed "stretcher strains" and are readily apparent on stamped surfaces. This impairs the surface appearance and reduces the utility of the component. If materials that do not have yield points are stamped, smooth surfaces are developed because the strain-hardening process spreads the deformation uniformly throughout the material. Uniform, continuous deformation is important in many processing and finishing operations; thus, it is important to select a combination of material-processing conditions that minimize the tendency toward localized yielding.

Grain-Size Effects on Yielding

The metals and alloys used in most structural applications are polycrystalline. The typical metallic object contains tens of thousands of microscopic crystals or grains. The size of the grains is difficult to define precisely because the 3-D shape of the grain is quite complex. If the grain is assumed to be spherical, the grain diameter, d, may be used to characterize size. More precise characterizations of grain size include the mean grain intercept, $\bar{l}$, and the ratio of grain-boundary surface to grain volume, S_v. These two parameters may be established through quantitative metallographic techniques. The grain structure of the metal or alloy of interest is examined at a magnification, X, and a line of a known length l is overlaid on the microstructure. The number of grain-boundary intersections with that line in measured, divided by the length of the line, and multiplied by the magnification. The resulting parameter, N_l, is the average number of grain boundaries intersected per unit length of line. This value for N_l is related to $\bar{l}$ and S_v through

$$\bar{l} = 1/N_l$$

and

$$S_v = 2N_l$$

Unfortunately, for historical reasons, the parameter d is the most common measure used to characterize the influence of grain size on the yield strengths of metals and alloys. This influence is frequently quantified through the Hall-Petch relationship whereby yield strength, σ_y, is related to grain size through the empirical equation

$$\sigma_y = \sigma_0 + kd^{-1/2}$$

The empirical constants σ_0 and k are the lattice friction stress and the Petch slope, respectively. A graphical representation of this relationship is shown in Fig. 12(a).

Grain boundaries act as barriers to dislocation motion, causing dislocations to pile up behind the boundaries. This pileup of dislocations concentrates stresses at the tip of the pileup, and when the stress is sufficient, additional dislocations may be nucleated in the adjacent grain. The magnitude of the stress at the tip of a dislocation pileup is dependent on the number of dislocations in the pileup. The number of dislocations that may be contained in a pileup increases with increasing grain size because of the larger grain volume. This difference in the number of dislocations in a pileup makes it easier for new dislocations to be nucleated in a large-grain metal than in a fine-grain metal of comparable purity, and this difference in the ease of dislocation nucleation extrapolates directly to a difference in yield strength. Based on this model for grain-size strengthening, the effects of grain size should exist even after the yield strength is exceeded.

Strain Hardening and the Effect of Cold Work

A stress-strain curve for relatively pure nickel (Fig. 13) shows that the 0.2% offset yield strength of this metal was approximately 235 MPa (34 ksi). The stress necessary to cause continued plastic deformation increased as the tensile strain increased. After a strain of approximately 1%, the stress necessary to produce continued deformation was 330 MPa (48 ksi), and after 10% strain the necessary stress had increased to approximately 415 MPa (60 ksi).

The stress necessary for continued deformation is frequently designated as the flow stress at that specific tensile strain. Thus, at 1% strain, the flow stress is 330 MPa (48 ksi), and the flow stress at 10% strain is 415 MPa (60 ksi). This increasing flow stress with increasing strain is the basis for increasing the strength of metals and alloys by cold working. The effects of grain size on the strength of the alloy are retained throughout the cold working process (Fig. 14).

The fact that the grain-size dependence of strength is retained throughout the strain-hardening process demonstrates the possibility for interactions among the various strengthening mechanisms in metals and alloys. For example, cold work causes strength increases through the interaction between point defects and dislocations, and these effects are additive to the effects of alloying. This is apparent in Fig. 15(a), where the incremental increase in strength resulting

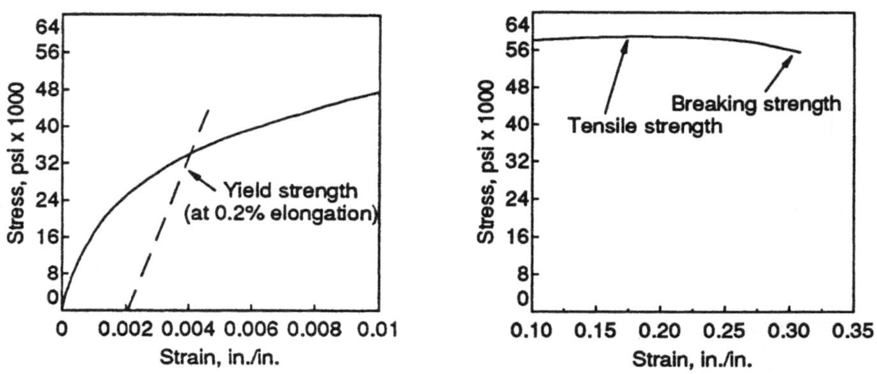

Fig. 13 Stress-strain curve for nickel

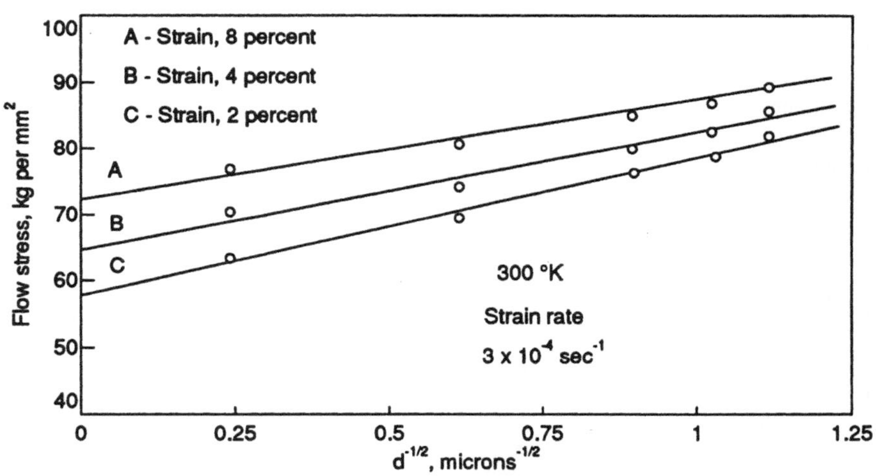

Fig. 14 Effects of grain size and cold work on the flow stress of titanium

from zinc additions to copper becomes larger when the alloy is cold worked.

Furthermore, strength is not the only tensile property affected by the cold working process. Ductility decreases with increasing cold work (Fig. 15b and c), and, if cold working is too extensive, metals and alloys will crack and fracture during the working operation. The over-all effects of cold work on strength and ductility are illustrated in Fig. 16, which compares the tensile behavior of steel rods that were cold drawn various amounts before being tested to fracture.

Note that the increase in strength and decrease in ductility cause the area under the stress-strain curve to decrease. This is significant because that area represents the work or energy required to fracture the steel bar, and the tensile-test results demonstrate that this energy decreases as the percentage of cold work increases.

Cold working, whether by rolling, drawing, stamping, or forging, changes the microstructure. The resulting grain shape is determined by the direction of metal flow during processing, as illustrated in Fig. 17. The grains in the cold rolled specimen were elongated and flattened, thus changing from the semispherical grains in Fig. 17(a) to the pancake-shape grains in Fig. 17(b). A rod-drawing process would have produced needle-shape grains in this same alloy. In addition to the changes in grain shape, the grain interior is distorted by cold forming operations. Bands of high dislocation density (deformation bands) develop, twin boundaries are bent, and grain boundaries become rough and distorted (Fig. 18). Because the deformation-induced changes in microstructure are anisotropic, the tensile properties of wrought metals and alloys frequently are anisotropic. The strain-hardened microstructures and the associated mechanical properties that result from cold work can be significantly altered by annealing. The microstructural changes that are introduced by heating to higher temperatures are dependent on both the time and temperature of the anneal. This temperature dependence is illustrated in Fig. 19 and results because atom motion is required for the anneal to be effective.

The sudden drop in hardness seen in the Cu-5%Zn alloy in Fig. 19 results from recrystallization, or the formation of new grains, in the alloy. Plastic deformation of metals and alloys at temperatures below the recrystallization temperature is cold work, and plastic deformation at temperatures above the recrystallization temperature is hot work. Metals and alloys, in tensile tests above the recrystallization temperature,

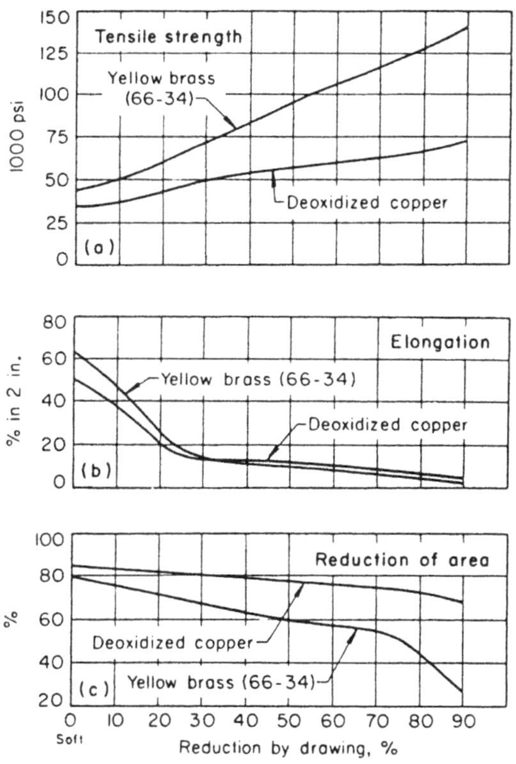

Fig. 15 Effects of cold work on the tensile properties of copper and yellow brass. (a) Tensile strength. (b) Elongation. (c) Reduction in area

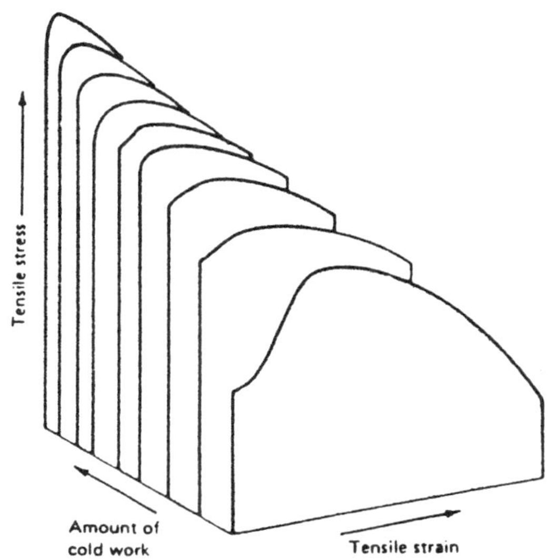

Fig. 16 Effects of cold work on the tensile stress-strain curves of low-carbon steel bars

do not show significant strain hardening, and the tensile yield strength becomes the maximum stress that the material can effectively support. See Chapter 13, "Hot Tensile Testing," for information on the effects of elevated temperatures on tensile properties.

Ultimate Strength

The ability to strain harden is one of the general characteristics in mechanical behavior that distinguish metals and alloys from most other engineering materials. Not all metallic materials

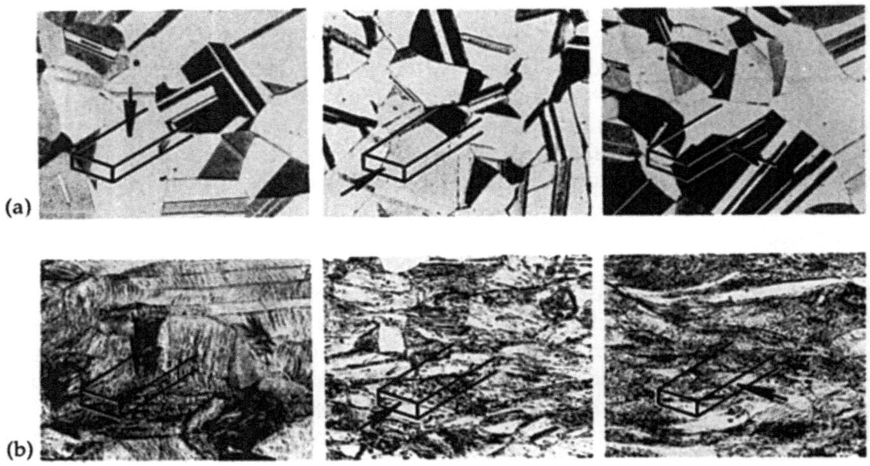

Fig. 17 Effect of cold rolling on grain shape in cartridge brass. (a) Grain structure in annealed bar. (b) Grain structure in same bar after 50% reduction by rolling. Diagram in the lower left of each micrograph indicates orientation of the view relative to the rolling plane of the sheet. 75×

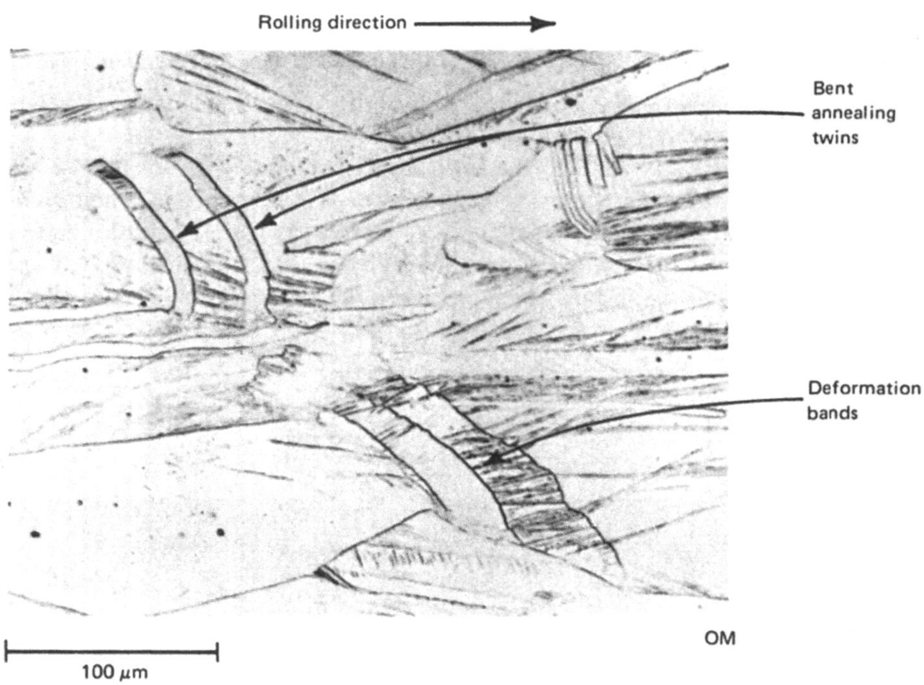

Fig. 18 Grain structure of severely deformed Cu-5%Zn alloy

exhibit this characteristic. Chromium, for example, is very brittle and fractures in a tensile test without evidence of strain hardening. The stress-strain curves for these brittle metals are similar to those of most ceramics (Fig. 20). Fracture occurs before significant plastic deformation takes place.

Such brittle materials have no real yield strength, and the fracture stress is the maximum stress that the material can support. Most metals and alloys, however, undergo plastic deformation prior to fracture, and the maximum stress that the metal can support is appreciably higher than the yield strength. This maximum stress (based on the original dimensions) is the ultimate or tensile strength of the material.

The margin between the yield strength and the tensile strength provides an operational safety factor for the use of many metals and alloys in structural systems. Other than this safety margin, the actual value of tensile strength has very little practical use. The ability of a structure to withstand complex service loads bears little relationship to tensile strength, and structural designs must be based on yielding.

Tensile strength is easy to measure and is frequently reported because it is the maximum stress on an engineering stress-strain curve. Engineering codes may even specify that a metal or alloy meet some tensile-strength requirement. Historically, tensile strengths, with experience-based reductions to avoid yielding, were used in design calculations. As the accuracy of measurement of stress-strain curves improved, utilization of tensile strength diminished, and by the 1940s most design guidelines were based on yielding. There is a large empirical database that correlates tensile strength with hardness, fatigue strength, stress rupture, and mechanical properties. These correlations, historical code requirements, and the fact that structural designs incorporating brittle materials must be based on tensile strength provide the technical basis for the continuing utilization of tensile strengths as design criteria.

Cold work and other strengthening mechanisms for metals and alloys do not increase tensile strength as rapidly as they increase yield strength. Therefore, as evident in Fig. 16, strengthening processes frequently are accompanied by a reduction in the ability to undergo plastic strain. This reduction decreases the ability of the material to absorb energy prior to fracture and, in many cases, is important to successful materials utilization. Analysis of the tensile behavior of metals and alloys can provide insight into the energy-absorbing abilities of the material.

Toughness

The ability to absorb energy without fracturing is related to the toughness of the material. Most, if not all, fractures of engineering materials are initiated at pre-existing flaws. These flaws may be small enough to be elements of the microstructure or, when slightly larger, may be macroscopic cracks in the material or, in the extreme, visually observable discontinuities in the

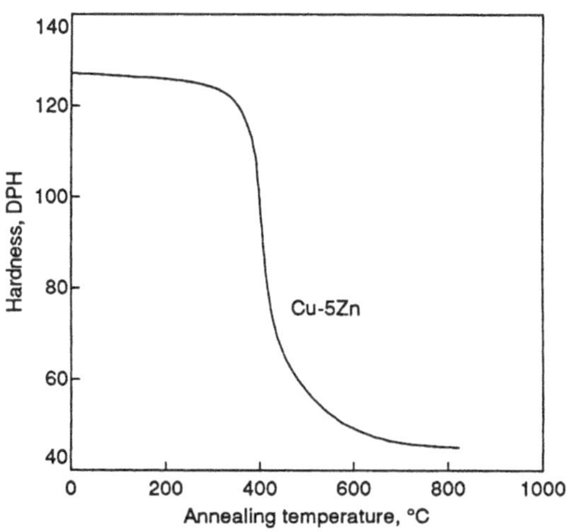

Fig. 19 Effect of annealing on hardness of cold rolled Cu-5%Zn brass. Hardness can be correlated with strength, and the strengths of this alloy would show similar annealing effects.

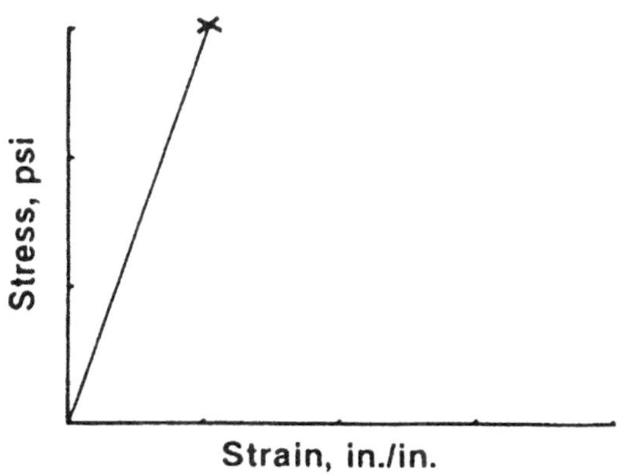

Fig. 20 Stress-strain curve for brittle material

structure. A tough material resists the propagation of flaws through processes such as yielding and plastic deformation. Most of this deformation takes place near the tip of the flaw. Because fracture involves both tensile stress and plastic deformation, or strain, the stress-strain curve can be used to estimate material toughness. However, there are specific tests designed to measure material toughness. Most of these tests are conducted with precracked specimens and include both impact and fracture-mechanics type studies (see *Mechanical Testing and Evaluation,* Volume 8 of *ASM Handbook,* for descriptions of these tests). Toughness calculations based on tensile behavior are estimates and should not be used for design.

The area under a stress-strain curve (normalized to specimen dimensions) is a measure of the energy absorbed by the material during a tensile test. From that standpoint, this area is a rough estimate of the toughness of the material. Because the plastic strain associated with tensile deformation of metals and alloys is typically several orders of magnitude greater than the accompanying elastic strain, plasticity or dislocation motion is very important to the development of toughness. This is illustrated by the stress-strain curves for a brittle, a semibrittle, and a ductile material shown schematically in Fig. 21.

Brittle fracture (see Fig. 21a), takes place with little or no plastic strain, and thus the area under the stress-strain curve, *A* is given by

$$A = (1/2)\sigma\varepsilon$$

and, because all the strain is elastic,

$$\sigma = E\varepsilon$$

Combining these equations gives

$$A = (1/2)(\sigma_f^2)E$$

where σ_f is the fracture stress. If the fracture stress for this material were 205 MPa (30 ksi) and Young's modulus were 205 GPa (30×10^6 psi), the fracture energy, estimated from the stress-strain curve, would be 1.2×10^{-3} J/mm^3 (15 lbf · in./in.^{3}) per cubic inch of gage section in the test specimen. If the test specimen were ductile (Fig. 21c), the area under the stress-strain curve could be estimated from

$$A = (\sigma_y + \sigma_t)(\varepsilon_f/2)$$

where σ_y is yield strength, σ_t is tensile strength, and ε_f is strain to fracture. Estimation of the fracture energy from the typical tensile properties of mild steel test specimens, σ_y = 205 MPa (30 ksi), σ_t = 415 MPa (60 ksi), and ε_f = 0.3, gives 1.12 J/mm^3 (13,500 lbf · in./in.^{3}) of gage section in the test specimen.

The ratio of the energy for ductile fracture to the energy for brittle fracture is 900. This ratio

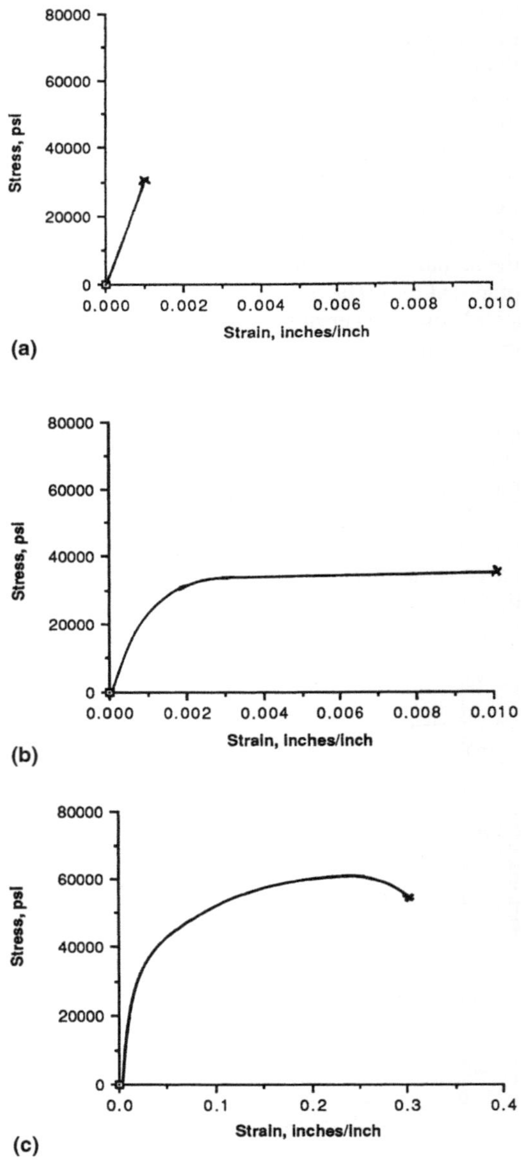

Fig. 21 Stress-strain curves for materials showing various degrees of plastic deformation or ductility. (a) Brittle material. (b) Semibrittle material. (c) Ductile material

is based on the calculations shown above and will increase with increasing strain to fracture and with increasing strain hardening. These area and energy relationships are only approximations. The stresses used in the calculations are based on the original dimensions of the test specimen. The utility of such toughness estimates is the ease with which testing can be accomplished and the insight that the estimates provide into the importance of plasticity to the prevention of fracture. This importance is illustrated by considering the area under the stress-strain curve shown in Fig. 21(b). Assuming that, for this semibrittle material, the yield strength and tensile strength are both 205 MPa (30 ksi) (no significant strain hardening) and that fracture takes place after a plastic strain of only 0.01 mm/mm (in./in.), the area under the stress-strain curve is 410 J (300 lbf · in.) per unit area. This area is 20 times higher than the area under the stress-strain curve for brittle fracture shown in Fig. 21(a). This calculation demonstrates that a plastic strain of only 0.01% can have a remarkable effect on the ability of a material to absorb energy without fracturing.

Toughness is a very important property for many structural applications. Ship hulls, crane arms, axles, gears, couplings, and airframes are all required to absorb energy during service. The ability to withstand earthquake loadings, system overpressures, and even minor accidents will also require material toughness. Increasing the strength of metals and alloys generally reduces ductility and, in many cases, reduces toughness. This observation illustrates that increasing the strength of a material may increase the probability of service-induced failure when material toughness is important for satisfactory service. This is seen by comparing the areas under the two stress-strain curves in Fig. 22.

The cross-hatched regions in Fig. 22 illustrate another tensile property—the modulus of resilience, which can be measured from tensile stress-strain curves. The ability of a metal or alloy to absorb energy through elastic processes is the resilience of the material. The modulus of resilience is defined as the area under the elastic portion of the stress-strain curve. This area is the strain energy per unit volume and is equal to

$$A = (1/2)(\sigma_y^2/E)$$

Increasing the yield strength and/or decreasing Young's modulus will increase the modulus of resilience and improve the ability of a metal or alloy to absorb energy without undergoing permanent deformation.

Ductility

Material ductility in a tensile test is generally established by measuring either the elongation to fracture or the reduction in area at fracture. In general, measurements of ductility are of interest in three ways:

1. To indicate the extent to which a metal can be deformed without fracture in metalworking operations such as rolling and extrusion.
2. To indicate to the designer, in a general way, the ability of the metal to flow plastically before fracture. A high ductility indicates that the material is 'forgiving' and likely to deform locally without fracture should the designer err in stress calculation or the prediction of severe loads.
3. To serve as an indicator of changes in impurity level or processing conditions. Ductility measurements may be specified to assess material quality even though no direct relationship exists between the ductility measurement and performance.

Tensile ductility is therefore a very useful measure in the assessment of material quality. Many codes and standards specify minimum values for tensile ductility. One reason for these specifications is the assurance of adequate toughness without the necessity of requiring a more costly toughness specification. Most changes in alloy composition and/or processing conditions will produce changes in tensile duc-

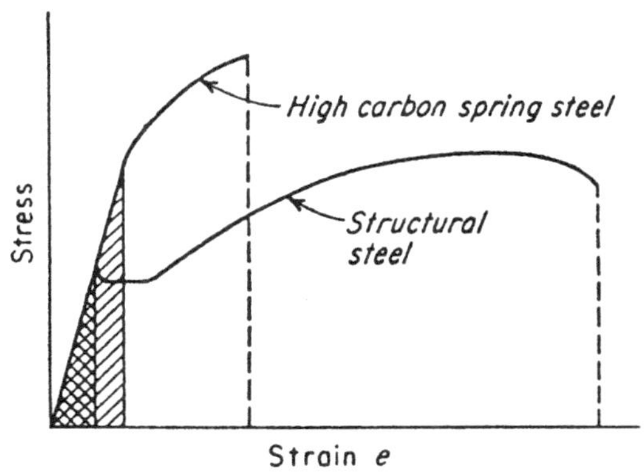

Fig. 22 Comparison of the stress-strain curves for high- and low-toughness steels

tility. The "forgiveness" found in many metals and alloys results from the ductility of these materials. Although there is some correspondence between tensile ductility and fabricability, the metalworking characteristics of metals and alloys are better correlated with the ability to strain harden than with the ductility of the material. The strain-hardening abilities of many engineering alloys have been quantified through the analysis of true stress-strain behavior.

True Stress-Strain Relationships

Conversion of engineering stress-strain behavior to true stress-strain relationships may be accomplished using the techniques represented by Eq 8 through 13 in Chapter 1, "Introduction to Tensile Testing." This conversion, summarized graphically in Fig. 23, demonstrates that the maximum in the engineering stress-strain curve results from tensile instability, not from a decrease in the strength of the material. The drop in the engineering stress-strain curve is artificial and occurs only because stress calculations are based on the original cross-sectional area. Both testing and analysis show that, for most metals and alloys, the tensile instability corresponds to the onset of necking in the test specimen. Necking results from strain localization; thus, once necking is initiated, true strain cannot be calculated from specimen elongation. Because of these and other analytical limitations of engineering stress-strain data, if tensile data are used to understand and predict metallurgical response during the deformation associated with fabrication processes, true stress-true strain relationships are preferred.

The deformation that may be accommodated, without fracture, in a deep drawing operation varies with the material. For example, austenitic stainless steels may be successfully drawn to 50% reductions in area whereas ferritic steel may fail after only 20 to 30% reductions in area in similar drawing operations. Both types of steel will undergo in excess of 50% reduction in area in a tensile test. This difference in drawability correlates with the strain-hardening exponent (n) and therefore is apparent from the slope of the true stress-strain curves for the two alloys (Fig. 24). A detailed discussion on the strain-hardening exponent, or coefficient, can be found in Chapter 6, "Tensile Testing for Determining Sheet Formability."

The strain-hardening exponents, or n values for ferritic and austenitic steels, are typically 0.25 and 0.5, respectively. A perfectly plastic material would have a strain-hardening exponent of zero and a completely elastic solid would have a strain-hardening exponent of one. Most metals and alloys have strain-hardening exponents between 0.1 and 0.5. Strain-hardening exponents correlate with the ability of dislocations to move around or over dislocations and other obstacles in their path. Such movement is termed "cross slip." When cross slip is easy, dislocations do not pile up behind each other and strain-hardening exponents are low. Mild steels, aluminum, and some nickel alloys are examples of materials that undergo cross slip easily. The value of n increases as cross slip becomes more difficult. Cross slip is very difficult in austenitic stainless steels, copper, and brass and the strain-hardening exponent for these alloys is approximately 0.5.

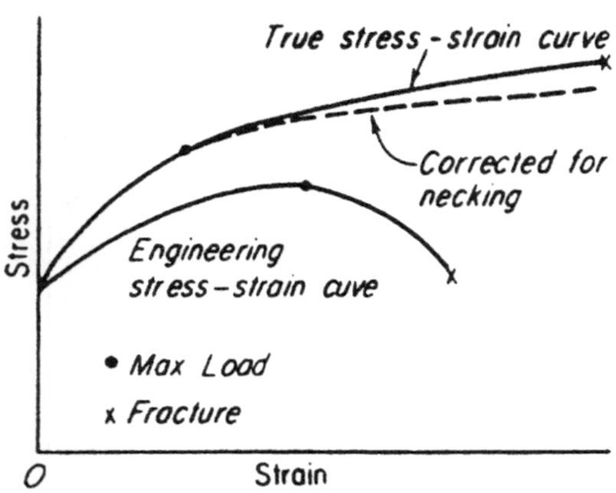

Fig. 23 Comparison of engineering and true stress-strain curves

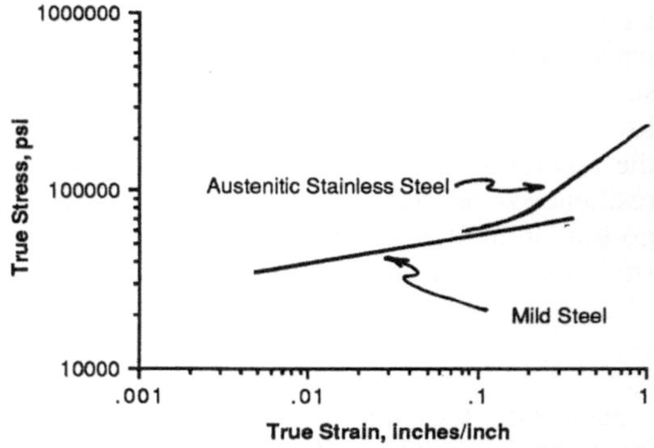

Fig. 24 True stress-strain curves for austenitic and mild steels

Tensile specimens, sheet or plate material, wires, rods, and metallic sections have spot-to-spot variations in section size, yield strength, and other microstructural and structural inhomogeneities. Plastic deformation of these materials initiates at the locally weak regions. In the absence of strain hardening, this initial plastic strain would reduce the net section size and focus continued deformation in the weak areas. Strain hardening, however, causes the flow stress in the deformed region to increase. This increase in flow stress increases the load necessary for continued plastic deformation in that area and causes the deformation to spread throughout the section. The higher the strain-hardening exponent, the greater the increase in flow stress and the greater the tendency for plastic deformation to become uniform. This tendency has a major impact on the fabricability of metals and alloys. For example, the maximum reduction in area that can be accommodated in a drawing operation is equal to the strain-hardening exponent as determined from the true stress-strain behavior of the material. Because of such correlations, the effects of process variables such as strain rate and temperature can be evaluated through tensile testing. This provides a basis to approximate the effects of process variables without direct, in-process assessment of the variables.

Temperature and Strain-Rate Effects

The yield strengths of most metals and alloys increase as the strain rate increases and decrease as the temperature increases. This strain-rate temperature dependence is illustrated in Fig. 25. These dependencies result from a combination of several metallurgical effects.

For example, dislocations are actually displacements and therefore cannot move faster than the speed of sound. Furthermore, as dislocation velocities approach the speed of sound, cross slip becomes increasingly difficult and the strain-hardening exponent increases. This increase in the strain-hardening exponent increases the flow stress at any given strain, thus increasing the yield strength of the material. A decrease in ductility and even a transition from ductile to brittle fracture may also be associated with strain-rate-induced increases in yield strength. In many respects, decreasing the temperature is similar to increasing the strain rate. The mobility of dislocations decreases as the temperature decreases, and thus, for most metals and alloys, the strength increases and the ductility decreases as the temperature is lowered. If the reduction in dislocation mobility is sufficient, the ductility may be reduced to the point of brittle fracture. Metals and alloys that show a transition from ductile to brittle when the temperature is lowered should not be used for structural applications at temperatures below this transition temperature.

Dislocation motion is inhibited by interactions between dislocations and alloy or impurity (foreign) atoms. The effects of these interactions are both time and temperature dependent. The interaction acts to increase the yield strength and limit ductility. These processes are most effective when there is sufficient time for foreign atoms to segregate to the dislocation and when dislocation velocities are approximately equal to the diffusion velocity of the foreign atoms. Therefore, at any given temperature, dislocation-foreign atom interactions will be at a maximum at some intermediate strain rate. At low strain rates, the foreign atoms can diffuse as rapidly as the dislocations move and there is little or no tendency for the deformation process to force a separation of dislocations from their solute atmospheres. At high strain rates, once separation has been effected, there is not sufficient time for the atmosphere to be re-established during the test. Atom movement increases with increasing temperature, thus the strain rates that allow dislocation-foreign atom interactions to occur are temperature dependent. Because these interactions limit ductility, the elongation in a tensile test may show a minimum at intermediate test temperatures where such interactions are most effective (Fig. 26).

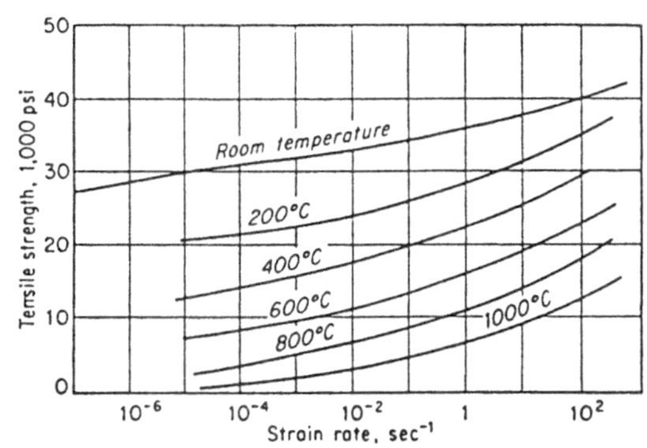

Fig. 25 Effects of temperature and strain rate on the strength of copper

The effects of time-dependent dislocation foreign atom interactions on the stress-strain curves of metals and alloys are termed "strain aging" and "dynamic strain aging." Strain aging is generally apparent when a tensile test, of a material that exhibits a sharp yield point, is interrupted. If the test specimen is unloaded after being strained past the yield point, through the Lüders strain region and into the strain-hardening portion of the stress-strain curve, either of two behaviors may be observed when the tensile test is resumed (Fig. 27). If the specimen is reloaded in a short period of time, the elastic portion of the reloading curve (line *d-c* in Fig. 27a) is parallel to the original elastic loading curve (line *a-b* in Fig. 27a) and plastic deformation resumes at the stress level (level *c*) that was reached just before the test was interrupted.

However, if the time between unloading and reloading is sufficient for segregation of foreign atoms to the dislocations, the yield point reappears (Fig. 27b) and plastic strain is not reinitiated when the unloading stress level (point *c*) is reached. This reappearance of the yield point is strain aging, and the strength of the strain-aging peak is dependent on both time and temperature because solute-atom diffusion and segregation to dislocations are required for the peak to develop. If tensile strain rates are in a range where solute segregation can occur during the test, dynamic strain aging is observed. Segregation pins the previously mobile dislocations and raises the flow stress, and when the new, higher flow stress is reached the dislocations are separated from the solute atmospheres and the flow stress decreases. This alternate increase and decrease in flow stress causes the stress-strain curve to be serrated (Fig. 28).

Serrated flow is common in mild steels, in some titanium and aluminum alloys, and in other

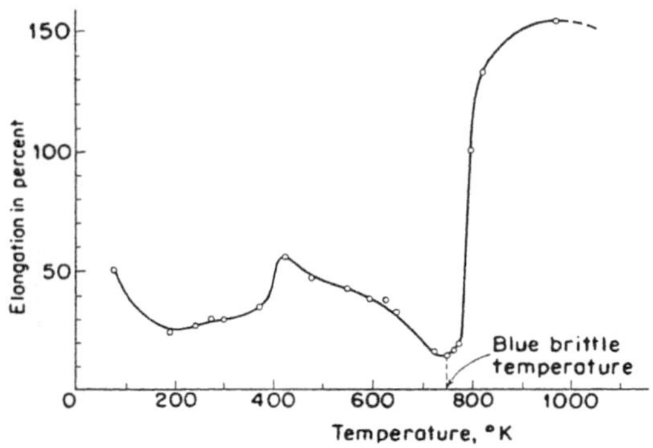

Fig. 26 An intermediate-temperature ductility minimum in titanium

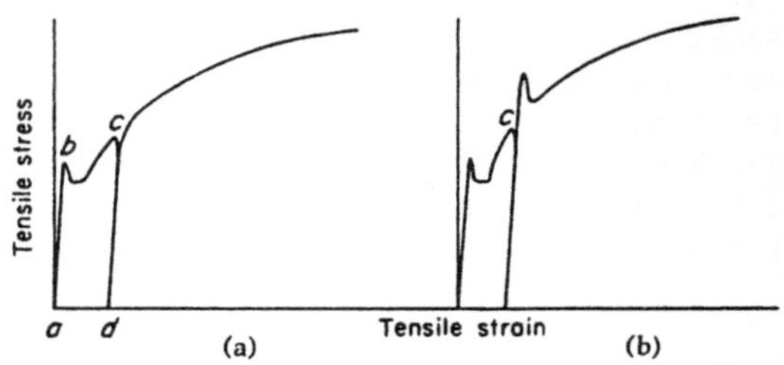

Fig. 27 Illustration of strain aging during an interrupted tensile test. (a) Specimen reloaded in a short period of time. (b) Time between loading and unloading is sufficient

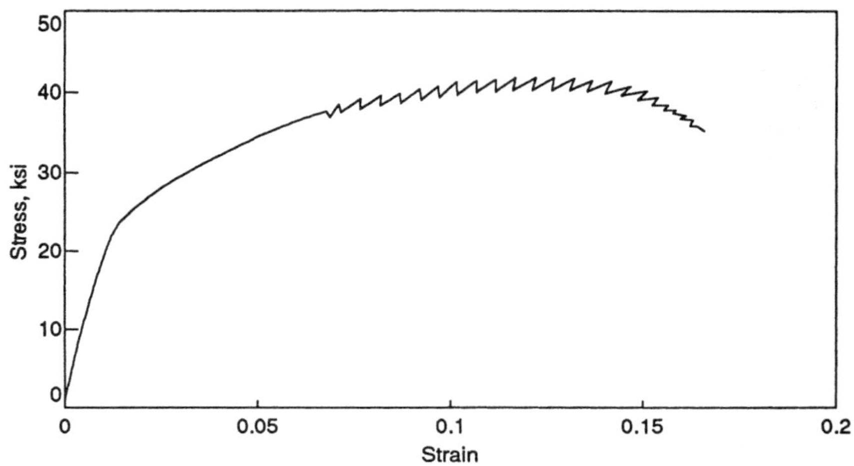

Fig. 28 Dynamic strain aging or serrated yielding in an aluminum alloy tested at room temperature

metals that contain mobile, alloy or impurity elements. This effect was initially studied in detail by Portevin and LeChatelier and is frequently called the Portevin-LeChatelier effect. Processing conditions must be selected to avoid strain-aging effects. This selection necessarily involves the control of processing strain rates and temperatures.

Special Tests

The tensile test provides basic information concerning the responses of metals and alloys to mechanical loadings. Test temperatures and strain rates (or loading rates) generally are controlled because of the effects of these variables on the metallurgical response of the specimen. The tensile test typically measures strength and ductility. These parameters are frequently sensitive to specimen configuration, test environment, and the manner in which the test is conducted. Special tensile tests have been developed to measure the effects of test/specimen conditions on the strengths and ductilities of metals and alloys. These tests include the notch tensile test and the slow-strain-rate tensile test.

Notch Tensile Test. Metals and alloys in engineering applications frequently are required to withstand multiaxial loadings and high stress concentrations owing to component configuration. A standard tensile test measures material performance in smooth bar specimens exposed to uniaxial loads. This difference between service and test specimens may reduce the ability of the standard tensile test to predict material response under anticipated service conditions. Furthermore the reductions in ductility generally induced by multiaxial loadings and stress concentrations may not be apparent in the test results. The notched tensile test therefore was developed to minimize this weakness in the standard tensile test and to investigate the behavior of materials in the presence of flaws, notches, and stress concentrations.

The notched tensile specimen generally contains a 60° notch that has a root radius of less than 0.025 mm (0.001 in.) (see Fig. 23 in Chapter 3). The stress state just below the notch tip approaches triaxial tension, and for ductile metals this stress state generally increases the yield strength and decreases the ductility. This increase in yield strength results from the effect of stress state on dislocation dynamics. Shear stresses are required for dislocation motion. Pure triaxial loads do no produce any shear stress; thus, dislocation motion at the notch tip is restricted and the yield strength is increased. This restriction in dislocation motion also reduces the ductility of the notched specimen. For low-ductility metals, the notch-induced reduction in ductility may be so severe that failure takes place before the 0.2% offset yield strength is reached.

The sensitivity of metals and alloys to notch effects is termed the "notch sensitivity." This sensitivity is quantified through the ratio of notch strength to smooth bar tensile strength. Metals and alloys that are notch sensitive have ratios less than one. Smooth bar tensile data for these materials are not satisfactory predictors of material behavior under service conditions. Tough, ductile metals and alloys frequently are notch-strengthened and have notch sensitivity ratios greater than one, thus the standard tensile test is a conservative predictor of performance for these materials.

Slow-Strain-Rate Testing. Test environments also may have adverse effects on the tensile behavior of metals and alloys. The characterization of environmental effects on material response may be accomplished by conducting the tensile test in the environment of interest (for example, sodium chloride solutions). Because the severity of environmental attack generally increases with increasing time, tensile tests designed to determine environmental effects frequently are conducted at very low strain rates. The low strain rate increases the test time and maximizes exposure to the test environment. This type of testing is termed either "slow-strain-rate testing" (SSRT) or "constant-extension-rate testing" (CERT). Exposure to the aggressive environment may reduce the strength and/or ductility of the test specimen. These reductions may be accompanied by the onset of surface cracking and/or a change in the fracture mode. A CERT or SSRT study that shows detrimental effects on the tensile behavior will establish that the test material is susceptible to environmental degradation (Fig. 29a).

This susceptibility may cause concern over the utilization of the material in that environment. Conversely, the test may show that the tensile behavior of the material is not influenced by the environment and is therefore suitable for service in that environment (Fig. 29b). CERT and SSRT may be used to screen materials for potential service exposures and/or investigate the effects of anticipated operational changes on

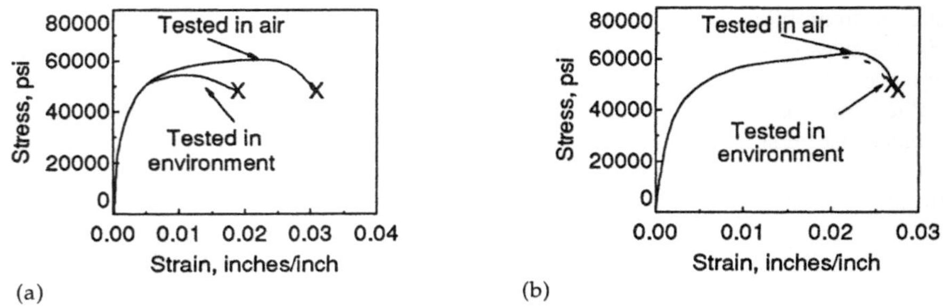

Fig. 29 Typical CERT and SSRT results showing (a) material susceptibility to environmental degradation and (b) material compatibility with the environment

the materials used in process systems. In either event, the intent is to avoid materials utilization under conditions that may degrade the strength and ductility and cause premature fracture. In addition to the tensile data *per se,* evidence of adverse environmental effects may also be found through examination of the fracture morphologies of CERT and SSRT test specimens.

Figure 30 shows a typical SSRT or CERT testing machine. Various types of corrosion cells may be required to control the test conditions for specific studies. Standard tensile specimens (ASTM E 8) are generally recommended for use with specified conditions of gage lengths, radii, and so on, unless specialized studies are being conducted. Notched or precracked specimens are also used for certain tests. More detailed information SSRT/CERT testing can be found in the article "Evaluating Stress-Corrosion Cracking" in *Corrosion: Fundamentals, Testing, and Protection,* Volume 13A of *ASM Handbook.*

Fracture Characterization

Tensile fracture of ductile metals and alloys generally initiates internally in the necked portion of the tensile bar. Particles such as inclusions, dispersed second phases, and/or precipitates may serve as the nucleation sites. The fracture process begins by the development of small holes, or microvoids, at the particle-matrix interface (Fig. 31).

Continued deformation enlarges the microvoids until, at some point in the testing process, the microvoids contact each other and coalesce. This process is termed "microvoid coalescence" and gives rise to the dimpled fracture surface topography characteristic of ductile failure processes (Fig. 32).

The surface topography of a brittle fracture differs significantly from that of microvoid coalescence. Brittle fracture generally initiates at imperfections on the external surface of the material and propagates either by transgranular cleavagelike processes or by separation along grain boundaries. The resultant surface topography is either faceted, perhaps with the riverlike patterns typical of cleavage (Fig. 33a), or intergranular, producing a "rock candy"-like appearance (Fig. 33b). The test material may be inherently brittle (such as chromium or tungsten), or

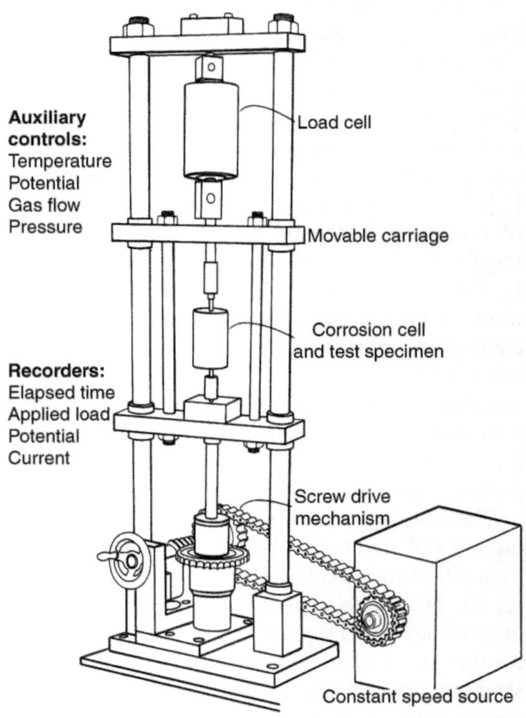

Fig. 30 Typical slow-strain-rate test apparatus

brittleness may be introduced by heat treatment, lowering the test temperature, the presence of an aggressive environment, and/or the presence of a sharp notch on the test specimen.

The temperature, strain rate, test environment, and other conditions, including specimen surface finish for a tensile test, are generally well established. An understanding of the effects of such test parameters on the fracture characteristics of the test specimen can be very useful in the determination of the susceptibility of metals and alloys to degradation fabrication and during

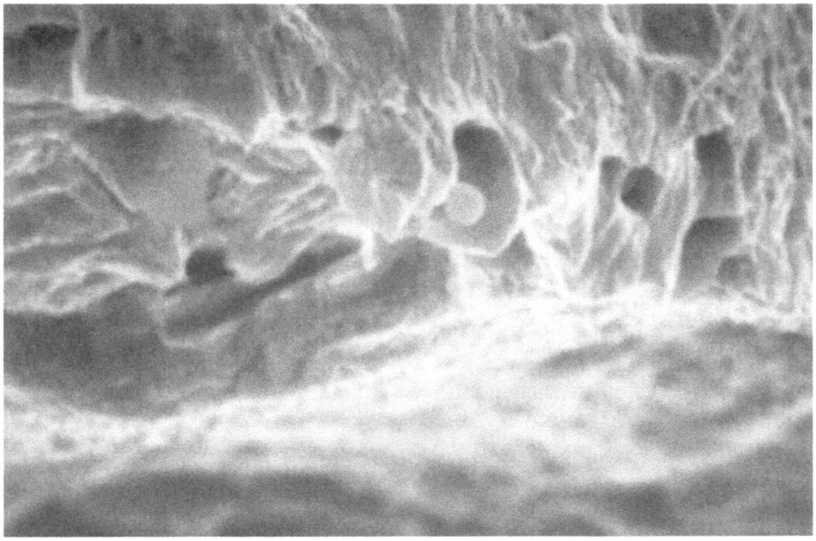

1.00 μ

Fig. 31 Photomicrograph illustrating fracture initiation at particles. Particle is small sphere near the center of the micrograph.

5.00 μ

Fig. 32 Scanning electron micrograph illustrating ductile fracture surface topography. This fracture topography is identified as microvoid coalescence.

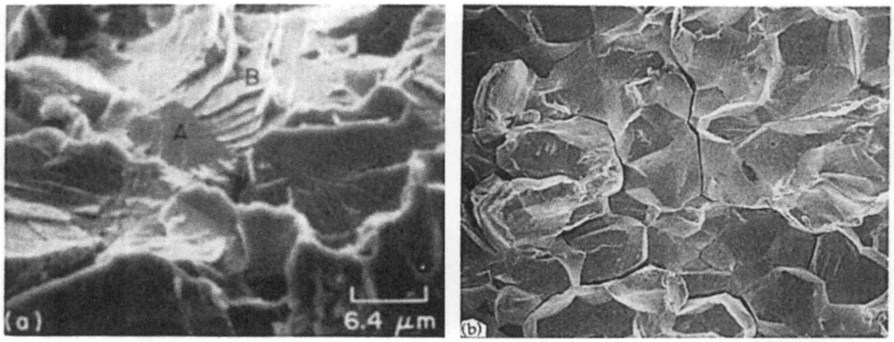

Fig. 33 Scanning electron micrographs illustrating transgranular and intergranular fracture topographies. (a) Transgranular cleavage-like fracture topography. Direction of crack propagation is from grain A through grain B. (b) Intergranular fracture topography

service. Typically, any heat treatment or test condition that causes the fracture process to change from microvoid coalescence to a more brittle fracture mode reduces the ductility and toughness of the material and may promote premature fracture under selected service conditions. Because the fracture process is very sensitive to both the metallurgical condition of the specimen and the conditions of the tensile test, characterization of the fracture surface is an important component of many tensile-test programs.

Summary

The mechanical properties of metals and alloys are frequently evaluated through tensile testing. The test technique is well standardized and can be conducted relatively inexpensively with a minimum of equipment. Many materials utilized in structural applications are required to have tensile properties that meet specific codes and standards. These requirements are generally minimum strength and ductility specifications. Because of this, information available from a tensile test is frequently under utilized. A rather straightforward investigation of many of the metallurgical interactions that influence the results of a tensile test can significantly improve the usefulness of test data. Investigation of these interactions, and correlation with metallurgical/material/service variables such as heat treatment, surface finish, test environment, stress state, and anticipated thermomechanical exposures, can lead to significant improvements in both the efficiency and the quality of materials utilization in engineering service.

ACKNOWLEDGMENT

This chapter was adapted from M.R. Louthan, Jr., Tensile Testing of Metals and Alloys, *Tensile Testing,* 1st ed., P. Han, Ed., ASM International, 1992, p 61–104

SELECTED REFERENCES

- C.R. Brooks, Plastic Deformation and Annealing, *Heat Treatment, Structure and Properties of Nonferrous Alloys,* American Society for Metals, 1982, p 1–73
- G.E. Dieter, *Mechanical Metallurgy,* 3rd ed., McGraw-Hill, New York, 1986
- T.M. Osman, Introduction to the Mechanical Behavior of Metals, *Mechanical Testing and Evaluation,* Vol 8, *ASM Handbook,* ASM International, 2000, p 3–12
- T.H. Courtney, Fundamental Structure-Property Relationships in Engineering Materials, *Materials Selection and Design,* Vol 20, *ASM Handbook,* ASM International, 1997, p 336–356

CHAPTER 8

Tensile Testing of Plastics

ENGINEERING PLASTICS are either thermoplastic resins (which can be repeatedly reheated and remelted) or thermosetting resins (which are cured resins with cross links that depolymerize upon exposure to elevated temperatures above the glass transition temperature). The glass transition temperature (T_g) is defined as the temperature at which an amorphous polymer (or the amorphous regions in a partially crystalline polymer) changes from a hard and relatively brittle condition to a viscous or rubbery condition.

The testing of plastics includes a wide variety of chemical, thermal, and mechanical tests (Table 1). This chapter reviews the tensile testing of plastics, which has been standardized in ASTM D 638, "Standard Test Method for Tensile Properties of Plastics," and other comparable standards. Tensile testing embraces various procedures by which modulus, strength, and ductility can be assessed. Tests specifically designed to measure phenomena as varied as creep, stress relaxation, stress rupture, fatigue, and impact resistance can all be classified as ten-

Table 1 ASTM and ISO mechanical test standards for plastics

ASTM standard	ISO standard	Topic area of standard
Specimen preparation		
D 618	291	Methods of specimen conditioning
D 955	294-4	Measuring shrinkage from mold dimensions of molded thermoplastics
D 3419	10724	In-line screw-injection molding of test specimens from thermosetting compounds
D 3641	294-1,2,3	Injection molding test specimens of thermoplastic molding and extrusion materials
D 4703	293	Compression molding thermoplastic materials into test specimens, plaques, or sheets
D 524	95	Compression molding test specimens of thermosetting molding compounds
D 6289	2577	Measuring shrinkage from mold dimensions of molded thermosetting plastics
Mechanical properties		
D 256	180	Determining the pendulum impact resistance of notched specimens of plastics
D 638	527-1,2	Tensile properties of plastics
D 695	604	Compressive properties of rigid plastics
D 785	2039-2	Rockwell hardness of plastics and electrical insulating materials
D 790	178	Flexural properties of unreinforced and reinforced plastics and insulating materials
D 882	527-3	Tensile properties of thin plastic sheeting
D 1043	458-1	Stiffness properties of plastics as a function of temperature by means of a torsion test
D 1044	9352	Resistance of transparent plastics to surface abrasion
D 1708	6239	Tensile properties of plastics by use of microtensile specimens
D 1822	8256	Tensile-impact energy to break plastics and electrical insulating materials
D 1894	6601	Static and kinetic coefficients of friction of plastic film and sheeting
D 1922	6383-2	Propagation tear resistance of plastic film and thin sheeting by pendulum method
D 1938	6383-1	Tear propagation resistance of plastic film and thin sheeting by a single tear method
D 2990	899-1,2	Tensile, compressive, and flexural creep and creep-rupture of plastics
D 3763	6603-2	High-speed puncture properties of plastics using load and displacement sensors
D 4065	6721-1	Determining and reporting dynamic mechanical properties of plastics
D 4092	6721	Dynamic mechanical measurements on plastics
D 4440	6721-10	Rheological measurement of polymer melts using dynamic mechanical procedures
D 5023	6721-3	Measuring the dynamic mechanical properties of plastics using three-point bending
D 5026	6721-5	Measuring the dynamic mechanical properties of plastics in tension
D 5045	572	Plane-strain fracture toughness and strain energy release rate of plastic materials
D 5083	3268	Tensile properties of reinforced thermosetting plastics using straight-sided specimens
D 5279	6721	Measuring the dynamic mechanical properties of plastics in torsion

sile tests provided that the stress system is predominantly tensile, but by common usage the term "tensile test" is usually taken to mean a test in which a specimen is extended uniaxially at a uniform rate. Ideally, the specimen should be slender, of constant cross section over a substantial gage length, and free to contract laterally as it extends; a tensile stress then develops over transverse plane sections lying within the gage region, and the specimen extends longitudinally and contracts laterally. A procedure was initially developed for tests on metals but was subsequently adopted and adapted for tests on rubbers, fibers, and plastics. In the case of plastics, their viscoelastic nature and the probable anisotropy of their end products (including test specimens) are factors that strongly influence both the conduct of the tests and the interpretation of the results.

Practical tensile testing often conforms to one (e.g., ASTM D 638) or another of several standard methods or to a code of practice, with variants dictated by local circumstances. Most of the stipulations set out in the standardized practices embody the collective wisdom of earlier tensile-test practitioners and fall into four distinct groupings:

- Stipulations relating to the specimen-machine system
- Stipulations relating to the derivation of excitation-response relationships from the raw data
- Stipulations relating to the precision of the data
- Stipulations relating to the physical interpretation of the data.

The stipulations in the first group are the primary ones, because, unless the specimen-machine system functions properly, no worthwhile data can be generated. The stipulations in the other three groups are supplementary but are nevertheless essential in that they enable the outcome of the machine-specimen interaction to be translated progressively into mechanical-properties data for the specimen under investigation.

Viscoelasticity and anisotropy cast their influences over all these groups. Viscoelasticity influences the excitation-response relationships, complicates the analysis of data, and affects some practical aspects of the test. Anisotropy does the same things, but also introduces an uncertainty about the utility of any specific datum because it varies from point to point in a specimen, and from specimen to specimen in a sample, depending on the processing conditions and other factors. These variations can be large, and therefore questions arise as to how such materials should be evaluated and whether or not results from tests on a particular specimen can ever be definitive. If a test has been properly executed, the properties data should be precise, but they may be precise without being accurate and may be accurate without being definitive.

In one particular respect, tensile testing suffers from a fundamental and inescapable deficiency that is common to many types of mechanical tests: the experimenter has no option but to measure force and deformation, whereas the physical characteristics of the specimen and the material should be expressed in terms of stress and strain. The translations of force into stress and deformation into strain are sources of errors and uncertainties, so much so that the transformed results may bear little relation to the strict truth, although this does not render them useless. Note No. 2 in Section 1, "Scope," of ASTM D 638 states, appropos of other factors but appropriate nevertheless, that "This test method is not intended to cover precise physical procedures. . . . Special additional tests should be used where more precise physical data are required."

Fundamental Factors that Affect Data from Tensile Tests

Viscoelasticity. Plastics are viscoelastic—that is, the relationships between the stress state and the strain state are functions also of time. *Linear* viscoelasticity, the simplest case, is represented by the relationship

$$\sum_{n=0}^{\infty} a_n \frac{\partial^n \sigma}{\partial t^n} = \sum_{m=0}^{\infty} b_m \frac{\partial^m \varepsilon}{\partial t^m}$$

where σ is stress, ε is strain, and t is time, and a and b are characterizing coefficients. When most of the coefficients are set to zero, the equation describes simple behavior. If only a_0 and b_0 differ from zero, the equation represents linear elastic behavior, and if only a_0 or b_1 differs from zero, it represents Newtonian viscosity, but as other coefficients differ from zero the differential terms progressively enter the equation and the relationships between stress and strain then become time-dependent. In simple cases, the viscoelasticity can be visualized as the mechanical behavior of assemblies of Hookean springs and Newtonian dashpots (which are representable by the same equation), the two simplest assemblies being a series combination and a parallel combination of one spring and one dashpot. The former, known as a Maxwell element, is used primarily to represent or demonstrate the time-dependence of the stress that arises when a strain is applied suddenly, and the latter, known as a Voigt element, demonstrates the time-dependence of the strain that develops when a stress is applied suddenly (see Fig. 1).

Due allowance must be made for viscoelasticity during both the practical execution of the test and the interpretation of the results, because the ramifications of viscoelasticity extend over virtually all the mechanical behavior. Thus, for example, after the specimen has been mounted in the grips, the clamping stresses may relax to the point where it is not held securely. There are several such specimen-machine interactions, but appropriate practical measures alleviate their consequences, and serious malfunctions generally can be avoided. In contrast, possibilities of misinterpretation of the results are not so easily circumvented, because viscoelasticity is an inescapable feature of almost every response curve. In general, the response of a viscoelastic body to an applied stress or strain is a function of the stress history or the strain history. Therefore, the moduli, which are defined in various ways depending on the time-form of the excitation, are functions of elapsed time and/or frequency (see Fig. 1). Furthermore, plastics are *nonlinearly* viscoelastic—that is, at constant time the relationship between stress and strain is nonlinear. The relaxation modulus, which is derived from an experiment in which a strain supposedly is applied instantaneously and held constant thereafter, is a function of the strain magnitude as well as of the elapsed time; similarly, the creep compliance is a function of the stress magnitude and the elapsed time. These two procedures are ideal in that they enable nonlinearity and time-dependence to be separated experimentally, but the apparently simple procedure of conventional tensile testing is not simple; the force or the stress that develops in the course of the test is governed by both the changes in strain and the passage of time.

A single tensile test provides merely one section across a relationship that for plastics is a complex one between stress, strain, and strain rate, and it follows that inferences drawn from

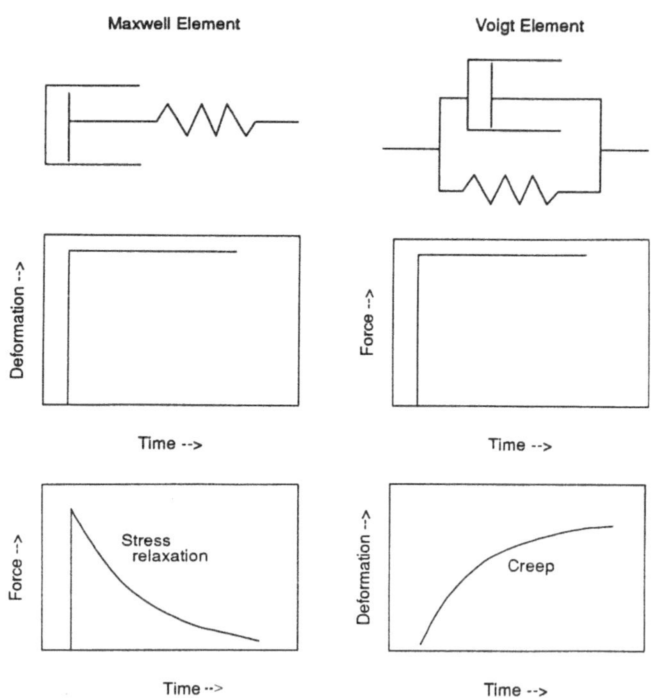

Fig. 1 Visualizations of simple viscoelastic systems

that single curve are correspondingly limited in their scope. For instance, such a curve contains no direct indication of load-bearing capability under loads sustained over any period greater than the duration of that particular tensile test. Tensile-testing practice accommodates this and related deficiencies pragmatically by regarding deformation rate as a critical variable. A comprehensive evaluation entails the use of several rates, which should range over several decades, although this raises certain practical issues. Very low rates may be prohibited on the grounds of uneconomical deployment of expensive apparatus, and very high rates pose technical demands on machine power and sensor response that may be resolved more effectively by use of impact tests.

The viscoelasticity, in combination with certain features of the test system itself, influences the choice of data for subsequent conversion into property values. Thus, the modulus, which is a multivalue property if the material is viscoelastic, must be qualified by specification of the current stress (or strain) and the stress (or strain) history up to a specific point in time. For ramp excitations—i.e., the constant deformation-rate conditions of a tensile test—modulus can be defined as the slope of either the tangent at, or the secant to, any desired point on a stress-strain curve. As such, each single datum is one point only in a viscoelastic function; it has no special merit, although, of the various options, the tangent at the origin is possibly the best in theory because the strain-dependence should be negligible there. However, mechanical inertias in the testing machine and finite response times of the sensors combine with the viscoelasticity to distort the observed force-deformation relationship. They reduce the initial slope, obscure the origin, and obscure or distort abrupt changes in slope that may signify structural changes in the deforming specimen, thereby detracting from the usefulness of the test and introducing the potential for errors in the measurements.

Strength is also a multivalue property, the viscoelasticity intruding both directly, as a time-dependence (rate-dependence) or the equivalent temperature-dependence, and indirectly, as a factor influencing the nature of the failure or fracture, through the sensitivity to strain rate and temperature of the ductile-brittle transition. This transition is usually a gradual one, with the ductility decreasing progressively as the deformation rate is increased or as the temperature is lowered. The practicalities of the evaluation of tensile properties are such that temperature usually is varied in preference to extension rate; Fig. 2 shows typical results from which it may be inferred that the shapes of the stress-strain curves of plastic materials are not uniquely characteristic, and it follows also that uncertainties can arise over the point at which a characterizing datum such as a strength or a yield strain should be extracted from the response curve.

In summary, the viscoelastic nature of plastics entails specific precautions concerning some practical aspects of the test and the analysis of the results. In the first category, mounting of the specimen in the grips and mounting of strain sensors on the specimen require special attention. In the second category, the response curve must be recognized as offering only a limited insight into the mechanical behavior of the sample under investigation, and the data must be used with appropriate caution. See also Fig. 3.

Anisotropy in Plastic Specimens. Test specimens, whether directly molded or cut from larger pieces, are often anisotropic—partly because plastics are viscoelastic in their molten state and very viscous, so that the shaping processes cause molecular alignments, and partly because ordered structural entities may develop during the cooling stage.

The property values derivable from such specimens often differ from what might be expected on the basis of isotropic idealizations, and, because of their limited range, the data usually generated are not definitive in that they do not adequately quantify the tensor array of modulus or strength and do not show how that array varies with processing conditions, flow geometry, and specimen geometry. Some of the ramifications have been troublesome in evaluation programs in the past, because certain consequential results have seemed to be anomalous. Two situations are particularly important: one relates to the position of the failure site, and the other relates to the strength of notched specimens.

The first situation involves tensile specimens of dumbbell or similar shape, which often are injection molded through an endgate (see Fig. 5 in Chapter 11, "Tensile Testing of Fiber-Reinforced Composites," for a typical tensile test specimen). The pattern of molecular and fiber orientation is then predominantly longitudinal in the outer layers of the parallel-sided section but is more complex in the core and at the ends of the specimen. At the end remote from the gate, the larger cross section causes diverging flow during the molding operation and therefore some lateral molecular orientation, which may lower the longitudinal strength locally to such a level that the specimen breaks there rather than at the smaller cross section in the gage region.

The second situation involves notched specimens. A molded notch may not affect strength to the same degree, or even in the same sense, that would be inferred from stress-concentration theory, because the local flow geometry near the crack tip may enhance the strength and thereby mitigate the effect of the stress concentration. On the other hand, the flow geometry may reduce the strength in the critical direction. A machined notch also interacts with the flow geometry in that the geometrical details govern where the tip lies in relation to the orientation pattern. Results are likely to be less ambiguous than those for molded notches but still at quantitative variance with predictions based on concepts of stress concentration or stress-field intensity.

A secondary consequence, but one of great practical importance, is that the essentially simple functional operation of the test machine is compromised, particularly in relation to the specimen-machine interaction. The force is transmitted to the specimen mainly by means of shear stresses at or near the grips, and the specimen is required to extend with lateral contraction but no extraneous distortion. However, a predominantly axial molecular orientation or fiber alignment confers a relatively high tensile strength but a relatively low shear strength along the longitudinal direction, with the result that shear failure near the grips may ensue before tensile failure occurs in the gage region. Modified grips, reinforcing plates attached to the ends of specimens, and changed specimen profiles can all reduce the risk of malfunction, but such steps may be detrimental in other respects. If the predominant orientation lies at some angle to the tensile axis, the specimen will distort into a sigmoid, the exact form of which will depend on whether or not the clamped ends are free to rotate; in either case, the observed force and extension will not convert into correct values of tensile modulus or tensile strength. In general, these effects are far more pronounced in specimens of continuous-fiber plastic-matrix composites than in simple plastic specimens (including those containing short fibers); but even if there is no gross malfunction in tests on plastics,

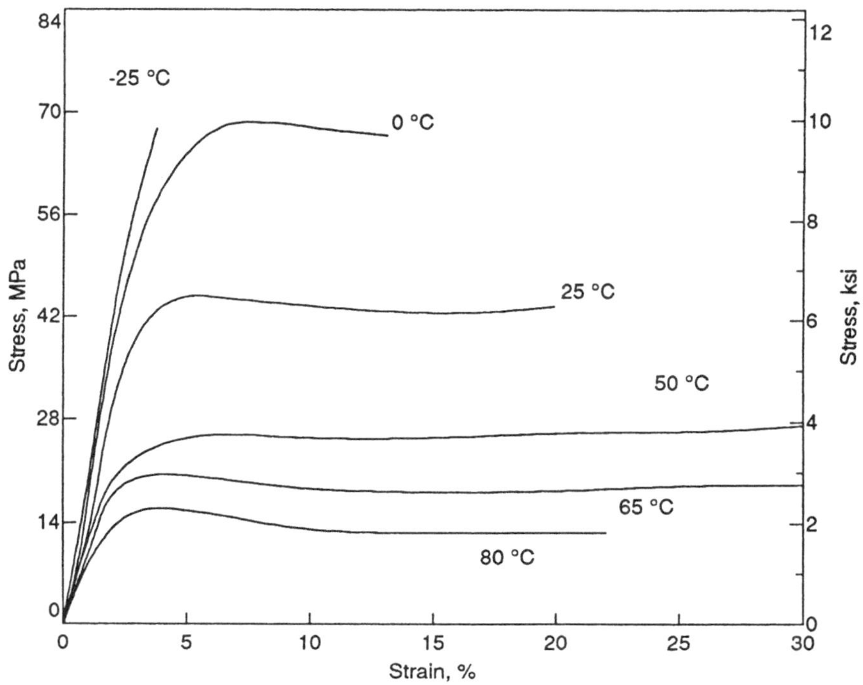

Fig. 2 Influence of temperature on the nature of the stress-strain relationship. Strain rate has a similar effect, with increasing rate being equivalent to decreasing temperature. Source: Ref 1

there is a high probability of mildly erroneous data being generated.

The influences of flow geometry and flow irregularities on derived property values are pervasive and can distort an investigator's perception of properties, trends, etc. Corrective action to avoid misconceptions entails expansion of the evaluation programs to cover samples with different flow geometries and, in some instances, modified test configurations—e.g., different specimen profiles. The choice of samples and specimens is a complex issue that has never been resolved adequately. Specimens machined from various judiciously chosen positions in larger items are possibly a wiser choice than the widely used injection-molded endgated bars. The latter are popular because they are economical in the use of material and manpower, but the pre-

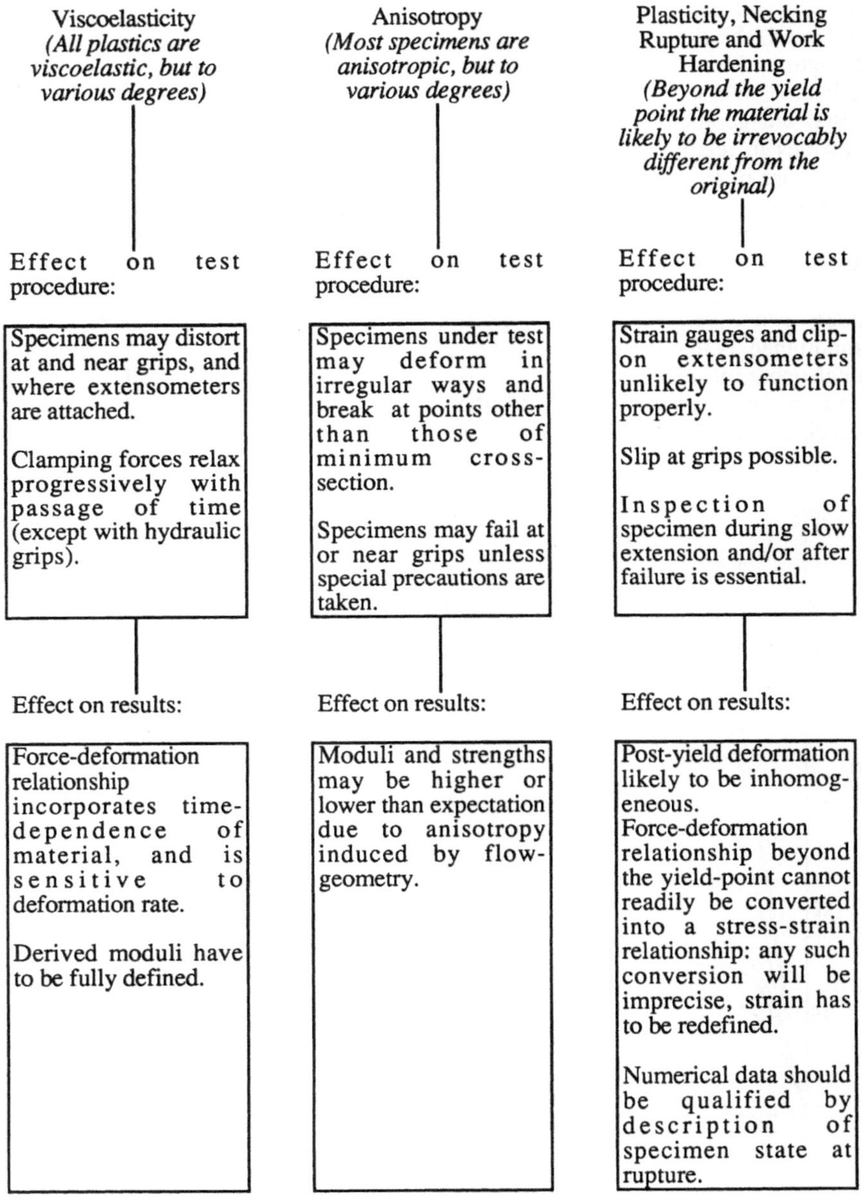

Fig. 3 Influence of the inherent nature of plastics on tensile-testing practice

dominantly axial molecular orientation of thin moldings confers higher tensile moduli and strengths than those exhibited by most end products. The pattern of orientation varies with the thickness of the bar; axial orientation arises mainly in the outer layers, and hence, as the thickness increases, the measured values of tensile modulus and strength decrease.

In summary, anisotropy can cause extraneous distortions in specimens under test, failure at or near the grips, unsuspected errors in data, and odd trends with respect to notch geometry, specimen profile, etc. (see also Fig. 3). In many instances, the evaluation program should be expanded, possibly with modified test procedures.

Plasticity, Necking Rupture, and Work Hardening. There is much experimental evidence, from creep studies and from tensile tests themselves, that with increasing strain the deformation processes become progressively dominated by molecular mechanisms that either are irreversible or are reversible but have very protracted recovery times. The over-all character of the deformation processes becomes "viscous" rather than "elastic," and the specimen then either extends uniformly or yields by means of a necking mechanism approximately in conformance with plasticity theory. Figure 4 gives a schematic impression of likely yielding and post-yield behavior. A material that has yielded is usually radically different in nature from what it was prior to yielding. The difference may be merely a reordered molecular state, but it also may be the presence of larger-scale discontinuities such as voids, crazes, or interphase cracks, all of which have various and different implications for the service performance of end products.

The yield stress, defined by some identifiable feature on the force-deformation curve, depends on the deformation rate, as does the probability that failure will occur before a neck is established. The higher the deformation rate, the higher the yield stress and the greater the chance that brittle or pseudobrittle failure will intervene. There are two principal reasons for the latter relationship. The temperature and the anisotropy may be such that the ductile-brittle transition is traversed as the deformation rate is increased, or the neck may form but fail to stabilize because of the particular microstructure of the polymer or composition of the plastic. The molecular architecture, the molecular weight, and the degree of branching all affect the propensity of the molecules to align in the neck and the consequential strength there. Similarly, the development of macroscopic discontinuities—e.g., microvoids, phase separations, and crazes—may be detrimental, although not necessarily, because the ligaments may be strengthened by favorably oriented molecules. Another factor is the local temperature, which will rise if the heat generated by virtue of the inherent loss processes exceeds what can be lost to the environment and which may reach a critical point at which the yield stress has fallen to such a level that the neck cannot support the prevailing force. If the neck stabilizes satisfactorily it will travel along the parallel-sided section of the specimen either at an approximately constant force or with a progressively increasing force if the molecular assembly is such that further orientation can occur.

The various features of time-dependent plasticity, necking rupture, inhomogeneous deformation, and work hardening affect the practical execution of tests in that a high extensibility imposes particular requirements on the grips and the deformation sensors, but, more importantly, these features affect the ways in which the derived data should be presented and interpreted (see Fig. 3). Thus, a yield stress identified by some features on the force-deformation curve should not be regarded as unambiguously definitive, because there is a zone in which the material is neither wholly viscoelastic nor wholly plastic and, additionally, the material in the neck is not necessarily a continuum. At a more mundane level, if a specimen has necked, the stress and the strain at failure are not readily calculable from the force and deformation data.

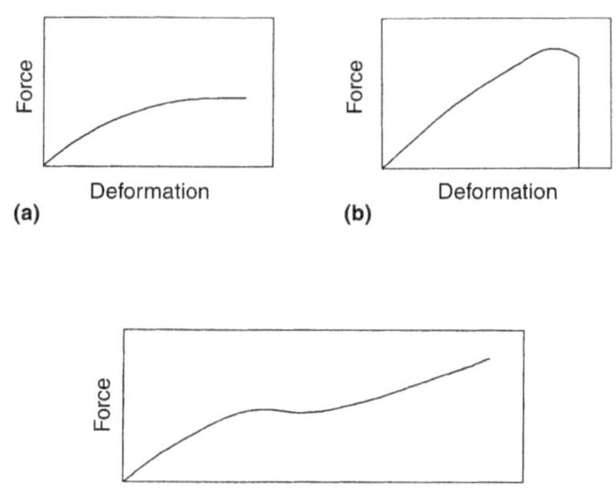

Fig. 4 Yielding and post-yield tensile behavior: (a) uniform extension; (b) yielding followed by necking rupture; (c) yielding followed by "cold drawing" and work hardening

Stipulations in Standardized Tensile Testing

The Specimen-Machine System. The superficially simple nature of the tensile test conceals a demanding mechanical requirement. The specimen must be extended uniformly at any one of several prescribed rates, which, when translated into a design specification, entails:

- Adequate power in a testing machine to ensure that the stiffest specimens can be extended at the designated rates
- Alignment of the line of action with the axis of symmetry of the specimen, to minimize the variation of stress across the specimen cross section
- Secure and balanced clamping of the specimen to ensure that it neither slips in the grips nor suffers extraneous forces
- High-quality specimens of the correct size and profile for the intended purpose and with a fine surface finish

These four design features are interconnected to some degree and are all influenced by the viscoelastic nature of the specimens.

The provision of adequate power poses no direct problem, but there may be secondary difficulties in that a powerful machine is likely to be massive and to have inertias and frictions in the actuator and the likages that are troublesome when the active forces are small—i.e., at low specimen strain or when the specimen has a low modulus or a low strength. The issue is whether a single machine is suitable for testing all classes of plastics at all conceivable strain rates and over the entire strain range to failure. If there is a range of machines at the investigator's disposal, the choice should be governed by the character of the specimen and should be such that the specimen is matched to the machine. The specimen should never dominate the machine, because in such an event the signals being extracted from the test would reflect a complex combination of machine and specimen characteristics. On the other hand, if the machine is excessively dominant, it may impose inadvertent and undesirable constraints on the specimen.

Accurate alignment of the specimen in the machine is not easily achieved, because the machine, the specimen, and the clamping of the one to the other are all prone to asymmetries that can cause misalignment. There are various design choices ranging from sufficient degrees of freedom to allow a misaligned specimen to settle into an aligned position as it begins to extend, at one extreme, to total constraint at the other. The former method relies on the specimen being sufficiently stiff to be essentially unaffected by the adjustment forces, which is unlikely to be the case for a plastic material. Similarly, however, the friction inherent in a fully constrained system may constitute a large error in the measured force.

Machine factors are largely outside the control of a user, but, to varying degrees, specimen-preparation procedures, choice of grips, and operational checks, all of which affect and/or control the axiality of the alignment, are discretionary. Specimens must be symmetrical about their longitudinal axes. One machined from a larger item can be very accurately symmetrical. A directly molded specimen can be similarly accurate, but inappropriate molding conditions or a badly designed mold can produce distorted specimens. Specimens molded from novel or newly developed materials, for which the processing conditions may not have been optimized, are prone to such distortion, but force of circumstance may dictate the data generated from such specimens must be used, despite the imperfections, as a basis for judgments crucial to the further progression of a research or development program. When this is the case, the judgments should be suitably circumspect.

Even if the specimen is satisfactorily symmetrical, it may be clamped unsymmetrically unless special precautions are taken to position it properly in the grips. Use of a hole in each specimen end and corresponding pins in the grips is the simplest solution, and has proved very satisfactory for tensile creep tests. The holes also facilitate the machining operation by defining the axis of symmetry. Ideally, the force should be transmitted to the specimen through the pins rather than through the faces of the grips, but this imposes special requirements on the specimen geometry to limit the chance of shear failure at the pins (see the subsection on anisotropy in plastic specimens), and the less ideal conventional clamping, ostensibly acting across the entire width of the specimen, is commonly preferred.

Misalignment is relatively unimportant if the strength of a ductile material is being measured, because limited plastic deformation suffices to correct the fault and the test progresses unimpaired thereafter. On the other hand, misalignment is a source of error if the strength of a brittle material or the modulus of any type of material is being measured, because the misalignment causes the specimen to bend or unbend, as the case may be, as it is extended in the test. The stress is then nonuniform over the cross section, and one face of the specimen bears a stress higher than the average stress; the measured strength is then likely to be an underestimate of the true strength. The error in the modulus measurement may be positive or negative, depending on the positioning of the strain sensor, and can even be eliminated if the strain on each face of the specimen is measured.

To some degree, there is a conflict of objectives in the design and operation of the grips. Secure clamping is desirable so that the specimen does not slip relative to the grips, or entirely out of the grips, during a test, but it simultaneously prevents self-aligning movement and thereby preserves any initial misalignment. On balance, total constraint is the preferred option. In this case, hydraulic grips are probably the most satisfactory because they exert a pressure that is uniform over the entire face and that remains constant as the specimen extends and correspondingly thins. Simple mechanical grips may have to be over-tightened initially, and consequently the specimen may be severely distorted. Such distortion can be reduced or eliminated by the use of reinforcing tabs on the ends of the specimens, but this is a tedious measure that is not widely used for tests on plastics.

The specimen-machine system cannot be expected to operate satisfactorily, however well designed it may be, unless the quality of the specimen is commensurate with the expectations. Methods of specimen production include direct molding, die cutting, and machining with a router or milling cutter. Certain procedures must be followed with each method if the symmetry required for axial stressing is to be attained: the cooling systems of molding cavities should be so designed that any residual strains are in equilibrium, specimens being machined should be supported so that they do not distort under the machining forces, etc. The surface finish is also important, because imperfections may act as stress concentrators and cause the specimen to fail prematurely. Unsuitable molding conditions can produce surface textures and imperfections ranging from the visually obvious to the submicroscopic. Die cutters are fast in operation but often produce specimens with poor edge faces. In general, milling cutters and routers produce better surface finishes than die cutters, but this depends on the cutting speed, which should be high but not so high that generated heat softens or melts the surface.

The various elements of the specimen-machine interaction that affect the over-all operation efficiency in the tensile testing of plastics are summarized in Fig. 5.

Derivation of Excitation-Response Relationships. Many investigators require only a single datum from a tensile test and naturally tend to regard the derivation procedure as a simple operation, which it may be when the testing machine is set up with a single objective in mind. The over-all operation, however, is a more complex matter, the single datum being only a small element in the total response of the specimen. The excitation-response relationship provides numerical values of various mechanical properties—e.g., modulus and yield strength; also, in its entirety it gives an over-all impression of "tensile characteristics" although, as was pointed out in the section on viscoelasticity, each curve provides only one section across a complex relationship between stress, strain, and strain rate. The particular type of excitation used in tensile testing was originally chosen for its mechanical simplicity; it loosely approximates a ramp function of strain versus time, which is not particularly tractable analytically even for a linear viscoelastic body and is even less tractable for a nonlinear one. Thus, because of both practical and theoretical limitations, it is unlikely that the observable response can ever be translated into fundamental quantities at the molecular level—for example, relaxation time spectra.

However, irrespective of the details, information can be obtained from a test only if there are suitable sensors to convert the excitation and response into numerical or analog data. These sensors, which are described in Chapter 4, "Tensile Testing Equipment and Strain Sensors," must have sensitivities and response times that are appropriate for the intended purpose of the test. The sensitivity should be such that the sensor discriminates at, say, 1% of full scale display, and the response time should be such that the fine structure of response is detected even though this generally entails the likelihood that extraneous vibrations in the machine will be incorporated as noise in the signal. The observable quantities are limited to force and deformation, and the former is actually measured as deformation in a transducer.

The force is always measured directly and accurately provided that the machine and the transducer are adequately stiff. The deformation up to the yield point may be measured directly by means of an extensometer attached to the specimen, strain gages bonded to it, or an independent optical device operating without physical contact. These methods entail careful and sometimes expensive subsidiary operations, and, furthermore, only the remote optical devices are practicable beyond the yield point. Consequently, for certain classes of test, they may be dispensed with, the deformation then being measured indirectly as actuator movement, with possible corrections for extraneous effects caused by clamping and the specimen profile.

Strain gages and clip-on extensometers have their respective advantages and disadvantages. The former are more troublesome to mount on the specimen and measure the strain over only a small zone, but, on the other hand, they can be so positioned as to measure strain along whichever direction is of interest. Clip-on gages provide an average strain over a larger span. They are less versatile in relation to strain axis but can measure transverse strains, and therefore the change in volume during a test can be determined by either type of sensor. Such information provides insight into pre-yield mechanisms.

In the case of modulus measurement, the strains involved are small and the over-all deformation is homogeneous; force translates easily into stress with only small errors, and deformation can be measured over a defined gage

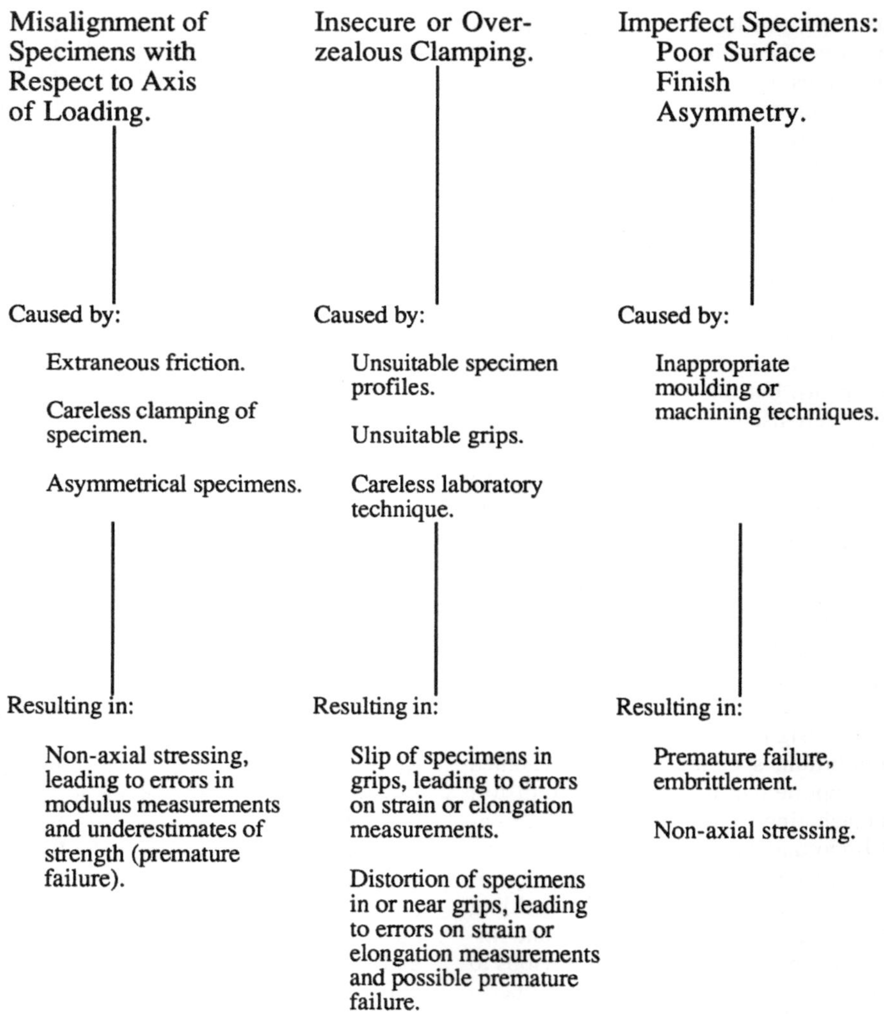

Fig. 5 Sources of experimental error in the specimen-machine system

length. In principle, deformation measurements should be accurate in such situations, but clip-on extensometers may slip if the retaining spring force is small or may indent the specimen if the spring force is large, and bonded strain gages may affect the surface strain that they are intended to measure if the stiffness of the specimen is low. Even so, with minor reservations, the modulus can be measured to a satisfactory precision. Coefficients of variation of about 0.03 are commonplace, and coefficients of 0.02 are attainable.

In the case of strength tests, the over-all precision of the measurements is lower—primarily because the calculation of stress is inevitably an approximation, and secondarily because extraneous defects in the specimen may promote failure or induce brittleness. If the failure is brittle, the calculated strength can be based on the initial cross-sectional area, but this measured quantity may be neither precise nor accurate because of nonaxial loading, defects in the specimen, or variable anisotropy. Coefficients of variation of 0.10 for the interspecimen variability are commonplace, and the values may be a substantial underestimate of the true strength. If the failure is ductile, the estimate of area is likely to be erroneous, and if the deformation is also inhomogeneous, as is common, the calculation of failure stress is further confounded. The nominal yield stress calculated on the basis of the initial cross section is likely to be precise, with a coefficient of variation of about 0.03, but not highly accurate because of the complexity of the associated phenomena. In contrast, the nominal breaking stress of a specimen that extends beyond the yield point is little more than a normalized breaking force and is physically meaningless.

The deformation or strain at failure is similarly a dubious quantity. It is usually inferred from the movement of the actuator, because strain gages and extensometers normally are not used in tests that are intended to progress to failure of the specimen. With brittle fracture, the error in the inferred deformation is usually large, because extraneous deformations at and near the grips constitute a relatively large proportion of the over-all movement of the actuator. This source of error is less influential when the failure is ductile, but the measured deformation usually does not then translate directly into strain. Even so, the commonly quoted extension to fracture is a useful quantity because it relates loosely to the stability of the neck, the propensity of the specimen for subsequent work hardening, and the incidence of defects in the specimen. There are no quantitative rules for underpinning of judgments on these matters, and the investigator must assess new results against a background of whichever accrued data are appropriate. The same is true, to varying degrees, of most of the data relating to failure; they are accommodated within a framework of comprehension that enables useful information to be extracted despite uncertainties about the physical credentials of the experimental data. This framework of comprehension is based on the collective experience of many previous investigators, accumulations of data, established correlations between test results and service performance, perceptions of quality, and other knowledge. It follows that the reliability of such rationalizations depends heavily on the quality of the database.

The principle sources of error that are encountered in this phase of the testing operation are summarized in Fig. 6. In combination, the various sources of error summarized in Fig. 3, 5, and 6 often lead to coefficients of variation of 0.10 or higher; at this level, the imprecision is such that ten nominally identical specimens should be tested for the derivation of a property value (most standard specifications stipulate a minimum of five).

Physical Interpretation of Data. The force-determination relationships of specimens are converted by calculation and inference into stress-strain relationships for the constituent material. At low strains, this stress-strain relationship defines various moduli, and, provided that appropriate procedural precautions are taken, the accuracy of the modulus data can be high. If the specimens are brittle, the precision of the measured strength may also be high, but the accuracy is likely to be low because of the deleterious effects of imperfections in the specimens. As the strain increases—beyond, say, 0.02—the conversions become progressively more approximate, and therefore, even though the original test results may have been precise, the final strength data are unlikely to be accurate. Even so, the approximations and oversimplifications entailed in this stage are minor impediments in comparison with those involved in the train of inference leading from the over-all bulk values derived from the test to the local values prevailing at the site of fracture of failure. The theories of fracture mechanics and plasticity, taken in conjunction with a mathematical model of the local situation, provide some conversion rules, and it is possible, therefore, for an investigator to gain an insight into micromechanical behavior from macromechanical data. Some procedures, however, are too cumbersome for routine use, and are also questionable to the extent that some doubt persists about the over-all quality of the data generated by them.

The features on a force-determination curve that are taken as identifying important events such as yielding or the onset of critical crack growth may have been chosen more for their macroscopic convenience than for their physical validity, one practical consequence being the enhancement of precision at the expense of accuracy or realism. The diagram in Fig. 7 summarizes the possible ambiguity over the identification of the "yield point" in even the simplest case—i.e., when the force-determination curve passes through a maximum. The possible error in the measured yield stress is likely to be small because of the shape of the force-deflection curve as the yield point is approached; on the other hand, for the same reason, an estimate of the yield strain is likely to be imprecise. Where there is no maximum, the characterizing point may be less easily identified and will almost certainly be associated with different physical manifestations; the derived yield stress may be as prone to error as the derived yield strain. Similarly with brittle failure, the error in

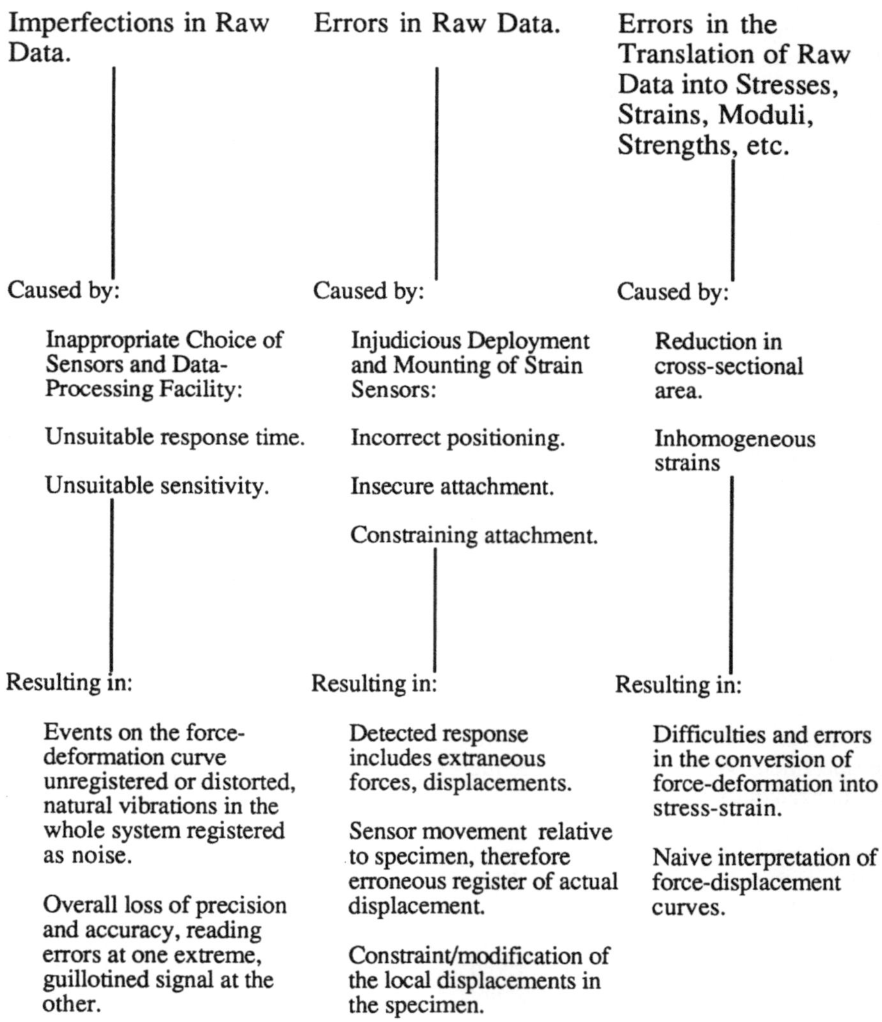

Fig. 6 Faulty techniques and errors in the derivation of raw data and property values

the critical stress-field intensity factor may be large because of the shape of the rising flank of the curve and because the selected feature may not mark the critical point; for instance, the dominant peak may denote the over-all collapse of the specimen as a load-bearing structure rather than the point at which the growth of the crack becomes critical.

If an investigator needs to clarify such points or to study the phenomena in greater detail, supplementary tests can be helpful. Photography of the specimen at specific moments or continually throughout the test enables correlations to be established between the features on the force-deformation curve and the physical events in the specimen. The simplest expedient is nothing more than a supplementary tensile test at an extension rate sufficiently low for the correlations to be established through visual inspection of the extending specimen; such a test can even be interrupted temporarily to permit a more intense scrutiny, although when an interrupted tensile test is resumed the subsequent force-deformation relationship will differ from that of an uninterrupted test because of viscoelastic relaxation during the static period.

As the extension progresses beyond the yield region, the link between the observed force-deformation relationship and the inferred stress-strain relationship becomes progressively more tenuous. The causes are the aforementioned approximations entailed in the translations of force into stress and deformation into strain, developing inhomogeneities in the specimen and molecular and structural rearrangements in the material. In the post-yield region, the measurable quantities are the ultimate strength, commonly defined as force divided by initial cross-sectional area; the elongation to fracture, derived from the actuator movement; the shape of the

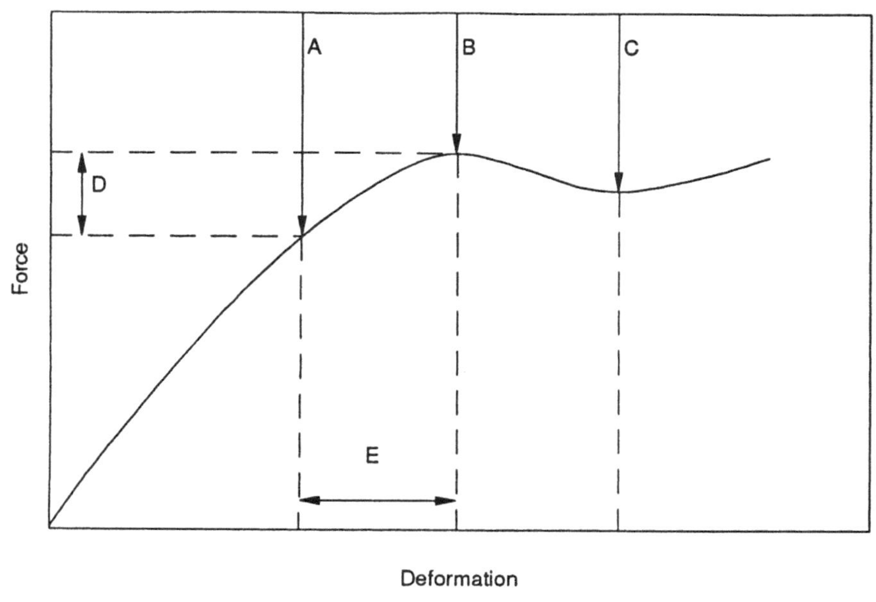

A. Point at which localized yielding may be observed

B. Point at which resistance to further deformation decreases

C. Band of uncertainty over yield stress

D. Band of uncertainty over yield stress

E. Band of uncertainty over yield strain

Fig. 7 Simple force-deformation curve. The maximum force usually is taken as signifying the onset of yielding, but it merely marks the point at which the specimen, as a structure, becomes less resistant to further deformation. Thus, the yield stress and the yield strain are not unambiguously quantifiable.

curve immediately after necking; and the overall slope of the curve. These are all characterizing quantities for the force-deformation curve, but it is important that they be regarded as nothing more. These quantities have to be transformed into characterizing quantities for the specimen as an engineering entity, and the reliability of this operation depends on the validity of the mathematical model that is chosen to simulate the mechanical behavior of the specimen. There must be a second transformation, into characterizing data for the material. This is the more difficult of the two, because the flow geometry and processing conditions inherent in the production of specimens impose particular states of molecular order, aggregation, etc., that govern the anisotropy and the levels of the property values. Thus, even though data may be precise and accurate, they may not be representative of the material properties as manifested in the majority of end products, and therefore they may be either unsuitable for some purposes or misleading.

Thermoplastics differ in their sensitivities to flow geometry and processing conditions. High molecular weights, discrete second phases, and large crystal entities tend to worsen the anisotropy, and the consequential ranges of property values can be large—for example, a factor of two for modulus and a factor of three for strength. However, such large ranges normally do not appear as overt variabilities, because the specimen-preparation routines have been standardized and restricted in the interests of reproducibility and operational economy rather than in the interests of practical relevance. Furthermore, the data so generated usually lie near the upper limit of attainable values and are therefore potentially misleading.

The nebulous nature of the post-yield data and the potential variation in all data do not detract unduly from the usefulness of the data, because there are many semiquantitative correlations between the characterizing features and property values on the one hand and certain attributes and properties of end products on the other. For example, even though elongation to fracture varies with the shape of the specimen and cannot be equated accurately with strain, a high value is generally a desirable attribute that is indicative of probable toughness in service items. Intersample differences often can be attributed to specific factors such as molecular weight and the incidence of flaws, contaminants, defects, etc., but results must always be judged in the context of the particular evaluation program and set against an established pattern of data. The overall success depends on the quality of the infrastructure and the database.

Utilization of Data from Tensile Tests

Materials Evaluation. Tensile tests are multipurpose, the data derived from them being commonly used for purposes ranging from quality control to research. Property tables, such as shown in Table 2, feature modulus, tensile strength, and elongation to fracture derived from tensile tests, but for only one standard deformation rate and one temperature (23 °C, or 73

Table 2 Room-temperature tensile properties of selected engineering plastics

Thermoplastic	Tensile strength(a) MPa	Tensile strength(a) ksi	Tensile elongation at break(a), %	Tensile modulus(a) kPa	Tensile modulus(a) psi
Styrene	46	6.7	2.2	320	46
Styrene-acrylonitrile (SAN)	72	10.5	3.0	390	56
Acrylonitrile-butadiene-styrene (ABS)	48	7	8.0	210	30
Flame-retardant ABS	40	5.8	5.1	240	35
Polypropylene (PP)	32	4.7	15.0	130	19
Glass-coupled PP	32	4.7	15.0	130	19
Polyethylene (PE)	30	4.3	9.0	100	15
Acetal (AC)	61	8.8	60.0	280	41
Polyester	55	8	200.0	280	40
Flame-retardant polyester	61	8.9	60.0	280	40
Nylon 6	81	11.8	200.0	280	40
Flame-retardant Nylon 6	85	12.3	60.0	290	42
Nylon 6/6	79	11.4	300.0	130	19
Flame-retardant Nylon 6/6	67	9.7	35.0	130	19
Nylon 6/12	61	8.8	150.0	200	29
Polycarbonate (PC)	62	9	110.0	240	34.5
Polysulfone (PSU)	70	10.2	75.0	250	36

(a) ASTM D 638 test method

°F), whereas an emerging body of opinion contends that data for other rates and temperatures should be provided by the data generators.

The current minimum evaluation scheme falls far short of what is now being called for formally, and even the latter calls for less than what could be derived from tensile tests, namely:

- The tensile modulus (tangent and secant) at various strains below the yield point
- The lateral contraction ratios
- The yield stress and, in some instances, the yield strain
- The "load drop" after yielding
- The slope of force versus deformation after the yield point
- The ultimate strength (based on initial cross-sectional area)
- The elongation to fracture.

As discussed in previous sections, these quantities are measurable to different levels of precision, have variously dubious claims to the status of physical properties, and all relate to the specimen rather than to the material from which it has been made. It follows that they should be interpreted with caution. Above all, an evaluator/investigator should bear two points in mind whenever the results/data are being communicated to others:

- Data at one deformation rate and one temperature may not be adequately representative of the tensile properties and fall short of prospective recommendations on data generation and presentation.
- Whatever the range of test conditions and whatever the information extracted from the test, the data relate to the specimen; the properties of the sample and of the material must be inferred.

Despite these reservations, the types of data presented in the seven-item list above serve a variety of purposes satisfactorily, although they are also subject to misinterpretation and misuse. Misinterpretation by the investigator can result from:

- Reliance on a single datum, and failure to make use of the entire force-deformation relationship
- Failure to impose independent checks on inferences drawn from the data.

Misuse by the investigator and others can result from:

- Disregard of the possible uniqueness of each sample
- Insufficient regard for the potentially deleterious effects of unfavorable flow geometries
- Disregard of the boundaries beyond which particular data are invalid or irrelevant.

Materials Comparisons and Selection. The elementary table of properties on which many comparisons are based features tensile modulus, tensile yield strength, ultimate tensile strength, and ultimate elongation. It is currently criticized for its various shortcomings, but it may owe its simple form to the fact that, for some purposes, many data on each of many properties or pseudoproperties may be confusing rather than enlightening. On the other hand, judgments in some areas require special or selective data, and judgments in other areas require data that extend far beyond the confines of "single-point" data. Thus, the criteria by which a material is chosen in preference to others vary with circumstance, in accordance with often subjective rules. Attempts have been made, and are being made, to automate the operation, which entails a pseudoquantification of the judgment processes, but this latter step is generally a difficult one because the specification for the end product often asks for a combination of property values or characteristics that are mutually exclusive.

One such dilemma arises regularly in the perpetual search for an optimum balance between modulus and ductility, which relate, respectively, to stiffness and toughness in an end product. Practical experience has provided the rough working rule, which also has a basis in theory, that the two properties are reciprocally related, and it follows, therefore, that an acceptable balance at one temperature and deformation rate may not be sustained under different conditions. Currently, the two requisite measurements often are made by independent techniques that use specimens of different shapes, but the tensile test offers the advantage of them being measured on one specimen in one operation.

Decisions about the data formats and logic pathways for materials comparison and selection lie generally outside the scope and influence of the evaluators/investigators, although these workers can nevertheless exert an indirect influence through the tactics and strategies that they adopt for testing and evaluation. It is desirable, in the longer term, that any such steps should be formalized by modifications of existing standard test methods, but this is always a protracted process because of the necessary consultation stage, and as an interim measure limited but useful enhancements of the data can be achieved by strict observance of those strictures of the standards that relate to the qualifying information that describes and specifies the test sample. This suggestion is not likely to recommend itself to people heavily engaged in testing, because the qualifying data can be more voluminous than the actual property data, although there is a growing realization that the latter are virtually useless without the former.

Design Data. The principle underlying design calculations is that the behavior of a structure under a system of forces can be deduced from a formula combining relevant material properties with an appropriate form factor. The property values used should be appropriate for the practical situation to which the design calculation relates—i.e., service temperature, pattern of loading, flow geometry, and other influential factors should all be considered. A distinction is drawn between "design data" and single-point property data, the implication being that the former have a higher status. However, this distinction is an artificial one because even a single datum, such as a modulus or a strength derived from a tensile test, may be used in a design calculation provided that adjustments are made to allow for the differences between the laboratory test conditions and the service/design situation.

At least some of the adjustment factors can be derived from other tensile tests. Thus, anisotropy and other consequences of specific processing conditions and flow geometries can be assessed by tests on appropriately chosen specimens and samples. On the other hand, adjustments that allow for long loading times, intermittent loading, or similar situations must be based on independent creep, creep-rupture, and fatigue tests that are specifically structured to identify and quantify the response of specimens to such loading patterns.

The degree of adjustment varies with the polymer architecture, the composition of the plastic, and the operative factor. If the strength of a standard injection-molded endgated tensile bar is taken as a reference point, an unfavorable flow geometry can reduce the strength to 50% of the reference value, and a long loading period can reduce it to 20% of the reference value, for example. To use unadjusted data in a design for service conditions radically different from those of the test would be to misuse them.

Summary

Tensile testing produces information about the mechanical behavior of specimens subjected to a predominantly tensile stress. The scope and quality of that information depend mainly on the degree of practical finesse that is deployed. The main factors that can affect the outcome of the test program are:

- Sample and specimen selection
- Machine design and function
- Specimen preparation
- Choice and mounting of sensors
- Specimen-machine interaction
- Translation of sensor signals into properties data
- Trains of inference.

However, the over-all test procedure and the strategy depend on the purpose of the test; comprehensive evaluations are expensive, and curtailed evaluations are relatively uninformative.

The over-all balance of the machine-specimen system affects the precision, and to some degree the accuracy, of the raw data. The choices of sample, specimen geometry, specimen position and alignment with respect to the sample, and number of specimens tested affect the precision, accuracy, and fitness-for-purpose of the derived data.

The raw data take the form of a relationship between force and deformation, which can be converted into approximations of a stress-strain relationship and other properties. The raw data and the transformed data relate only to the specific conditions of the test.

The mean value of a measured quantity and the standard deviation as derived from a small subset of nominally identical specimens are approximations of the true mean and standard deviation of a large set of nominally identical specimens.

A tensile test provides data relating to the specimen tested. Tensile-property values derived from one type of specimen drawn from a sample do not fully characterize the tensile properties of the entire sample, and the tensile property values of one sample do not usually suffice to characterize the tensile properties of the constituent material. Finally, tensile properties alone do not characterize the mechanical behavior of a specimen, sample, or material, although they constitute invaluable indicators.

ACKNOWLEDGMENT

This chapter was adapted from S. Turner, Tensile Testing of Plastics, *Tensile Testing,* 1st ed., P. Han, Ed., ASM International, 1992, p 105–133

REFERENCE

1. S. Turner, Mechanical Testing, *Engineering Plastics,* Vol 2, *Engineered Materials Handbook,* ASM International, 1988, p 544–558

SELECTED REFERENCES

- A.-M.M. Baker and C.M.F. Barry, Effects of Composition, Processing, and Structure on Properties of Engineering Plastics, *Materials Selection and Design,* Vol 20, *ASM Handbook,* ASM International, 1997, p 434–456
- J. Rietveld, Viscoelasticity, *Engineering Plastics,* Vol 2, *Engineered Materials Handbook,* ASM International, 1988, p 412–422
- M.L. Weaver and M.E. Stevenson, Introduction to the Mechanical Behavior of Nonmetallic Materials, *Mechanical Testing and Evaluation,* Vol 8, *ASM Handbook,* ASM International, 2000, p 13–25
- Mechanical Testing of Polymers and Ceramics, *Mechanical Testing and Evaluation,* Vol 8, *ASM Handbook,* ASM International, 2000, p 26–48

CHAPTER 9

Tensile Testing of Elastomers

ELASTOMERS comprise a subclass of the larger group of materials, based on very large molecules, called polymers. Various common plastics such as polystyrene and polyethylene, and other materials such as household films and wraps, are polymer-base materials but are not called elastomers because of their limited capacity for reversible stretching. Elastomers must display the ability to stretch and recover that is typical of a rubber band.

Although the terms "elastomer" (from "elastic polymer") and "rubber" at one time had slightly different meanings, they have become synonymous for all practical purposes. These terms are used to designate the mixture of polymers and other ingredients that makeup a rubber formulation. Each unique formulation is called a "compound," much as a mixture of metals is known as an "alloy."

Manufacturing of Elastomers (Ref 1)

The manufacture of rubbers or elastomers involves three major processing steps: mixing or compounding, shaping, and vulcanizing or crosslinking.

Compounding. The properties of elastomers are typically adjusted by compounding, that is, the incorporation of additives that improve properties, aid processing, or reduce cost. A typical formulation might include the elastomer base itself; fillers for reinforcement, hardness control, or cost reduction; a plasticizer to improve low-temperature properties; antioxidants; and the crosslinking system.

The actual mixing process depends on the type of elastomer. A high-viscosity elastomer such as natural rubber requires the use of a powerful mixer such as a Banbury mixer or rubber mill, whereas a more liquid material can be processed using ordinary rotary mixers.

Shaping. The compounded elastomer can be shaped using molding, extrusion, or calendering. Compression molding, transfer molding, and injection molding can be used to produce forms ranging from cable connectors and champagne stoppers to tires. Hoses are the major example of elastomers formed by extrusion. Calendering is used to produce sheet rubber products such as conveyor belts, protective liners, and floor tiles.

Vulcanization is generally carried out after the elastomer is in its final shape, frequently while it is in a mold, at temperatures between 135 and 200 °C (275 and 390 °F). Vulcanization is necessary to transform the raw elastomer into a useful material by providing crosslinks between the long chains of the polymer molecules. As described in the section "Factors Influencing Elastomer Properties" in this chapter, vulcanization has a profound effect on the properties of elastomers.

Properties of Interest

A test of the tensile strength of an elastomer can yield readings of several different properties. In some cases, these properties are totally independent of each other. In other cases, they are interrelated. At times, some will be of more interest than others, depending on what is being investigated or controlled. Typical properties of some of the more common elastomers are listed in Table 1.

Ultimate Tensile Strength. Naturally, the first property of interest determined in a tensile test is the ultimate tensile strength. For elastomers, a class of materials that contain substantial numbers of very different polymers, tensile strength can range from as low as 3.5 MPa (500 psi) to as high as 55.2 MPa (8.0 ksi); however, the great majority of common elastomers tend to fall in the range from 6.9 to 20.7 MPa (1.0 to 3.0 ksi).

Ultimate Elongation. The second property noted is ultimate elongation, which is the property that defines elastomeric materials. Any material that can be reversibly elongated to twice its unstressed length falls within the formal ASTM definition of an elastomer. The upper end of the range for rubber compounds is about 800%, and although the lower end is supposed to be 100% (a 100% increase of the unstressed reference dimension), some special compounds that fall slightly below 100% elongation still are accepted as elastomers.

Modulus. The third characteristic that may be of interest is referred to in the rubber industry as the modulus of the compound, but a specific designation such as 100% modulus or 300% modulus is used. That is due to the fact that the number generated is not an engineering modulus in the normal sense of the term, but rather is the stress required to obtain a given strain. Therefore, the "100% modulus," also referred to as M-100, is simply the stress required to elongate the rubber to twice its reference length.

Tension Set. A final characteristic that can be measured, but that is used less often than the other three, is called "tension set." Often, when a piece of rubber is stretched to final rupture, the recovery in length of the two sections resulting from the break is less than complete. It is possible to measure the total length of the original reference dimension and calculate how much longer the total length of the two separate sections is. This is expressed as a percentage. Some elastomers will exhibit almost total recovery, whereas others may display tension set as high as 10% or more. Tension set may also be measured on specimens stretched to less than breaking elongation.

Factors Influencing Elastomer Properties

Because elastomers are enormously different in molecular structure from other materials such as metals, and in fact are complex organic composites of numerous ingredients of very differing characteristics, it is not surprising that they tend to exhibit a wide range of characteristics. Some of the important factors that influence elastomer properties include:

- Structuring of the molecular matrix
- Compounding
- Specimen preparation
- Specimen type
- Vulcanization parameters
- Temperature

Molecular Structure. Very often the processing of the mixture that makes up the elastomer results in some level of orientation of the molecules involved. This structuring of the molecular matrix is commonly referred to as the

Table 1 Properties of common elastomers

		Mechanical properties				Service temperature (continuous use)	
Common name	Specific gravity	Shore Durometer hardness	Tensile strength, MPa (ksi)	Modulus, 100%, MPa (psi)	Elongation, %	min, °C (°F)	max, °C (°F)
Butadiene rubber	0.91	45A–80A	13.8–17.2 (2.0–2.5)	2.1–10.3 (300–1500)	450	−100 (−150)	95 (200)
Natural rubber, isoprene rubber	0.92–1.037	30A–100A	17.2–31.7 (2.5–4.6)	3.3–5.9 (480–850)	300–800	−60 (−75)	70 (160)
Chloroprene rubber	1.23–1.25	30A–95A	3.4–24.1 (0.5–3.5)	0.7–20.7 (100–3000)	100–800	−50 (−60)	107 (225)
Styrene-butadiene rubber	0.94	30A–90D	12.4–20.7 (1.8–3.0)	2.1–10.3 (300–1500)	450–500	−60 (−75)	120 (250)
Acrylonitrile-butadiene (nitrile) rubber	0.98	30A–100A	6.9–24.1 (1.0–3.5)	3.4 (490)	400–600	−50 (−60)	120 (250)
Isobutylene-isoprene (butyl) rubber	0.92	30A–100A	>13.8 (>2.0)	0.3–3.4 (50–500)	300–800	−45 (−50)	150 (300)
Ethylene-propylene (-diene) rubber	0.86	30A–90A	3.4–24.1 (0.5–3.5)	0.7–20.7 (100–3000)	100–700	−55 (−70)	150 (300)
Silicone rubber	1.1–1.6	20A–90A	10.3 (1.5)	. . .	100–800	−117 (−178)	260 (500)
Fluoroelastomer	1.8–1.9	55A–95A	10.3–13.8 (1.5–2.0)	1.4–13.8 (200–2000)	150–250	−50 (−60)	260 (500)

Source: Ref 1

"grain" of the rubber, and tensile properties usually differ to a detectable degree with and across the grain. This anisotropy may not be significant or even exist in actual elastomeric components, depending on both the specific compound and its processing history. When the grain direction can be determined from knowledge of the processing, tensile testing is done parallel to the grain.

Compounding. Over 20 different types of polymers can be used as bases for elastomeric compounds, and each type can have a significant number of contrasting subtypes within it. Properties of different polymers can be markedly different: for instance, urethanes seldom have tensile strengths below 20.7 MPa (3.0 ksi) whereas silicones rarely exceed 8.3 MPa (1.2 ksi). Natural rubber is known for high elongation, 500 to 800%, whereas fluoroelastomers typically have elongation values ranging from 100 to 250%.

Literally hundreds of compounding ingredients are available, including major classes such as powders (carbon black, clays, silicas), plasticizers (petroleum-base, vegetable, synthetic), and curatives (reactive chemicals that change the gummy mixture into a firm, stable elastomer). A rubber formulation can contain from four or five ingredients to 20 or more. The number, type, and level of ingredients can be used to change dramatically the properties of the resulting compound, even if the polymer base remains exactly the same.

Thus, the same base material—polychloroprene (widely known as neoprene), for example—can be used by the rubber chemist to make compounds as soft as a baby-bottle nipple or as hard as a hockey puck, with tensile strengths ranging from less than 6.9 MPa (1.0 ksi) to more than 20.7 MPa (3.0 ksi) and elongation values from 150 to 600%. Considering the wide varieties of starting polymers and ingredient choices, it is understandable that extremely broad contrasts in properties are found among elastomers.

Specimen Preparation. In addition, tensile properties of elastomers are sensitive to factors involved in specimen preparation. The majority of the time, specimens are cut from molded sheets of rubber. This is done using sharpened dies of a specific dumbbell shape, and the smoothness and sharpness of the die are important. Any nick or tiny tear along the edge of the cut specimen can act as a crack initiator and lead to premature failure of the specimen. Inappropriately low levels of ultimate tensile strength and elongation can be observed in such instances.

Similarly, lack of thoroughness in mixing of the ingredients can lead to poor dispersion, and careless mixing can cause incorporation of small foreign particles in the rubber. Either case will again lead to lower and less precise test results.

Specimen Type. Use of specimens other than the standard type called for in the ASTM procedures (see below) is sometimes necessary. Pieces from large moldings can be cut out and ground to reasonable flatness and appropriate thickness, or strips of small tubing can be tested. Correlation between such specimens and standard types is not always precise. Ground specimens do not have the smooth, molded surfaces of laboratory specimens, and therefore it is very likely that cracks will propagate from surface imperfections in the early stages of strain, leading to tensile rupture at lower elongations. Because the stress-strain curve is terminated at a lower strain, the associated tensile force is automatically lower as well, and thus nonstandard specimens seem to display lower values of elongation and tensile strength than lab specimens of identical material.

Vulcanization. Differences in test results between lab specimens and specimens cut from actual parts may also be caused partly by another variable—the level of vulcanization of the elastomer, also called its "state of cure." It is difficult to determine whether or not the state of cure for a lab specimen is truly the same as that for a specimen cut from a large article.

Vulcanization, which is the formation of chemical crosslinks between the long chains of the polymer molecules, is usually accomplished through exposure to some level of heat over time. Although different thermal cycles may yield rubber articles that appear and feel the same, their properties can vary appreciably. The various tensile properties will change in different degrees with increasing thermal treatment, so that there is seldom an optimum state of cure in the sense of all of the compound properties reaching their ideal levels simultaneously. For instance, tensile strength may reach a maximum following some particular curing cycle, whereas elongation at that point is well along a steeply decreasing curve.

Thus, the optimum curing cycle for molding of a given compound must be determined through various means too diverse to be explained here, and that curing cycle must then be used consistently for test specimens made of that compound. Otherwise, differences in tensile properties that do not truly relate to any real difference in the formulation will very likely be observed.

At times, a compound will be tested at its normal cure level, and then a second set of samples not only will be molded with the standard curing cycle, but will then undergo an additional phase of high-temperature exposure prior to thermal testing. This thermal aging, usually done in an oven at a combination of temperature and time appropriate to the particular type of elastomer, will result in definite changes in the polymer matrix.

Such changes are reflected in alteration of the tensile-test results. Reduction in elongation is typical, but ultimate tensile strength may increase or decrease. The degree of change of tensile properties resulting from thermal aging is frequently used as an indicator of the compound's ability to withstand aging and/or lower thermal exposure over long time periods. One rule of thumb is that the time required at a given temperature for a compound's tensile strength to drop to about half its original level represents the functional life of the compound at that temperature.

A more subtle effect on standard test results is the effect of time delay between vulcanization and testing of the elastomer. Various complex processes continue to take place in the polymeric matrix for some time after molding is completed, which can affect tensile properties. Therefore, normal procedures call for a minimum delay of 8 h between molding and testing. However, in certain production situations for which such a delay is not tolerable, a correlation could be developed between "warm testing" results—i.e., from tests run within a short time of the sample being vulcanized—and those from standard procedures.

Test Temperature. Aside from the types of specimen-preparation effects mentioned above, there are also significant effects from differing test conditions. The great bulk of testing is done at room temperature and a standard rate of elongation, but occasionally special conditions will be called for. For instance, knowledge of tensile strength at some elevated temperature is sometimes desired. Raising or lowering test temperature usually has an inverse effect on tensile strength that can be very substantial, changing it by a factor of two or more.

ASTM Standard D 412

The official standard for tensile testing of elastomers is ASTM D 412 (Ref 2). It specifies two principal varieties of specimens: the more commonly used dumbbell-type die cut from a standard test slab 150 by 150 by 20 mm (6 by 6 by 0.8 in.), and actual molded rings of rubber. The second type was standardized for use by the O-ring industry. For both varieties, several possible sizes are permitted, although, again, more tests are run on one of the dumbbell specimens (cut using the Die C shape described in ASTM D 412) than on all other types combined. Straight specimens are also permitted, but their use is discouraged because of a pronounced tendency to break at the grip points, which makes the results less reliable. Unless otherwise specified, the standard temperature for testing elastomer specimens is 23 ± 2 °C (73.4 ± 3.6 °F).

The power-driven equipment used for testing is described, including details such as the jaws used to grip the specimen, temperature-controlled test chambers when needed, and the crosshead speed of 500 mm/min (20 in./min). The testing machine must be capable of measuring the applied force within 2%, and a calibration procedure is described. Various other details, such as die-cutting procedures and descriptions of fixtures, are also provided.

The method for determining actual elongation can be visual, mechanical, or optical, but is required to be accurate within 10% increments. In the original visual technique, the machine operator simply held a scale behind or alongside the specimen as it was being stretched and noted the progressive change in the distance between two lines marked on the center length of the dogbone shape. The degree of precision that could be attained using a hand-held ruler behind a piece of rubber being stretched at a rate of over 75 mm/s (3 in./s) was always open to question, with 10% being an optimistic estimate.

More recent technology employs extensometers, which are comprised of pairs of very light grips that are clamped onto the specimen and whose motion is then measured to determine actual material elongation. The newest technology involves optical methods, in which highly contrasting marks on the specimen are tracked by scanning devices, with the material elongation again being determined by the relative changes in the reference marks.

Normal procedure calls for three specimens to be tested from each compound, with the median figure being reported. Provision is also made for use of five specimens on some occasions, with the median again being used.

Techniques for calculating the tensile stress, tensile strength, and elongation are described for the different types of test specimens. The common practice of using the unstressed cross-sectional area for calculation of tensile strength is used for elastomers as it is for many other materials. It is interesting to note that if the actual cross-sectional area at fracture is used to calculate true tensile strength of an elastomer, values that are higher by orders of magnitude are obtained.

Test Method Precision. In recent years, attention has been given to estimating the precision and reproducibility of the data generated in this type of testing. Interlaboratory test comparisons involving up to ten different facilities have been run, and the later versions of ASTM D 412 contain the information gathered.

Variability of the data for any given compound is to some degree related to that particular formulation. When testing was performed on three different compounds of very divergent types and property levels, the pooled value for repeatability of tensile-strength determinations within labs was about 6%, whereas reproducibility between labs was much less precise, at about 18%. Comparable figures for ultimate elongation were approximately 9% (intralab) and 14% (interlab).

Surprisingly, the same comparisons for M-100 (100% modulus) showed much less precision, with intralab variation of almost 20% and interlab variation of over 31%. The theory had been held for some time that, because tensile strength and ultimate elongation are failure properties, and as such are profoundly affected by details of specimen preparation, tensile modulus figures would be more narrowly distributed. Because the data given above clearly do not support such a theory, some other factor must be at work. Possibly it is the lack of precision with which the 100% strain point is observed, but in any case it was important to determine the actual relationship between the precision levels of the different property measurements.

Significance and Use of Tensile-Testing Data

Tensile Strength. The meaning of tensile strength of elastomers must not be confused with the meaning of tensile strength of other materials such as metals. Whereas tensile strength of a metal may be validly and directly used for a variety of design purposes, this is not true for elastomers. As stated early in ASTM D 412, "Tensile properties may or may not be directly related to the end use performance of the product because of the wide range of performance requirements in actual use." In fact, it is very seldom if ever that a given high level of tensile strength of a compound can be used as evidence that the compound is fit for some particular application.

It is important to note that the tensile properties of elastomers are determined by a single application of progressive strain to a previously unstressed specimen to the point of rupture, which results in a stress-strain curve of some particular shape. The degree of nonlinearity and in fact complexity of that curve will vary substantially from compound to compound.

In Fig. 1, tensile-test curves from five very different compounds, covering a range of base polymer types and hardnesses, are displayed. The contrasts in properties are clearly visible, such as the high elongation (>700%) of the soft natural rubber compound compared with the much lower (about 275%) elongation of a soft fluorosilicone compound. Tensile strengths as low as 2.4 MPa (350 psi) and as high as 15.5 MPa (2.25 ksi) are observed. Different shapes in the curves can be seen, most noticeably in the pronounced curvature of the natural rubber compound.

Figure 2 demonstrates that, even within a single elastomer type, contrasting tensile-property responses will exist. All four of the compounds tested were based on polychloroprene, covering a reasonably broad range of hardnesses, 40 to 70 Shore A Durometer. Contrasts are again seen, but more in elongation levels than in final tensile strength. Two of the compounds are at the same Durometer level, and still display a noticeable difference between their respective stress-strain curves. This shows how the use of differing ingredients in similar formulas can result in some properties being the same or nearly the same whereas others vary substantially.

It should be noted that successive strains to points just short of rupture for any given compound will yield a series of progressively different stress-strain curves; therefore, the tensile-strength rating of a compound would certainly change depending on how it was flexed prior to final fracture.

Thus, the real meaning of rubber tensile strength as determined using the official procedures is open to some question. However, some minimum level of tensile strength is often used as a criterion of basic compound quality, because the excessive use of inexpensive ingredients to fill out a formulation and lower the cost of the compound will dilute the polymer to the point that tensile strength decreases noticeably. For example, neoprene compounds are capable of achieving tensile strengths up to 20.7 MPa (3.0 ksi) or higher when compounded using good technical practice.

In many cases, use of legitimate compounding techniques to optimize specific performance characteristics will result in neoprene compounds whose tensile strengths range from 10.3 to 17.2 MPa (1.5 to 2.5 ksi). The fact that the range has a lower end well below 20.7 MPa (3.0 ksi) does not in any way imply that the com-

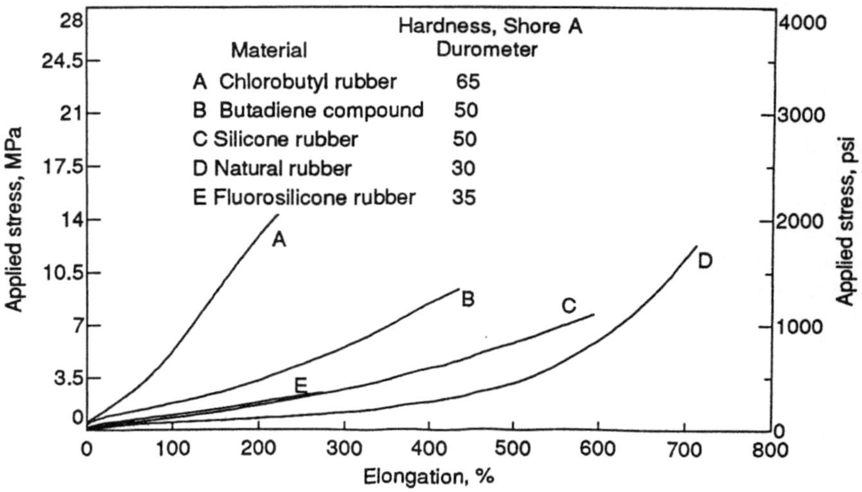

Fig. 1 Tensile-test curves for five different elastomer compounds

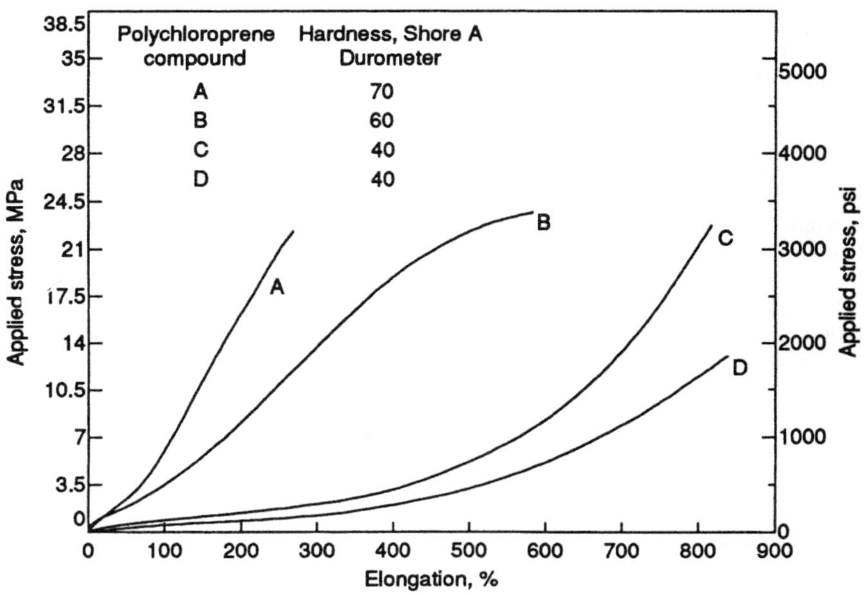

Fig. 2 Tensile-test curves for four polychloroprene compounds

pounds are deficient in some sense, but it is generally accepted that a tensile strength of a neoprene compound below 10.3 MPa (1.5 ksi) is evidence that the compound is low in polymer content and therefore its ability to provide good performance over time is questionable.

Various specifications on elastomers, including government and industrial standards, call for minimum tensile strengths at different levels for different types of polymers. Such minima range from perhaps 4.8 MPa (700 psi) for silicones to over 21 MPa (3.0 ksi) for urethanes.

Because elastomeric elements are hardly ever used in tension, tensile strength of compounds is not a useful property measurement for predicting performance. Also, because tensile strength does not correlate with other important characteristics such as stress relaxation and fatigue resistance, it is principally used as a quality-control parameter relating to consistency.

Elongation is the unique defining property of elastomers, and its meaning is somewhat more applicable to end uses. However, because service conditions normally do not require the rubber to stretch to any significant fraction of its ultimate elongative capacity, ultimate elongation still does not provide a precise indication of serviceability.

It is commonly accepted that as the elongation of a compound declines, that material's ability to tolerate strain, including repetitive strain, generally decreases. Thus, if two compounds based on the same elastomer but having quite contrasting elongation values are compared in fatigue properties when both are subjected to equal strain levels, the formula with the higher elongation might well be expected to have the longer life.

Just as with tensile strength, certain minimum levels of ultimate elongation are often called out in specifications for elastomers. The particular elongation required will relate to the type of polymer being used and the stiffness of the compound. For example, a comparatively hard (80 Durometer) fluoroelastomer might have a requirement of only 125% elongation, whereas a soft (30 Durometer) natural rubber might have a minimum required elongation of at least 400%.

Tensile modulus, better described as the stress required to achieve a defined strain, is a measurement of a compound's stiffness. When the stress-strain curve of an elastomer is drawn, it can be seen that the tensile modulus is actually a secant modulus—that is, a line drawn from the graph's origin straight to the point of the specific strain. However, if an engineer really needs to understand what forces will be required to deform the elastomer in a small region about that strain, he or she would be better off drawing a line tangent to the curve at the specific level of strain, and using the slope of that line to determine the approximate ratio of stress to strain in that region. This technique can be utilized in regard to actual elastomeric components as well as lab specimens.

Tension set is used as a rough measurement of the compound's tolerance of high strain. This property is not tested very often, but for some particular applications such a test is considered useful. It could also be used as a quality-control measure or compound development tool, but most of the types of changes it will detect in a compound will also show up in tests of tensile strength, elongation, and other properties, and so its use remains infrequent.

Summary

Tensile properties of elastomers vary widely, depending on the particular formulation, and scatter both within and between laboratories is appreciable compared with scatter in tensile testing of metal alloys. ASTM D 412 is the defining specification, and presents detailed instructions on specimen preparation, equipment, test conditions, etc. The meaning of the data is comparatively limited in regard to the utility of any compound for a specific application. Tensile-test data are used effectively as quality-control parameters and general development tools for the rubber technologist.

ACKNOWLEDGMENT

This chapter was adapted from R.J. Del Vecchio, Tensile Testing of Elastomers, *Tensile Testing,* P. Han, Ed., ASM International, 1992, p 135–146

REFERENCES

1. R. Tuszynski, Elastomers, *Engineered Materials Handbook, Desk Edition,* ASM International, 1995, p 282–286
2. ASTM D 412, "Standard Test Methods for Vulcanized Rubber and Thermoplastic Elas-

tomers—Tension," *Annual Book of ASTM Standards,* Vol 09.01, ASTM International

SELECTED REFERENCES

- A.K. Bhowmick and H.L. Stephens, Ed., *Handbook of Elastomers,* 2nd ed., Marcel Dekker, 2000
- A.K. Bhowmick, M.M. Hall, and H.A. Benarey, Ed., *Rubber Products Manufacturing Technology,* Marcel Dekker, 1994
- A.N. Gent, Ed., *Engineering with Rubber,* Hanser Publishers, 1992
- W.F. Harrington, Elastomeric Adhesives, *Engineered Materials Handbook,* Vol 3, *Adhesives and Sealants,* ASM International, 1990, p 143–150
- J.E. Mark, B. Erman, and F.R. Eirich, Ed., *Science and Technology of Rubber,* 2nd ed., Academic Press, 1994
- B.M. Walker and C.P. Rader, Ed., *Handbook of Thermoplastic Elastomers,* 2nd ed., Van Nostrand Reinhold, 1988

CHAPTER 10

Tensile Testing of Ceramics and Ceramic-Matrix Composites

THE ADVANCED CERAMIC MATERIALS described in this chapter include both noncomposite, or monolithic, ceramics (for example, oxides, carbides, nitrides, and borides) and ceramic-matrix composites (CMCs). Ceramic-matrix composites can be broadly classified into two types: discontinuously reinforced CMCs (for example, particulate- or whisker-reinforced materials) and continuous fiber-reinforced materials. These advanced ceramic materials exhibit superior mechanical properties, corrosion/oxidation resistance, or electrical, optical, and/or magnetic properties when compared to traditional ceramics (ceramics products that use clay or have a significant clay component in the batch).

Rationale for Use of Ceramics

Advanced ceramics have been shown to have significant potential as structural materials. This is especially true for various specialized applications—particularly those involving high use temperatures. Ceramic materials have several real or potential advantages for such specialized applications that make them very appealing and possibly very competitive with existing structural materials. These advantages include the fact that ceramics can be made from noncritical raw materials (for example, aluminum, boron, carbon, nitrogen, oxygen, silica, and so on), in contrast to the scarce materials (nickel, cobalt, chromium, niobium, and so on) required for high-temperature superalloys. Another advantage is a potential for low cost, based in part on low-cost raw materials. Other advantages are based on the intrinsic properties characteristic of ceramics, including high stiffness (elastic modulus), high hardness, low thermal expansion, low density, chemical stability, thermal stability, and good electromagnetic properties (which are important for electromagnetic windows and electronic materials). The combination of low density, high stiffness, high strength and toughness (in composites), high use temperature, and chemical stability make some ceramics and CMCs most appealing as high-temperature structural materials. In such applications, these materials can be expected to have properties such as stiffness-to-weight and strength-to-weight ratios that far surpass those achievable with competitive materials such as superalloys or intermetallics (for example, NiAl).

Intrinsic Limitations of Ceramics

Unfortunately, some of the desirable intrinsic properties of ceramics also lead to some highly undesirable characteristics. The most significant of these derives from the ionic/covalent bonding typical of most ceramics, which severely limits plastic deformation. This limited plasticity greatly reduces the energy absorbed during fracture. The fracture energy then approaches the very low values of the cleavage energy. The low fracture energy or fracture toughness further results in several undesirable traits. Monolithic ceramics are typically flaw-sensitive, failing as a result of defects that are undetectable by conventional NDE techniques. The same flaw sensitivity also gives rise to great variability in strength, as a result of variations in the flaw population, and thus very low values of design strength. The low fracture energy also implies that monolithic ceramics will typically fail catastrophically—i.e., they will exhibit no stable crack propagation below the critical stress-intensity value, K_{Ic}. The effect here is most severe with respect to the tensile properties of ceramics. Ceramics typically are much higher in compressive strength than in tensile strength, and do not fail in shear modes, because K_{IIc} and K_{IIIc} are much higher than K_{Ic}.

As a result of the severe flaw sensitivity, lack of plastic deformation and relatively high stiffness of ceramics, the tensile strengths of ceramics are typically measured indirectly, rather than in direct tensile tests, as is common for other engineering materials. The results of direct tensile tests are relatively clear, assuming that failure occurs in appropriate locations and modes. In that case, the strength value derived from a direct uniaxial tensile test reflects the true tensile strength of the material. For most ceramics, however, "tensile" strength is measured indirectly by one of two types of flexural or bending tests. In these tests, the specimen is subjected to a complex stress state including tension, compression, shear, and significant stress gradients. In interpreting the results of these flexure tests, the maximum tensile stress present in the specimen at failure is usually reported as the "tensile" strength of the ceramic. Although such testing is straightforward, and calculation of the failure stress simple, many complications are involved. This is particularly true with fiber-reinforced CMCs, for which the results can be very misleading in terms of the true tensile strength of the material tested.

In addition to the widely used flexure tests (three-point, or modulus of rupture, and four-point), there are also other indirect tensile tests, each with its advantages and disadvantages, as will be discussed. Most of these tests have been developed with the intention of overcoming some of the difficulties associated with direct tensile tests or the complications inherent in flexure tests. In addition, especially in recent years, some modifications of tensile-test fixtures and specimens have become available, which make direct tensile testing of some ceramics more tractable.

Overview of Important Considerations for Tensile Testing of Advanced Ceramics

There are four key considerations that must be taken into account when carrying out tensile tests on advanced monolithic ceramics and CMCs. These include:

- Effects of flaw type and location on tensile tests
- Separation of flaw populations
- Design strength and scale effects
- Lifetime predictions and environmental effects

Effects of Flaw Type and Location on Tensile Tests

One of the complications of tensile testing is the physical location of the flaws that lead to failure. Most ceramics (and other materials) contain both surface and volume flaws. Surface flaws typically result from finishing operations and/or damage during service (for example, damage by foreign objects). Volume flaws typically are intrinsic to the material microstructure or are processing defects (voids, inclusions, etc.). It is important that any "tensile" test characterize the effects of all of these defects (or at least the most severe in terms of performance) on strength. Unfortunately, many of the indirect tensile tests, including flexure tests, produce severe stress gradients that may bias failure toward one type of flaw, most typically toward surface defects. Thus a flexure test on a ceramic material may detect primarily the flaws associated with the machining required to produce the test specimen, rather than the volume flaws associated with the processing of the material. It is quite important here, in trying to assess the "tensile" strength of a material, to be aware of these different flaw types and locations, and their effects on the results of different test procedures.

Separation of Flaw Populations

Assessment of the importance of different types and locations of flaws ideally is based on identification of the actual flaw types using fractography (Ref 1, 2). This is generally a time-consuming and sometimes very difficult task, especially if scanning electron microscopy is required. An alternative although less deterministic approach is to use data-analysis procedures suitable for separating multimodal distributions of strength data into their constituent parts. In some cases, this can be done effectively, although some uncertainties are always associated with this purely mathematical approach to separating the effects of different flaw populations in a material.

Fractography, as performed on ceramics and some ceramic composites, is typically done using reflected light microscopy for the larger flaws, but more often requires scanning electron microscopy for resolution of the small flaws (10 to 30 μm, or 0.39 to 1.2 mils) that are typical of monolithic ceramics. Recommended procedures for fractographic analysis are outlined in ASTM C 1322, "Standard Practice for Fractography and Characterization of Fracture Origins in Advanced Ceramics."

Many data-analysis procedures for characterizing strength distributions can be found in the applied mathematics and statistics literature. Commercial computer programs that perform some types of data analysis are widely available, although there are some pitfalls here as well. Different techniques for fitting the same distribution function to a set of data can produce different results for both the function's parameters and the errors in the parameters. These differences can then lead to problems with the use of the strength data, such as with lifetime predictions, predictions of failure probabilities, or estimates of scale effects on strength.

Design Strength and Scale Effects

For ceramics, determination of design strength and prediction of scale effects are two of the most important uses of strength data and thus two of the most important reasons for performing some type of tensile testing and the associated data analysis. For the designer, one of the key requirements is the specification of design strength as a function of service conditions (temperature and environment) and time. Presumably, the designer can specify quite accurately these service conditions (stress, temperature, and so on) as well as the desired lifetime of the component. Thus, accurate and hopefully conservative design-strength values can be incorporated into design codes to help ensure that components will perform as desired.

One aspect of the design process that is more significant for ceramics than for other, less brittle materials is the effect of specimen or component size on strength. The qualitative effect here is that larger specimens or components, on average, will have lower strengths and less scatter in strength values than small specimens. This results from presence in the larger components of greater numbers of flaws and a greater probability of the presence of more severe flaws. If design-strength data based on testing of relatively small specimens are to be used for predicting the performance of larger components, it is necessary to account for the scale effect on strength. This is typically done through the use of Weibull strength distributions, which were developed in the 1940s (Ref 3–5) and have since been widely used for characterizing a variety of material and component properties. Note that variations in size between laboratory test specimens and actual components can be quite large, with very large effects on design strengths. The difference in stressed volume between a metal tensile-test specimen and a solid-fuel rocket-motor casing, and the difference in gage length between a laboratory tensile-test specimen of an optical fiber and a transatlantic communication cable, both may be on the order of 10^6. Because testing of actual components in these and other cases is clearly impractical, accurate and conservative techniques for predicting such scale effects on strength and other significant properties are essential.

Lifetime Predictions and Environmental Effects

An issue that is also related to the nature of flaw and strength distributions is the prediction of component lifetimes from initial strength distributions and knowledge of service conditions. This relies even more heavily on accurate knowledge of the nature of the initial flaw distribution, because the nature of subsequent delayed failure depends strongly on the type and location of the initial flaw that leads to failure. Surface flaws can easily react with the environment, leading to delayed failure in modes such as stress-corrosion cracking. Volume flaws may be stable and may not lead to delayed failure under long-term loading. However, such flaws may also react with the remainder of the material—for instance, with an inclusion that differs chemically from the rest of the material—or may react with the environment diffusing into the bulk of the material. Such changes in volume flaws may subsequently lead to failure of the material. It is clearly important to have detailed knowledge of the nature of the initial flaw population, the manner in which the flaws evolve during service, how they interact with the service environment and the applied loads, and which of them control the service life of the material.

Tensile Testing Techniques

Tensile testing techniques, as applied to ceramics and CMCs, fall into four basic categories, each of which has its own advantages, problems, and complications. These categories are:

- True direct uniaxial tensile tests at ambient temperatures
- Indirect tensile tests (for example, three- and four-point flexural tests)
- Other tests where failure is presumed to result from tensile stresses
- High-temperature tensile tests

Applicable standards for some of these tests include:

- ASTM C 1273, "Standard Test Method for Tensile Strength of Monolithic Advanced Ceramics at Ambient Temperatures"
- ASTM C 1275, "Standard Test Method for Monotonic Tensile Behavior of Continuous Fiber-Reinforced Advanced Ceramics with Solid Rectangular Cross-Section Test Specimens at Ambient Temperature"
- ASTM C 1161, "Standard Test Method of Flexural Strength of Advanced Ceramics at Ambient Temperature"
- ASTM C 1211, "Standard Test Method for Flexural Strength of Advanced Ceramics at Elevated Temperatures"

Direct Tensile Tests

In terms of analysis of test results, the most straightforward tests are the direct tensile tests covered in ASTM C 1273 and C 1275. In these tests, the gage length of the specimen is nominally in a state of uniaxial tensile stress. Consequently, both the volume and surface of the gage length are subject to the same simple stress state, which is assumed to be constant throughout the gage volume; that is, it is normally assumed that both the surface and the volume of the gage section of the test specimen are subjected to a state of uniform uniaxial tension.

Test Specimen Geometries. There are two basic types of tensile specimen geometries. One type of specimen that can be prepared using readily available machine tools is the flat or "dog-bone" specimen shown in Fig. 1(a) and Fig. 2. Such specimens can be prepared readily using milling machines with carbide tooling for some materials and diamond tooling for others. It is also feasible, in some cases, to mold specimens directly into the desired shape (for example, by injection molding), which permits testing of materials with as-fabricated surfaces. These may be preferable to the machined surfaces typical of specimens prepared by grinding, where actual components are not surface finished.

The other type of specimen normally used is a cylindrical specimen (Fig. 1b and 2), typically with a reduced gage section and ends machined to suit some gripping arrangement. Such specimens are typically prepared (in the case of metals and polymers) by machining to the desired shape on a template-controlled profile lathe. In the case of ceramics and CMCs, the analogous procedure uses diamond grinding in the same mode to produce a cylindrical specimen of the desired shape. Again, it is possible, and sometimes desirable, to produce such specimens directly by a molding process, or by machining in the green state prior to firing, when an as-fired surface finish is appropriate for testing.

Gripping and Load Transfer in Direct Tensile Tests. Gripping of both flat and cylindrical specimens can be accomplished in various manners, depending on the particular material being tested. Success in using various gripping techniques will depend on the relative values of tensile strength, shear strength, hardness, and so on, of the material being tested. The dog-bone specimens can be gripped in conventional mechanical grips (Fig. 3a) or hydraulic or pneumatic grips (Fig. 3b), using friction alone to transmit the load to the specimen. Conventional mechanical wedge-action grips (Fig. 3c) can also be used successfully in some cases, although the high and uncontrollable clamping pressure may result in crushing or shear failure in the grip section for some materials. Pneumatic or hydraulic grips are generally preferable, because the gripping pressure can be controlled precisely, and because deformation of the specimen does not produce any change in the gripping pressure.

The success in load transfer through friction depends on achieving a reasonable friction coefficient between the specimen and the grip faces without causing the specimen to fail in compression. As an illustration of this, consider gripping a cylindrical aluminum oxide specimen with a 6.4 mm (0.25 in.) diameter in the gage section and a 12.7 mm (0.5 in.) diameter smooth shank. If the tensile strength is assumed to be approximately 350 MPa (50 ksi), a tensile test will require a load of 10,900 N (2450 lbf) to fracture the specimen. With a coefficient of friction of 0.13 between the specimen and the grip faces, the lateral clamping force would have to be 83,980 N, or 18,880 lbf. This clamping force is easily achievable with commercially available hydraulic grips.

The compressive stress on the shank of the specimen is based on the area of the specimen surface inserted into the grip. For this example, if the specimen is inserted into the grip to a depth of 25 mm (1 in.), the compressive stress is about 83 MPa (12 ksi), or well within the capability of the material.

It is very important to verify that the specimen geometry of the material being tested is appropriate for that material's strength. A combination of reducing the cross-sectional area of the gage section and increasing the length of insertion into the grips may be necessary to allow frictional gripping on some ceramic materials. If these specimen geometry enhancements are not possible because of limitations in the material, the use of frictional gripping may not be appropriate.

The problems of frictional gripping are generally severe for most ceramics, which typically have high hardnesses and low friction coefficients against other hard materials. This gripping technique is also particularly difficult with some fiber-reinforced CMCs, which combine high tensile strength, high hardness, and low shear strength. The problems are doubly complicated for the CMCs because the low shear strength limits the load transfer, as well as providing the possibility of shear failure in the grip section at high gripping pressures. There is a relatively

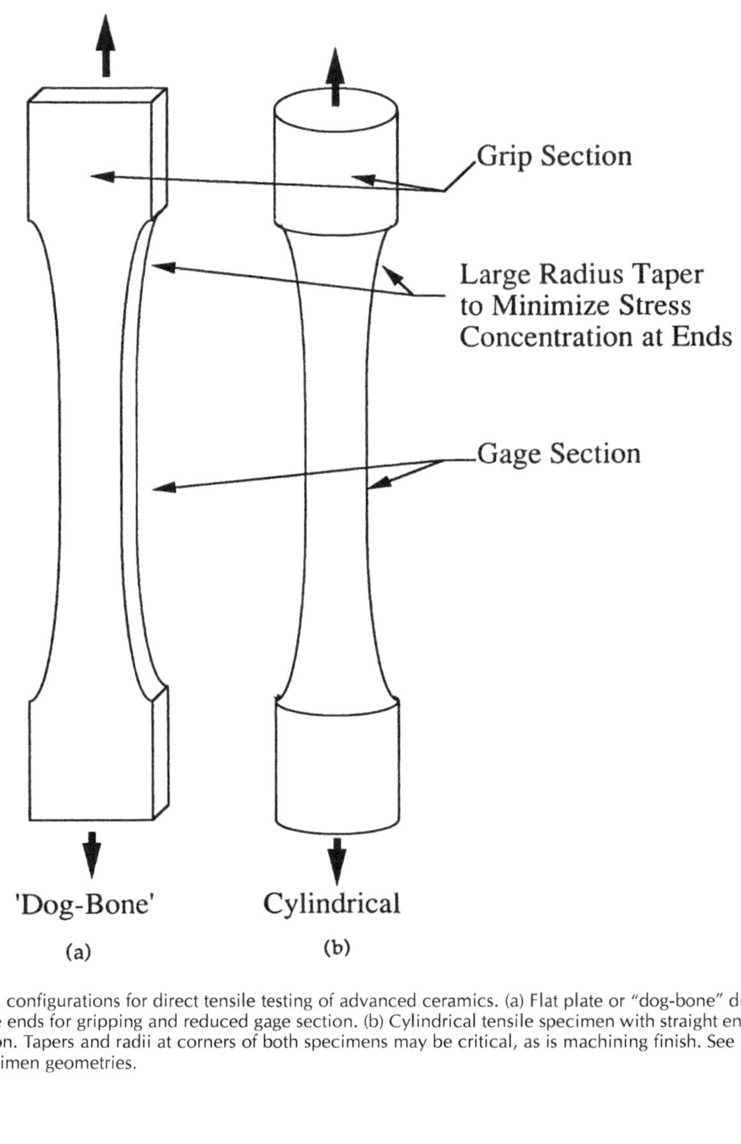

Fig. 1 Specimen configurations for direct tensile testing of advanced ceramics. (a) Flat plate or "dog-bone" direct tensile specimen with large ends for gripping and reduced gage section. (b) Cylindrical tensile specimen with straight ends for collet grips and reduced gage section. Tapers and radii at corners of both specimens may be critical, as is machining finish. See Fig. 2 for examples of more complex specimen geometries.

simple technique for minimizing these problems with CMCs, namely the use of large ratios of grip area to gage section cross-sectional area; however, this technique introduces other problems as well, primarily in terms of the effects of machining damage on the relatively large surface area of the gage section versus the intrinsic flaws in the relatively small volume of a highly reduced gage section.

Gripping of cylindrical specimens can also be done by means of friction, using wedge-type or collet grips, but this involves the same problems as those detailed above, plus the additional difficulties of requiring precise machining of specimen ends to mate with collets, and strict requirements in regard to specimen straightness. In the case of tapered specimen ends, which are used to increase load transfer, or in the case of the buttonhead specimen discussed below, machining can be even more critical.

Load transfer for flat plate or dog-bone specimens can also be effected by means of pins inserted through the grip section of the specimen (Fig. 3d), or such pinned ends can be combined with frictional gripping. Load transfer through pins requires, again, a balance between the load that can be transmitted through the bearing area, $\sigma_b A_b$, and the load required to produce tensile failure, $\sigma \cdot A_{gage}$. In most cases, this requires the use of multiple pins for load transfer. The use of multiple pins requires great precision both in the test apparatus and in machining of the specimen (precise hole location and diameter to ensure equal distribution of loading).

One approach sometimes taken to overcome some of these difficulties in specimen gripping and load transfer is bonding of the ceramic or composite specimen to grips of a more forgiving material. A low-shear-strength, high-tensile-strength, unidirectional CMC specimen can be

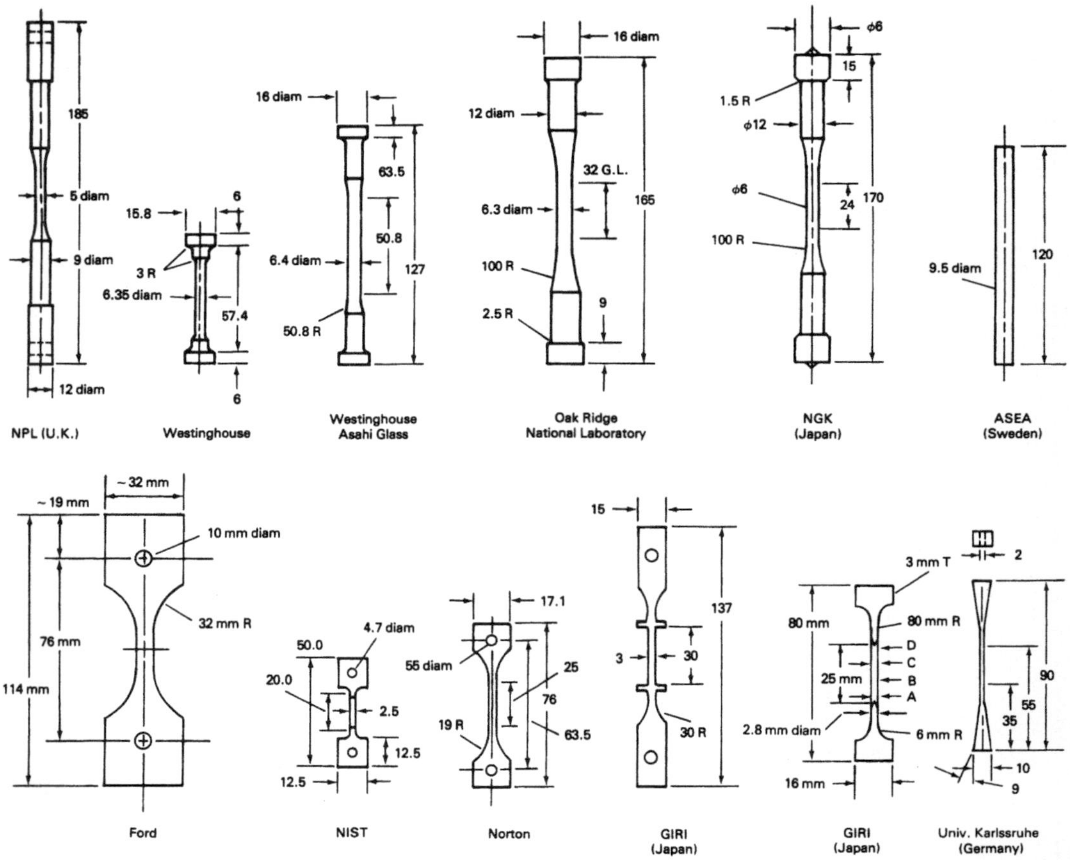

Fig. 2 Tensile specimens used for monolithic ceramics (each is in correct proportion to the others); all dimensions in mm. Upper row for round specimens; lower row for flat specimens. Source: G.D. Quinn, NIST

bonded to metallic grips that are a good match for the CMC in terms of Young's modulus (to minimize stress concentration). Provided that sufficient gripping area is available for load transfer through the adhesive, there is then little difficulty in applying load by conventional means to the now-metallic gripping area of the specimen (note that conventional epoxy adhesives have shear strengths that exceed those of some continuous-fiber CMCs). This procedure, which works very well, unfortunately is not useful for the more important high-temperature tensile tests, as will be discussed later.

The last technique to be discussed here is one that has come into use in commercial test fixtures for tensile testing of ceramics, based on a system developed by personnel at Oak Ridge National Laboratory (Ref 6–8). These test fixtures utilize complex systems for eliminating some of the major sources of errors in tensile testing of ceramics with low strains to failure. Both use what is referred to as a "buttonhead" specimen (see Fig. 3e), to which the load is transferred through enlarged regions on the specimen ends. Although these specimens have operational advantages, such as minimal re-

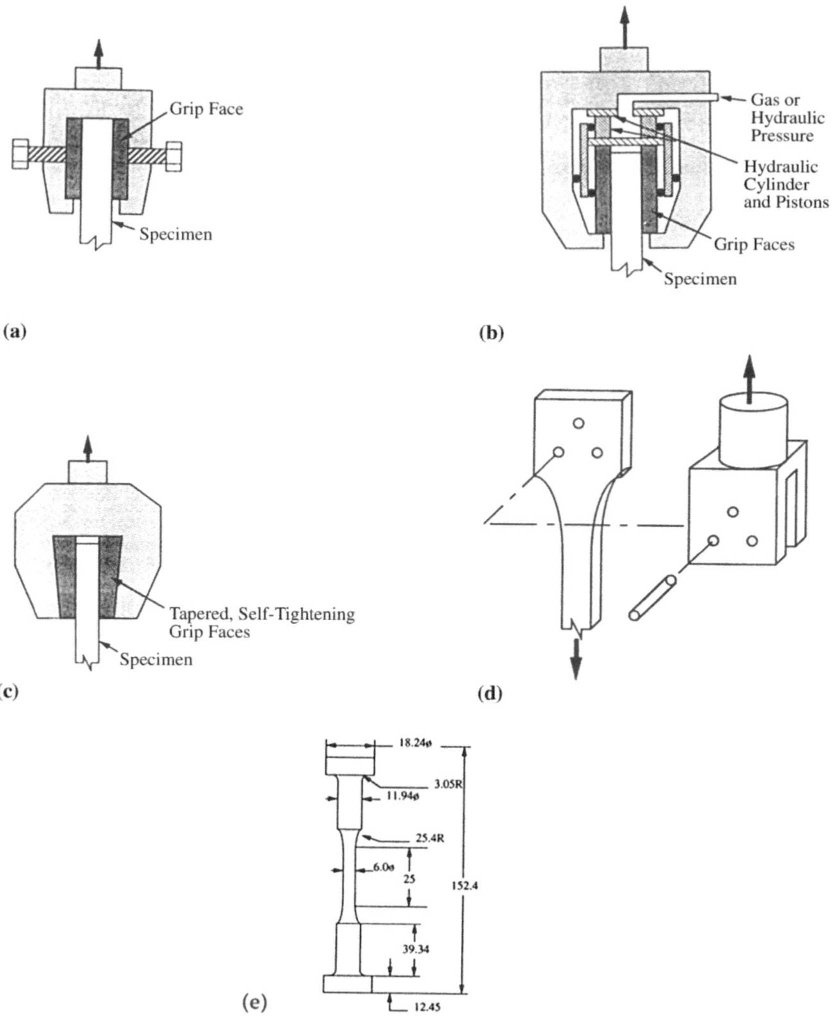

Fig. 3 Gripping systems for direct tensile tests. (a) Mechanical grips with screw clamping. (b) Pneumatic (or hydraulic) grips with force applied through lever arrangement and pneumatic pressure, ensuring constant clamping force. (c) Wedge-action, self-tightening mechanical grips; clamping pressure is roughly proportional to the tensile load in the specimen. (d) Pinned grips with load transfer by means of pins through grip and specimen. (e) Specimen configuration (buttonhead) for self-aligning commercial grip systems (all dimensions in millimeters; ground surface finish, 2 to 3 μm).

quirements for specimen alignment in the test fixtures, there are severe restrictions on the amount of load that can be transmitted through the buttonhead. The result has been that this type of gripping/load transfer has been very successful with materials of moderate tensile strength, and with long-term, low-stress tests such as creep and stress-rupture tests, but tends to fail for materials with high tensile strengths. Theoretical analysis of the requirements and limitations of this test are extremely difficult, as a result of the complex contact-stress problem at the buttonhead/grip interface. Thus, little guidance, aside from practical experience, can be utilized for determining when this type of test will be successful, and when the large investment in the grips themselves is appropriate.

Experimental Problems and Errors. One major source of error that is inherent in direct tensile tests has been eliminated to a major extent by the introduction of self-aligning grip systems. This error is associated with eccentricities in load application (see Fig. 4), which lead to a combined state of tension and bending in the test specimen. The large magnitudes of the parasitic bending stresses, even for small degrees of misalignment, result in significant errors in the calculated tensile stress (based on a state of pure tension). However, the use of various types of self-aligning grips, together with appropriate specimen geometries and careful specimen preparation, have largely eliminated these errors (Ref 8). The current self-aligning grips, available from the two major testing-machine manufacturers, use compact hydraulic systems to accomplish the same effect previously achieved through large and costly gas-bearing tensile-test fixtures. The only difficulties with these grip systems are noted above, involving specimen preparation, testing of high-strength materials, and the relatively high cost of the grips.

To some extent, the testing problems for certain continuous-fiber CMCs have been alleviated. This is particularly true for those CMCs

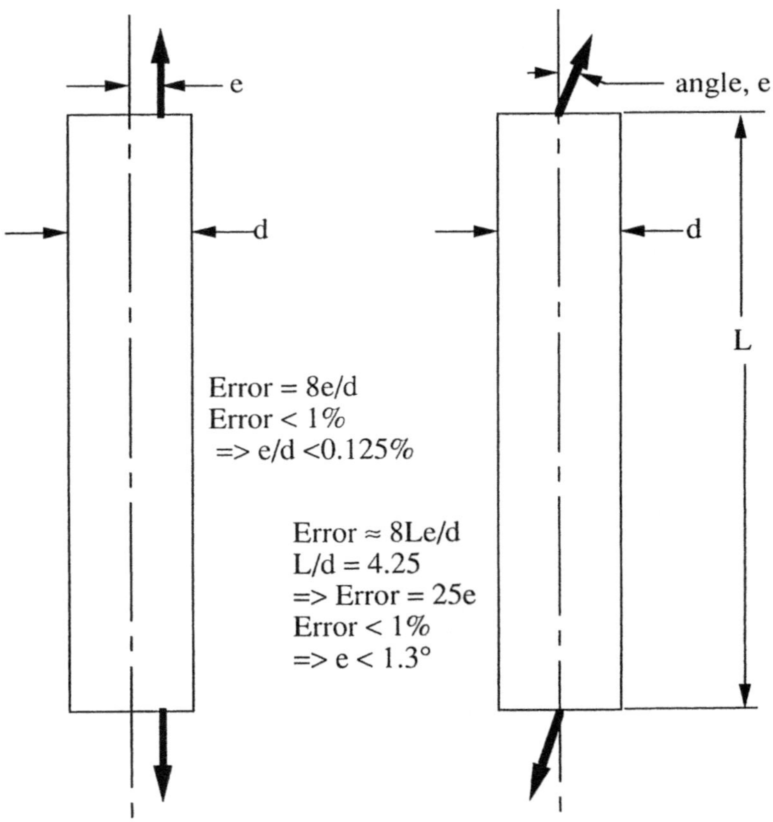

Fig. 4 Errors in tensile testing derived from load applied off-center and at angle to centerline of gage section; errors for two effects combined are roughly additive.

that have relatively high strain to failure (for ceramics) and relatively low modulus. In many cases, the simple gripping techniques used for metals and polymers will suffice for such CMCs, and few special precautions need to be taken, aside from ensuring sufficient gripping area relative to the cross-sectional area of the gage section (see the discussion above on gripping). The author has, without great difficulty, performed tensile tests on conventional dog-bone specimens of CMCs, using ordinary pneumatic grips with smooth grip faces made of materials slightly less hard than the CMC itself (for example, aluminum, copper, or silver) and appropriately sized grip and gage areas. Such results suggest that direct tensile testing of advanced CMCs may be far less difficult than testing of monolithic ceramics, and may not require the specialized test fixtures and specimens needed for testing of monolithic ceramics.

Summary of Direct Tensile Tests. The advantages and limitations of direct tensile testing of ceramics and ceramic composites are very clear. The advantages are:

- Direct measurement of the tensile strength in a known and simple stress state
- Stressing of the entire gage-section volume and surface, sampling both surface and volume flaws in the material being tested

The disadvantages and limitations include:

- The need for large specimens (because of the need for large gripping areas)
- Complex and precise specimen machining requirements for collet grips and especially for buttonhead specimens
- The need for relatively expensive (and bulky) test fixtures and grips

Indirect Tensile Tests

Indirect tensile tests are quite similar, typically involving some complex specimen geometry that induces a state of uniaxial tension in a portion of a specimen loaded in a fairly simple manner. Two examples are the theta specimen test, which is a variant of the diametral compression test discussed below, and the trussed beam test, which is similar to the theta specimen test but involves loading in flexure rather than in compression (see Fig. 5a and b). Both of these tests provide the capability for performing what is very close to a direct tensile test, but without the need for expensive tensile-test fixtures. Both are also amenable to use at high temperatures, without the great complications that accompany the use of conventional tensile-testing fixtures and procedures. The primary disadvantages of the theta and trussed beam specimens are the difficulty of machining them, especially with respect to the cutouts, and the problem of flaws introduced through such machining. In some cases, direct molding of specimens in these configurations may be possible, eliminating the machining problem altogether, as well as providing sintering, rather than machined, external surfaces—a possible advantage if actual components are prepared to net shape with no external surface finishing.

It should be noted that a great many other similar tests are possible, limited only by the creativity and ability of the experimenter to fabricate the test specimen and analyze the stress state produced. One such example is shown in Fig. 5(c), where a thin layer of material to be tested is used as the skin on the tensile side of a sandwich beam. The only requirement for determining the tensile stress at failure is knowledge of the elastic properties of the skin and core materials, and assurance that failure occurs first in the face sheet loaded in tension. The face sheet on the compressive side can be of virtually any high-strength, high-modulus material with known properties.

Flexure and Other "Tensile" Tests. There are a great variety of other tests used to characterize the tensile strengths of ceramics and ceramic composites, where the gage section of the specimen is not in a state of pure, uniaxial tension, but rather in some combined stress state. Such tests include the three-point and four-point flexure tests commonly used for ceramics, diametral compression tests, C-ring tests, combined-stress-state tests on cylindrical specimens, and various biaxial tests such as ball-on-ring and ring-on-ring tests. When these tests are used to measure tensile strength, it is presumed that there is no effect of combined stresses on failure and that the specimen fails from the largest tensile stress present—that is, the principal tensile stress. Historically, this has been a very good assumption for many monolithic ceramics with low toughness, identical elastic behavior in tension and compression, and essentially linear behavior to failure. However, in the case of many of the tougher ceramic composites, these assumptions are frequently incorrect. Note, however, that the biaxial tests, in some cases, have been used to evaluate the possible dependence of strength on stress state in ceramics.

For many toughening mechanisms present in CMCs, such as phase-transformation toughening, crack bridging, and fiber pullout, the behavior may be stress-state-dependent. In addition, for many such materials, the behavior in tension and the behavior in compression are not equal. The worst case of the latter occurs with some continuous ceramic-fiber composites, in which the compressive failure stress, as a result of fiber buckling, may be substantially lower than the tensile strength. The continuous-fiber CMCs also exhibit, for unidirectional materials, extremely low values of shear strength. This poses the additional problem of possible shear failure in tests where significant shear stresses are present, such as the three-point flexure test and the C-ring test. At present, the only solution to this problem is the careful monitoring of tests to determine the actual mode of failure (for example, compression, shear, or tension). This has been accomplished by means of video and telemicroscopic recording of specimen failure processes.

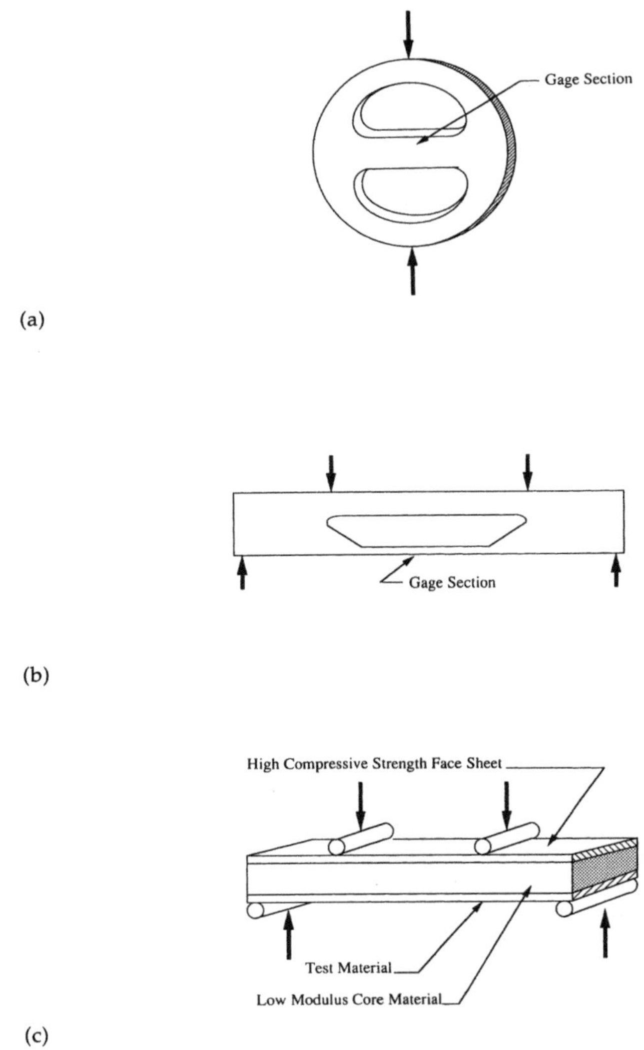

Fig. 5 Specimens for indirect tensile tests. (a) Theta specimen, which provides uniaxial tension for central member when specimen is loaded in diametral compression. (b) Trussed beam specimen, which provides approximately uniaxial tension in lower portion when beam is loaded in four-point bending. (c) Sandwich beam specimen, which loads lower skin in approximately pure bending with four-point flexural loading of beam.

In addition to the difficulties encountered in testing of fiber CMCs, there is the problem of the effects of shear stresses and combined stress states on phenomena such as the martensitic phase transformation used to toughen zirconia and zirconia-containing composites. This phase transformation is primarily a shear transformation, with substantial volume increase as well. Thus, a stress state with a high dilatational stress and high shear stresses may result in a high degree of phase transformation, with consequent effects on the measured "tensile" stress, in contrast to the behavior that might be seen in a direct tensile test with lower dilatational stress and no shear.

Of these various "tensile" tests, by far the most commonly used are the three- and four-point flexure tests. A detailed analysis of the errors that occur in the four-point flexure test (the preferred test; see Fig. 6a and 7) has been performed (Ref 9, 10), and standards have been developed (Ref 11) for the use of these tests for monolithic ceramics, together with recommendations for both test-specimen geometry and test fixturing. These will not be repeated here, but experience has shown that use of the recom-

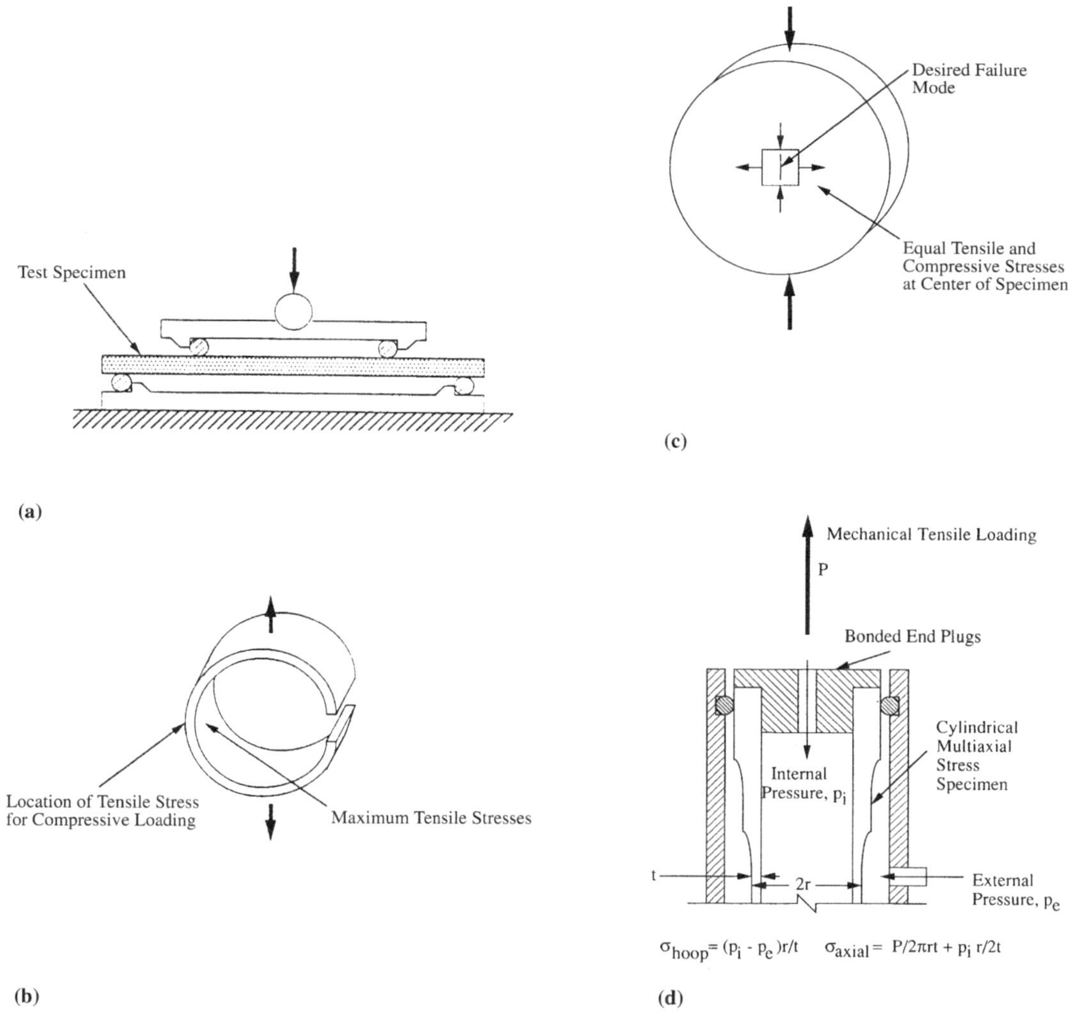

Fig. 6 Other "tensile" tests. (a) Four-point flexure test, which loads lower part of central portion of beam in tension, with a stress gradient in the vertical direction. (b) C-ring test, which provides flexural loading of a segment of a tubular component. (c) Diametral compression, or "Brazilian," test, which produces equal tensile and compressive stresses at the center of the specimen loaded in diametral compression. (d) Cylindrical specimen internally and externally pressurized and mechanically loaded in tension and compression, which can produce any desired combination of tensile and compressive stresses in the hoop and axial directions.

mended specimen geometry and test fixtures provides very good characterization of the tensile strengths of monolithics in which strength is controlled by surface flaws. It is also feasible, as with some of the other tests noted, to conduct such tests at high temperature, using appropriate materials for the test fixtures, although many other complications then arise, as will be discussed subsequently. One difficulty with these flexure tests occurs in the presence of stress gradients, with maximum stress occurring at the surface, leading to preferential failure from surface flaws. Another is the presence of shear stresses in regions of the specimen, which is a problem with some materials relatively weak in shear. A third is the presence of compressive stresses as well, which constitute an additional problem for materials, as noted, that fail in compression first. A last problem, which may be handled analytically if sufficient information is available about material response, is the problem of different stress-strain behavior in tension and compression. In the case of the flexure testing of fiber CMCs, matrix microcracking at a low stress level leads to an effective decrease in modulus in a portion of the tensile region of the specimen. This in turn leads to a shift in the neutral axis away from the tensile surface and a redistribution of stresses. In this particular case, use of the conventional beam-bending equations for maximum tensile stress may produce significant errors in the calculated stresses (Ref 12).

Another test, which has been used to a lesser extent, is the C-ring test (Fig. 6b), which is especially convenient for testing of materials produced in the form of thin-wall tubes, such as ceramic heat exchangers. In such cases, a slice is taken from the tube, with a portion removed as shown in Fig. 6(b), and is tested in either tension or compression. Testing in tension produces bending and tensile stresses in the interior of the specimen, as shown, whereas compressive testing similarly stresses the exterior of the specimen in tension. Relatively simple test fixturing suffices to load the specimen in either case, and extension to high temperatures is also relatively simple. This test has been analyzed theoretically (Ref 13–15), and the results presumably are accurate except for the same limitations of the other flexure tests. These include, as above, the problems of stress gradients, failure from surface flaws, and the presence of significant shear stresses. Other tests that have been used for measuring the "tensile" strengths of ceramics in-

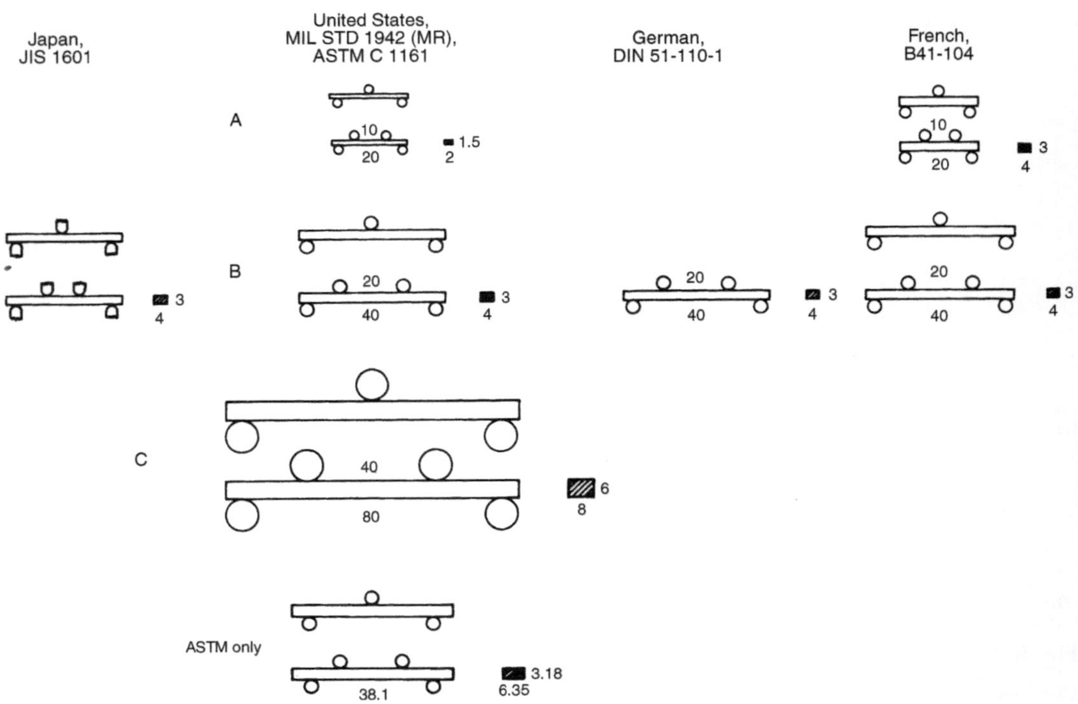

Fig. 7 Flexure strength standard test methods; all dimensions in mm. Source: S. Lampman, ASM International

clude various biaxial flexure tests (ball-on-ring, ring-on-ring) (Ref 16–18) that are equivalent to the three- and four-point flexure tests. These tests are convenient for materials that normally are available in the appropriate geometries—for example, thin plates or disks. These tests are very similar in most ways to the other flexure tests, except that the stress state is roughly equibiaxial, thus stressing flaws of all orientations, rather than only those oriented in the worst direction relative to the maximum tensile stress, as in a conventional flexure test.

Another type of test that is not widely used in the technical ceramics community, but more so in the geological area and with building materials, is the diametral compression, or "Brazilian," test, which uses a disk or short cylinder loaded in compression across its diameter (see Fig. 6c). In this test, the maximum tensile stresses are developed at the center of the specimen, where equal tensile and compressive stresses are present as shown. In a successful test of this type, the specimen fails by splitting vertically at its center. This test is particularly useful for materials such as cores from rock sampling, test cylinders of concrete, and similar materials. Typically, exact interpretation of the results in terms of the tensile strength of the material is difficult, because of the difficulty of determining the exact source of failure (from the machined surfaces or from the bulk of the material). There is also the problem, for some materials, of the presence of an equal compressive stress at the center of the specimen, which leads to the development of very large shear stresses at this site. Materials with relatively low shear strengths may thus fail first in shear, rather than in tension.

Combined-Stress-State Tests Using Multiaxial Cylindrical Specimens. The last type of "tensile" test to be discussed in this section is the combined-stress-state test employing multiaxial cylindrical specimens. These specimens (Fig. 6d), which can be loaded by various combinations of internal pressure, external pressure, axial tension or compression, and (when desired) torsion, are well suited to production of almost any desired stress state in the cylinder wall. As such, they have been used to address the problem of the failure criteria for brittle materials through systematic variation of the relative proportions and signs of the principal stresses. However, the major difficulties that are inherent in both preparation and use of such specimens have precluded their wide application. These tests require large amounts of material, extensive machining of specimens—typically with a profile lathe and diamond toolpost grinders for ceramics—and elaborate test fixturing. The extensive machining that is required, in addition to greatly increasing the cost of testing, introduces the potential for failure to be initiated by machining-induced flaws, rather than by volume flaws produced during processing. Such tests also have severe limitations with regard to high-temperature testing, as a consequence of the required loading arrangements.

Summary of the Advantages and Limitations of Flexure and Other "Tensile" Tests. The flexural and other indirect "tensile" tests described above provide several advantages over direct tensile tests for ceramic and ceramic composite specimens. These include:

- Simple specimen geometries, minimal specimen machining and simple test fixturing (flexure, biaxial, diametral compression, and C-ring tests)
- Use of as-fabricated materials (C-ring test)
- Capability for testing various stress states (flexure for tension, shear; biaxial flexure and cylindrical multiaxial specimens for combined stress states)

The particular disadvantages of these indirect tensile tests include:

- Extensive specimen preparation for multiaxial cylindrical specimens
- Stress gradients and combined stress states that may affect failure modes, especially in ceramic-matrix composites, or in other materials that are relatively weak in shear or exhibit different stress-strain behaviors in tension and compression

High-Temperature Tensile Tests

High-temperature tensile tests pose several specific difficulties and involve several specific requirements for both specimens and test fixtures. The particular difficulties depend on the temperature range involved and the atmosphere in which the test is to be conducted. Depending on the test particulars, suitable types of tests may include direct tensile tests, four-point flexure tests, and C-ring tests, the last two of which are subject to complications resulting from the stress states involved. Successful use has also been made of the theta specimen test, although this test has not become particularly popular because of the specimen machining involved.

Hot Grip Tests. The complications that involve the test temperature range are associated with the fixture materials available for transfer of load to the specimen (assuming that these fixtures are in the hot zone of the furnace). The alternative, which poses its own set of problems, is the use of large specimens and grips outside the test furnace. Typical ferrous materials for grips, pullrods, pushrods, loading anvils, and so on, are limited to approximately 1000 to 1200 °C (1830 to 2190 °F) because of severe strength loss at higher temperatures, as well as chemical problems (reaction, oxidation, etc.). The fixture materials suitable for higher-temperature use include various superalloys, which can be used at temperatures up to about 1200 °C (2190 °F), but may be expensive, difficult to machine, and subject to oxidation. Other, even more exotic materials include molybdenum, TZM (Mo-0.5Ti-0.08Zr-0.03C) thoria-dispersed nickel, and carbon or carbon-carbon composites. Some of these materials—for example, molybdenum and carbon/carbon—can be used at extremely high temperatures (up to about 2000 °C, or 3630 °F), but only in vacuum or inert atmospheres. Ceramics have also been used for high-temperature fixtures and grips, and may generally be used in a range of atmospheres. Unfortunately, some of the applications (for example, pullrods) are limited by the relatively low tensile strengths (200 to 400 MPa, or 30 to 60 ksi) of most of the available ceramics. The use of ceramics is also limited to temperatures of about 1500 to 1700 °C (2730 to 3090 °F) by the ceramics available in suitable forms for test fixtures and grips, such as aluminum oxide, silicon carbide, and silicon nitride. Finally, the use of ceramic grips and fixturing is severely limited by the difficulties and very high cost associated with machining test fixtures from suitable ceramic materials, which are hard and brittle.

Direct Tensile Tests. Assuming the desire to work with hot grips, or hot fixtures, to avoid some of the difficulties associated with cold grips, the selection of tests is very limited. Direct tensile tests can be performed only up to the temperature limitations of the grip materials, assuming that high-temperature-material analogs of one of the grip types have been acquired. This translates into a temperature limitation of about 1000 °C (1830 °F) for typical commercial metallic grips available at reasonable cost. Testing at somewhat higher temperatures can be performed, albeit at great cost, with ceramic analogs of these grips, and testing at temperatures of approximately 2000 °C (3630 °F) is possible with molybdenum grips in an inert atmosphere.

Four-Point Flexure Tests. Relatively appealing alternatives to direct tensile tests include C-ring and four-point flexure tests. Such tests can be readily performed with ceramic fixtures and pushrods, permitting testing in a variety of atmospheres at temperatures up to perhaps 1700 °C (3090 °F). The MTL four-point test fixture, depicted in Fig. 6(a) and 7, can be duplicated in a variety of ceramics (for example, alumina for the top and bottom anvil supports and pushrods, and sapphire for the loading-anvil rollers) at relatively low cost. Four-point flexure tests of this type can be used quite successfully, provided that some of the complications noted previously (for example, differing stress-strain behavior in compression and tension, and significant effects of shear stresses) do not occur. Another complication that may also arise in high-temperature flexure testing of ceramics is the presence of large strains and deflections resulting from increases in ductility or other flow processes that are operative at high temperatures. Such large strains may produce significant errors in stress values calculated by the use of beam-bending theories based on infinitesimal strains. As mentioned earlier in this chapter, elevated temperature flexure tests have been standardized in ASTM C 1211.

The C-ring test can also be readily used at high temperatures—particularly if the ring is loaded in compression by means of appropriate ceramic anvils and pushrods (Ref 13, 15). Loading in tension with ceramic attachments and pullrods is also possible, because of the relatively low loads required to cause failure via the bending stresses in this test. This particular test has, in fact, been used quite successfully in the development of ceramics and CMCs for high-temperature heat exchangers, which are fabricated from relatively thin-wall tubes. Attempts to characterize the tensile strengths of such tubes by testing of machined specimens would lead to very misleading results, because in this case specimen strength would be controlled primarily by machining damage, whereas the strength of the actual components, with their as-fabricated surfaces, is controlled by intrinsic defects.

Cold Grip Tests. In the event that it is feasible to work with either cooled grips (inside the furnace), or cold grips (outside the furnace), the tests that are most suitable are quite different. In this case, as shown in Fig. 8, any number of grip arrangements can be used, in conjunction with a long specimen, with the gage section contained within the hot zone of the test furnace. Several commercial vendors now offer systems that combine small test furnaces, some with hot zones as short as 2.5 to 5 cm (1 to 2 in.), with self-aligning grips (in some cases, water cooled) for the buttonhead specimens. Similarly, a small furnace around the gage section of a long, rectangular CMC tensile specimen gripped on aluminum tabs epoxy bonded to the end of the specimen has also been used.

This technique is not without its disadvantages. It requires large amounts of material for test specimens, which are typically more than 15 cm (5.9 in.) in length, and rather expensive test fixtures and furnaces (assuming that commercial equipment is used). Another unavoidable problem with this cold grip technique, and with the use of cooled grips in the furnace hot zone, is that of thermal gradients in the specimen, and increased requirements for power in the test furnace, because of the transfer of heat out of the furnace through the specimen and into the grips. The cold grip technique also poses some problems with control of the atmosphere inside the furnace, because seals must be provided around the test specimen where it passes into the furnace. This is not a major problem with hot grip tests, where very effective seals (for example, high-temperature bellows) can be provided at the points where the pullrods enter the furnace.

Strain Measurement. Historically, measurement of strains has been one of the major problems with high-temperature tensile testing of ceramics by either direct tensile tests or any of the indirect methods. One of the factors contributing to the difficulty of measuring strains in a high-temperature ceramic tensile specimen is the relatively low strain-to-failure in ceramics and CMCs. Frequently the maximum tensile strain achieved in monolithics is less than 0.1%, and even in the tough-fiber CMCs, the maximum strain may be only 2 to 3%. Measurement of such small strains is in general a very challenging task, and more so inside a high-temperature test furnace. In the past, the typical techniques used for "strain" measurements have involved measurement of the over-all travel of the load train outside the test furnace or measurement of the elongation or deformation of the specimen by means of displacement transducers coupled to the specimen by refractory rods (Ref 19) (see Fig. 9). Also available were dual-channel optical tracking systems capable of tracking two marks or flags on the specimen, thus providing a noncontact and highly precise method of measuring

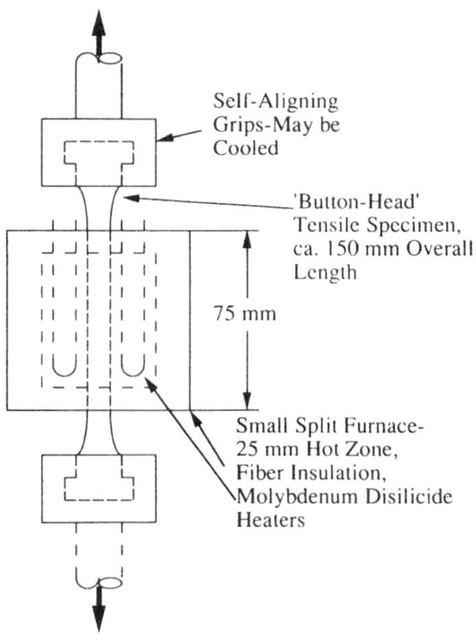

Fig. 8 Schematic illustration of cold grip tensile-testing arrangement with long specimen gripped outside of compact test furnace; commercial systems in this configuration are available for testing in air at temperatures up to about 1700 °C (3100 °F).

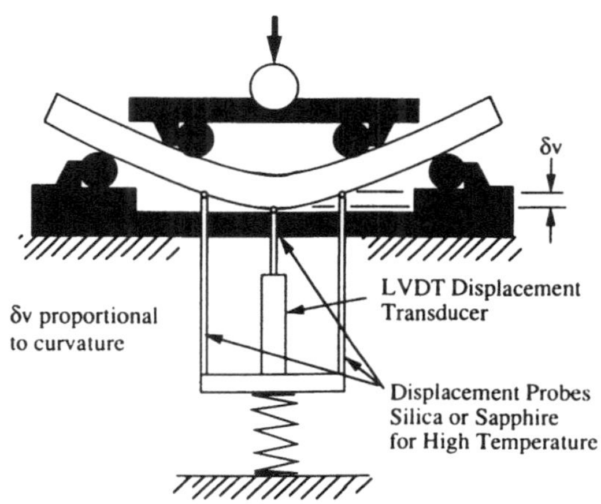

Fig. 9 Schematic diagram illustrating three-probe linear variable differential transformer (LVDT) measurement of curvature of central portion of four-point flexure system. The usual assumption of pure bending between inner load points implies that the strain is proportional to the curvature of the beam. The curvature is proportional to the difference in displacements as sensed directly by the LVDT (or other displacement transducer).

the strain in the gage section of the specimen. However, such optical trackers were extremely expensive, rivaling the cost of a complete test machine, and thus were not used extensively.

The situation with regard to strain measurement has improved dramatically in recent years, and several reasonably priced commercial systems for strain measurement inside high-temperature furnaces are now available (Ref 7, 20). One such system employs suitable extensions (silica, sapphire, silicon carbide, and so on) to the clip gages commonly used to measure strain in ambient-temperature tensile tests. These high-temperature clip gages permit accurate measurement of strain in a chosen portion of the test specimen, requiring only two ports in the side of the furnace for the extension rods. These direct-contact extensometers are available at moderate cost and are capable of measuring displacements and strains with extremely high accuracy.

Also available are various laser-based strain-measurement devices that can be used easily at high temperatures, requiring only a window in the side of the furnace through which the specimen can be sighted. These laser systems work in several distinct ways. One commercial system tracks two flags, as did the optical tracking systems previously mentioned, but offers laser technology and modern electronics at a cost comparable to that of the high-temperature clip gages cited above. The laser systems have the advantage that the radiation from the hot furnace interior does not interfere with the measurement, as it would with an optical tracking system following two marks on a specimen inside a hot furnace. The normal effect at temperatures above approximately 1000 °C (1830 °F) is that everything in the furnace looks the same (color differences are only a function of emissivity). With the use of lasers, the sensors can be equipped with narrow band filters that pass only the laser wavelength. Additionally, the laser signal can be modulated, with the sensors detecting only the modulated, ac signal, and not the dc background from the thermal radiation inside the furnace (helium-neon lasers are roughly the same color as the inside of a furnace at 800 to 900 °C, or 1470 to 1650 °F).

Another system that is amenable to use with a great variety of test specimens, even with extremely small-diameter (10 μm, or 0.4 mil) ceramic fibers, uses the speckle pattern generated by the reflection of a coherent laser beam from the surface of the specimen. As the specimen deforms, the speckle pattern deforms in a similar manner, and measurement of the changes in the speckle pattern permit accurate measurement of the strain in any direction on the surface of the specimen. These speckle interferometric strain gages are also reasonable in cost, easy to use, and require, again, only a sight port or small opening in the test furnace.

With the two types of laser strain gages and the high-temperature clip gage, there is now little difficulty in making direct and precise measurements of strain in high-temperature tensile specimens. With some of the other, indirect tensile tests, there are also relatively convenient ways of measuring strain. For example, for the four-point flexure test, a convenient and very accurate way of measuring strain in the central portion of the test specimen is the use of a three-probe displacement transducer system (see Fig. 9), which effectively measures the curvature of the central portion of the beam (which is normally assumed to be in pure bending where the strain is proportional to the curvature). Accordingly, strain measurement is not now considered to be a significant problem in tensile testing of ceramics.

Atmosphere Control. Control of the atmosphere in high-temperature tensile tests of ceramics and CMCs continues to be a significant problem. The situation for test temperatures below 1000 to 1200 °C (1830 to 2190 °F) is tractable, in that hot grips, or cooled grips inside the furnace, can be used, with effective seals on the pullrods and little restriction of atmosphere imposed by the grip materials (for example, oxidation of metal grips). However, for temperatures above 1200 °C (2190 °F), the problems are severe. The higher-temperature metallic grips (molybdenum) must be used only in inert or reducing conditions, and grips fabricated from graphite or carbon-carbon composites must be used under inert conditions (vacuum, argon, and so on). If the application requires testing in oxidizing conditions, as would be the case for gas turbine or hypersonic airframe materials, such tests may give very misleading results. High-temperature tests under oxidizing conditions (for example, in air or in simulated gas turbine combustion products) require either ceramic fixtures, which limit the type of test that can be performed and the loads that can be achieved in tensile tests, or the use of cold grips outside the furnace. Use of cold grips requires extremely large specimens (for experimental materials) and is complicated by the problem of sealing the furnace to provide effective atmosphere control. An appealing alternative, in many cases, is the use of four-point flexure tests with ceramic fixtures and pushrods, in which it is possible to test to quite high temperatures (about 1700 °C, or 3090 °F) in a variety of atmospheres ranging from reducing, through inert, to oxidizing conditions. Materials such as aluminum oxide and sapphire (for load points) will survive atmospheres such as forming gas, argon, nitrogen, vacuum, air, and oxygen, with little effect on the test fixturing, even at very high temperatures.

Recommendations for High-Temperature Tensile Testing of Ceramics. There are some clear choices for high-temperature tensile testing of ceramics, provided that appropriate test equipment and fixturing are affordable. The clear choice for most monolithic ceramics is the use of precisely aligned hydraulic grips or self-aligning grip systems, with straight-shank or buttonhead specimens, a small furnace system, and direct-contact extensometers or optical measurement of the specimen strain. Note that the buttonhead specimens are limited in load levels, as are pinned dog-bone specimens, and may be more suitable for lower stress level tests such as creep and fatigue tests.

Some modification of the gripping arrangement and grips (and some additional expense) may be necessary for testing of high-strength monolithics or CMCs. If test temperatures are always below 1000 °C (1830 °F), it is possible to use a much less expensive system, employing hot grips and a large furnace.

For situations where neither true tensile-testing system is practical, the most reasonable alternative is the use of the four-point flexure test with displacement transducer measurement of the strain in the central (gage) portion of the specimen. Use of some of the other tests described should be limited to the special cases applications for which they are appropriate (for example, use of the C-ring test for tube segments and the diametral compression tests for cylindrical specimens). The biaxial tests (ball-on-ring and ring-on-ring) may have some limited usefulness in situations where actual loading is biaxial and effects of combined stresses are expected to be significant.

Summary

The recommended procedures for ambient- and elevated-temperature tensile testing of advanced monolithic and CMCs are summarized in the following paragraphs. In addition, a brief discussion of data analysis for interpretation of uniaxial strength is also included.

Recommended Procedures for Ambient-Temperature Tensile Testing of Ceramics and CMCs

Monolithic Ceramics and Low-Toughness CMCs.

1. Direct tensile tests using the currently available commercial self-aligning grip systems and strain-measurement techniques. These tests require relatively expensive gripping systems, strain-measurement techniques, and large specimens with complex machining requirements. Specimen geometry has been established for these gripping systems to minimize failure in the gripping or transition regions.

2. Where material availability or economic constraints prevent such testing, four-point flexure testing following the ASTM standard C 1161; strain measurement preferably is done by measuring the displacement in the central portion of the test specimen at three points.

High-Toughness CMCs and other Ceramics with High Strains to Failure.

1. Direct tensile tests using either the self-aligning grip systems or simpler grip systems typically used for metals or polymers; strain measurement by conventional techniques (clip gages) may be adequate. With the use of the more conventional gripping systems, it may be possible to work with flat plate specimens, which may be easier to fabricate.

2. Four-point flexure tests in which the details of the fracture process are observed carefully, to ensure that failure does in fact occur first in a tensile mode, and with corrections for neutral axis shifts resulting from differing tensile and compressive stress-strain behavior.

Specialized Materials (Such as Heat-Exchanger Tubes).

1. Direct tensile tests if sufficiently large specimens can be obtained from components to minimize the effects of surface machining damage.

2. Otherwise, C-ring or other similar tests, with the same careful observation and corrections recommended for the four-point bend test.

Recommended Procedures for High-Temperature Tensile Testing of Ceramics and CMCs

Monolithic Ceramics and Low-Toughness CMCs.

1. Direct tensile tests using the currently available commercial self-aligning grip systems, with grips outside a compact furnace, and com-

mercial high-temperature strain-measurement techniques. These tests require relatively expensive gripping systems, strain-measurement techniques, furnace systems, and large specimens with complex machining requirements.

2. Where availability of material or financial limitations make the procedure above impractical, the alternative is four-point flexure with appropriate measurement of strain, as above.

High-Toughness CMCs and other Ceramics with High Strains to Failure.

1. Direct tensile tests using either self-aligning cold grip systems or simpler hot grip systems typically used for metals or polymers, with optical or capacitance (clip) gage measurement of strain; again, conventional grips may make it possible to work with the more easily fabricated flat plate or dog-bone specimen.

2. Four-point flexure tests in which the details of the fracture process are observed carefully (this is far more difficult in the confines of a high-temperature furnace), to ensure that failure does in fact occur first in a tensile mode, and with corrections for neutral axis shifts resulting from differing tensile and compressive stress-strain behavior.

Specialized Materials (Such as Heat-Exchanger Tubes).

1. Direct tensile tests if sufficiently large specimens can be obtained from components to minimize the effects of surface machining damage.

2. Otherwise, C-ring or other similar tests, with the same careful observation and corrections recommended for the four-point bend test. These observations and corrections are difficult to make in a high-temperature test, although C-ring and other indirect tensile tests are otherwise relatively easy to translate to high-temperature tests.

Recommended Procedures for Data Analysis

The recommended procedures for data analysis and reporting are partly covered in the ASTM standards for flexure and tensile testing. Another important source of information for data analysis is ASTM C 1239, "Standard Practice for Reporting Uniaxial Strength Data and Estimating Distribution Parameters for Advanced Ceramics." The failure strength of advanced ceramics is treated as a continuous random variable using this practice. Typically, a number of test specimens with well-defined geometries are failed under isothermal loading conditions. The load at which each specimen fails is recorded. The resulting failure stresses are used to obtain parameter estimates associated with the underlying population distribution. ASTM C 1239 is restricted to the assumption that the distribution underlying the failure strengths is the two-parameter Weibull distribution with size scaling (see also the discussion of "Design Strength and Scale Effects" earlier in this chapter). Furthermore, C 1239 is restricted to test specimens that are primarily subjected to uniaxial tensile stresses. This practice also outlines methods to correct for bias errors in the estimated Weibull parameters and to calculate confidence bounds on those estimates from data sets where all failures originate from a single flaw population (that is, a single failure mode). The methods outlined in C 1239 are not applicable to samples that fail due to multiple independent flaw populations (for example, competing failure modes).

Measurements of the strength at failure are taken for one of two reasons: either for a comparison of the relative quality of two materials, or the prediction of the probability of failure (or, alternatively, the fracture strength) for a structure of interest. ASTM C 1239 estimates the distribution parameters that are needed for either. In addition, this practice encourages the integration of mechanical property data and fractographic analysis (refer to ASTM C 1322 mentioned earlier in this chapter).

ACKNOWLEDGMENT

This chapter was adapted from D. Lewis III, Tensile Testing of Ceramics and Ceramic-Matrix Composites, *Tensile Testing,* P. Han, Ed., ASM International, 1992, p 147–181

REFERENCES

1. J.R. Varner, Descriptive Fractography, *Ceramics and Glasses,* Vol 4, *Engineered Materials Handbook,* ASM International, 1991, p 635–644.
2. R.W. Rice, Ceramic Fracture Features, Observations, Mechanisms and Uses, *Fractography of Ceramic and Metal Failures,* STP 827, ASTM, 1984, p 5–103.
3. S.B. Batdorf, Fundamentals of the Statistical Theory of Failure, *Fracture Mechanics of Ceramics,* Vol 3, R.C. Bradt, D.P.H. Has-

selman, A.G. Evans, and F.F. Lange, Ed., Plenum Press, 1978, p 1–29.
4. D. Lewis, Curve-Fitting Techniques and Ceramics, *Am. Ceram. Soc. Bull.,* Vol 57 (No. 4), 1978, p 434–437.
5. W. Weibull, A Statistical Distribution Function of Wide Applicability, *J. Appl. Mech.,* Vol 18, 1951, p 293–297.
6. D.F. Baxter, Jr., Tensile Testing at Extreme Temperatures, *Adv. Mater. Proc.,* Vol 139 (No. 2), 1991, p 22–32.
7. J.C. Bittence, New Emphasis on Automation, *Adv. Mater. Proc.,* Vol 136 (No. 5), 1989, p 45–56.
8. K.C. Liu and C.R. Brinkman, Tensile Cyclic Fatigue of Structural Ceramics, *Proc. 23rd Automotive Technology Development Contractor's Coordination Meeting,* Vol 165, Society of Automotive Engineers, Oct 1985, p 279–284.
9. F.I. Baratta and W.T. Matthews, "Errors Associated with Flexure Testing of Brittle Materials," U.S. Army Materials Technology Laboratory Report MTL TR 87-35, 1987.
10. F.I. Baratta, Requirements for Flexure Testing of Brittle Materials, *Methods for Assessing the Structural Reliability of Brittle Materials,* STP 844, ASTM, 1984, p 194–222.
11. G. Quinn, "Flexural Strength of High Performance Ceramics at Ambient Temperature," Department of the Army, MIL-STD-1942(MR), 1984.
12. D.B. Marshall and A.G. Evans, Failure Mechanisms in Ceramic Fiber-Ceramic Matrix Composites, *J. Am. Ceram. Soc.,* Vol 68 (No. 5), 1985, p 225–231.
13. M.K. Ferber, V.J. Tennery, S. Waters, and J.C. Ogle, Fracture Strength Characterization of Tubular Ceramics Using a Simple C-Ring Geometry, *J. Mater. Sci.,* Vol 8, 1986, p 2628–2632.
14. O.M. Jadaan, D.L. Shelleman, J.C. Conway, Jr., J.J. Mecholsky, and R.E. Tressler, Prediction of the Strength of Ceramic Tubular Components: Part I—Analysis, *J. Test. Eval.,* Vol 19 (No. 3), 1991, p 181–191.
15. D.L. Shelleman, O.M. Jadaan, J.C. Conway, Jr., and J.J. Mecholsky, Jr., Prediction of the Strength of Ceramic Tubular Components: Part II—Experimental Verification, *J. Test. Eval.,* Vol 19 (No. 3), 1991, p 192–201.
16. G. de With and H.H.H. Wagemens, Ball-on-Ring Test Revisited, *J. Am. Ceram. Soc.,* Vol 72 (No. 8), 1989, p 1538–1541.
17. H. Fessler and D.C. Fricker, A Theoretical Analysis of the Ring-on-Ring Loading Disk Test, *J. Am. Ceram. Soc.,* Vol 67 (No. 9), 1984, p 582–588.
18. D.K. Shetty, A.R. Rosenfield, and W.H. Duckworth, Statistical Analysis of Size and Stress State Effects on the Strength of An Alumina Ceramic, *Methods for Assessing the Structural Reliability of Brittle Materials,* STP 844, ASTM, 1984, p 57–80.
19. S.A. Bortz and T.B. Wade, Analysis and Review of Mechanical Testing Procedure for Brittle Materials, *Structural Ceramics and Testing of Brittle Materials,* S.J. Acquaviva and S.A. Bortz, Ed., Gordon and Breach, 1968, p 47–139.
20. Laser Gages Creep of Ceramics, *Adv. Mater. Proc.,* Vol 138 (No. 5), 1990, p 75–76.

SELECTED REFERENCES

- J.E. Amaral and C.N. Pollock, Machine Design Requirements for Uniaxial Testing of Ceramics Materials, *Mechanical Testing of Engineering Ceramics at High Temperatures,* B.F. Dyson, R.D. Lohr, and R. Morrell, Ed., 1989, p 51–68.
- H.C. Cao, E. Bischoff, O. Sbaizero, M. Ruhle, A.G. Evans, D.B. Marshall, and J. Brennan, Effects of Interfaces on the Mechanical Properties of Fiber-Reinforced Brittle Materials, *J. Am. Ceram. Soc.,* Vol 73 (No. 6), 1990, p 1691–1699.
- H. Cao and M.D. Thouless, Tensile Tests of Ceramic-Matrix Composites: Theory and Experiment, *J. Am. Ceram. Soc.,* Vol 73 (No. 7), 1990, p 2091–2094.
- D.F. Carroll, S.M. Wiederhorn, and D.E. Roberts, Technique for Tensile Testing Ceramics, *J. Am. Ceram. Soc.,* Vol 72 (No. 9), 1989, p 1610–1614.
- M.G. Jenkins, M.K. Ferber, R.L. Martin, V.T. Jenkins, and V.J. Tennery, "Study and Analysis of the Stress State in a Ceramic, Button-Head, Tensile Specimen," *ORNL/TM-11767,* Oak Ridge National Laboratory Technical Memorandum, Sept 1991.
- C.G. Larsen, *Ceramics Tensile Grip,* STP 1080, J.M. Kennedy, H.H. Moeller, and W.W. Johnson, Ed., ASTM, 1990, p 235–246.
- J.J. Mecholsky, Evaluation of Mechanical Property Testing Methods for Ceramic Matrix Composites, *Am. Ceram. Soc. Bull.,* Vol 65 (No. 2), 1986, p 315–322.

- L.C. Meija, High Temperature Tensile Testing of Advanced Ceramics, *Ceramic Engineering and Science Proceedings,* Vol 10 (No. 7–8), 1989, p 668–681.
- L.G. Mosiman, T.L. Wallenfelt, and C.G. Larsen, Tension/Compression Grips for Monolithic Ceramics and Ceramic Matrix Composites, *Ceramic Engineering and Science Proceedings,* Vol 12 (No. 7–8), 1991.
- T. Ohji, Towards Routine Tensile Testing, *Int. J. High. Technol. Ceram.,* Vol 4, 1988, p 211–225.
- G.D. Quinn, Strength and Proof Testing, *Ceramics and Glasses,* Vol 4, *Engineered Materials Handbook,* ASM International, 1991, p 599–609
- S.G. Seshadri and K.-Y. Chia, Tensile Testing Ceramics, *J. Am. Ceram. Soc.,* Vol 70 (No. 10), 1987, p C242–C244.

CHAPTER 11

Tensile Testing of Fiber-Reinforced Composites

THE CHARACTERIZATION of engineering properties is a complex issue for fiber-reinforced composites due to their inherent anisotropy and inhomogeneity. In terms of mechanical properties, advanced composite materials are evaluated by a number of specially designed test methods. These test methods are mechanically simple in concept but extremely sensitive to specimen preparation and test-execution procedures. They include:

- Tensile tests
- Compression tests
- Shear tests
- Flexural tests
- Fracture tests
- Fatigue tests

These test methods are covered by standards developed by ASTM, the International Standards Organization (ISO), and the Suppliers of Advanced Composite Materials Association (SACMA).

This chapter is limited to tensile property test methods. Tensile testing of fiber-reinforced composite materials is performed for the purpose of determining uniaxial tensile strength, Young's modulus, and Poisson's ratio relative to principal material directions. The unidirectional lamina provides the basic building block of the multidirectional laminate. Therefore, characterization of lamina material properties allows predictions of the properties of laminates. In actual practice, considerable success has been demonstrated in predicting laminate effective modulus or Poisson's ratio from ply properties. However, prediction of laminate strength properties from lamina strength data has proved more difficult, and therefore it is often necessary to resort to characterization of laminate strength properties. Thus, basic tensile testing is divided into lamina and laminate testing. There also are specimen differences between polymeric-matrix and metal-matrix composites that require separate discussions. Basic tensile-test methods for both polymeric-matrix and metal-matrix composites are confined to those materials that behave on the macroscale as orthotropic bodies.

Fundamentals of Tensile Testing of Composite Materials

Unlike homogeneous, isotropic materials, fiber-reinforced composites are characterized by properties that are direction-dependent. Advanced composites, whether of the polymeric-matrix class or the metal-matrix class, often are utilized in the form of a laminate. The lamina, or unidirectionally reinforced ply (Fig. 1), is the basic building block of the laminate. In order to perform engineering analysis, the heterogeneous lamina consisting of a fiber phase and a matrix phase is treated as a homogeneous, orthotropic material. In addition, laminate modeling assumes that plies are in a state of plane stress.

Stress-Strain Relationships for an Orthotropic Material. Development of stress-strain

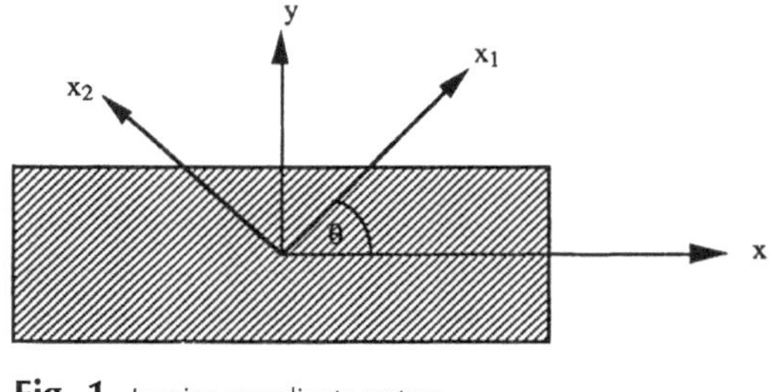

Fig. 1 Lamina coordinate system

relationships for an orthotropic material requires the definition of engineering constants. Using Fig. 1, the unidirectional material is orthotropic with respect to the x_1-x_2 axes. The stress-strain relationships for plane stress are of the forms

$$\varepsilon_1 = \frac{1}{E_1}\sigma_1 - \frac{\nu_{12}}{E_1} s_2 \quad \text{(Eq 1a)}$$

$$\varepsilon_2 = -\frac{\nu_{12}}{E_1}\sigma + \frac{1}{E_1}\sigma_2 \quad \text{(Eq 1b)}$$

$$\gamma_{12} = \frac{1}{G_{12}}\tau_{12} \quad \text{(Eq 1c)}$$

where, in the usual manner, the normal stresses and strains in the x_1 and x_2 directions are denoted by σ_1, ε_1, σ_2, and ε_2, respectively, whereas the shear stress and strain are denoted by τ_{12} and γ_{12}, respectively. In addition, E_1, E_2, and G_{12} are the Young's modulus parallel to the fibers, the Young's modulus transverse to the fibers, and the shear modulus relative to the x_1-x_2 plane, respectively. The major Poisson's ratio, as determined from contraction transverse to the fibers during a uniaxial test parallel to the fibers, is denoted by ν_{12}. For laminates in which the macroscopic stress-strain relationships are orthotropic, Eq 1 is valid, with the subscripts 1 and 2 replaced by x and y, respectively.

Shear Coupling Phenomenon. Components of stress and strain can be transformed from one coordinate system to another. Thus, it is possible to establish the stress-strain relationship in any coordinate system. For the unidirectional composite in Fig. 1, the constitutive relationships relative to the x-y coordinate system can be written in the forms

$$\varepsilon_x = \frac{1}{E_x}\sigma_x - \frac{\nu_{12}}{E_x}\sigma_y + \frac{\eta_x}{E_x}\tau_{xy} \quad \text{(Eq 2a)}$$

$$\varepsilon_y = -\frac{\nu_{xy}}{E_y}\sigma_x + \frac{1}{E_y}\sigma_y + \frac{\eta_y}{E_y}\tau_{xy} \quad \text{(Eq 2b)}$$

$$\gamma_{xy} = \frac{\eta_x}{E_y}\sigma_x + \frac{\eta_y}{E_x}\sigma_y + \frac{1}{G_{xy}}\tau_{xy} \quad \text{(Eq 2c)}$$

Equations 2a, b, and c correspond to the stress-strain relationships of an anisotropic material subjected to plane stress. Of particular significance is the fact that the normal strains are coupled to the shear stress and the shear strain is coupled to the normal stresses. Such behavior is referred to as the "shear coupling phenomenon" and requires the definition of two additional elastic properties. In particular, the elastic constants η_x and η_y are shear coupling coefficients determined from uniaxial tensile tests in the x and y directions, respectively—i.e.,

$$\eta_x = \frac{\gamma_{xy}}{\varepsilon_x} \text{ (uniaxial tension in the } x\text{-direction)} \quad \text{(Eq 3a)}$$

$$\eta_y = \frac{\gamma_{xy}}{\varepsilon_y} \text{ (uniaxial tension in the } y\text{-direction)} \quad \text{(Eq 3b)}$$

Symmetric Laminates and Laminate Notation. As shown in Fig. 1, the principal material directions within each ply of a laminate are denoted by an x_1-x_2 axis system. Laminate stacking sequences can be easily described for composites composed of layers of the same material with equal ply thickness by simply listing the ply orientations from the top of the laminate to the bottom. Thus, the notation [0°/90°/0°] uniquely defines a three-layer laminate. The angle denotes the orientation of the principal material axis, x_1, within each ply. If a ply were repeated, a subscript would be used to denote the number of repeating plies. Thus, $[0°/90°_3/0°]$ indicates that the 90° ply is repeated three times.

Any laminate in which the ply stacking sequence below the midplane is a mirror image of the stacking sequence above the midplane is referred to as a symmetric laminate. For a symmetric laminate, such as a $[0°/90°_2/0°]$ plate, the notation can be abbreviated by using $[0°/90°]_s$, where the subscript s denotes that the stacking sequence is repeated symmetrically. Angle-ply laminates are denoted by $[0°/+45°/-45°]_s$, which can be abbreviated as $[0°/\pm 45°]_s$. For laminates with repeating sets of plies—e.g., $[0°/\pm 45°/0°/\pm 45°]_s$, the abbreviated notation is of the form $[0°/\pm 45°]_{2s}$. If a symmetric laminate contains a ply that is split at the centerline, a bar is used to denote the split. Thus, the laminate [0°/90°/0°] can be abbreviated as $[0°/\overline{90°}]_s$. For unsymmetric laminates, a subscript T is often used to denote total laminate. For example, the laminate [0°/90°] can be written as $[0°/90°]_T$. This assures the reader that the laminate is indeed unsymmetric and that a subscript s was not inadvertently omitted.

Balanced Laminates. Laminates in which each ply oriented at an angle of $+\theta$ ($\theta \neq 0°$ or 90°) also contains a ply at $-\theta$ are referred to as balanced. Such composites are orthotropic relative to the *x-y* coordinate of the laminate. Thus, Eq 1a, b, and c with the subscripts 1 and 2 replaced by *x* and *y*, respectively, are applicable to balanced laminates.

Tensile Testing of Single Filaments and Tows

Although emphasis in this chapter has been placed on tensile testing of laminates, other constituent materials are also tested. These include single filaments and tows (untwisted bundles of continuous filaments).

Single-filament tensile strength can be determined using ASTM D 3379 (Ref 1), which can be summarized as a random selection of single filaments made from the material to be tested. Filaments are centerline-mounted on special slotted tabs (Fig. 2). The tabs are gripped so that the test specimen is aligned axially in the jaws of a constant-speed movable-crosshead test machine. The filaments are then stressed to failure at a constant strain rate. For this test method, filament cross-sectional areas are determined by planimeter measurements of a representative number of filament cross sections as displayed on highly magnified photomicrographs. Alternative methods of area determination include the use of optical gages, an image-splitting microscope, or the linear weight-density method.

Tensile strength and Young's modulus of elasticity are calculated from the load/elongation records and the cross-sectional area measurements. Note that a system compliance adjustment may be necessary for single-filament tensile modulus.

Tow tensile testing is carried out using ASTM D 4018 (Ref 2) or an equivalent test method. This is summarized as finding the tensile properties of continuous filament carbon and graphite yarns, strands, rovings, and tows by the tensile loading to failure of the resin-impregnated fiber forms. This technique loses accuracy as the filament count increases. Strain and Young's modulus are measured by an extensometer.

The purpose of using impregnating resin is to provide the fiber forms, when cured, with enough mechanical strength to produce a rigid test specimen capable of sustaining uniform loading of the individual filaments in the specimen.

To minimize the effect of the impregnating resin on the tensile properties of the fiber forms, the resin should be compatible with the fiber, the resin content in the cured specimen should be limited to the minimum amount required to produce a useful test specimen, the individual filaments of the fiber forms should be well collimated, and the strain capability of the resin should be significantly greater than the strain capability of the filaments.

ASTM D 4018 method I test specimens require a special cast-resin end tab and grip design to prevent grip slippage under high loads. Alternative methods of specimen mounting to end tabs are acceptable, provided that test specimens maintain axial alignment on the test machine centerline and that they do not slip in the grips at high loads. ASTM D 4018 method II test specimens require no special gripping mechanisms. Standard rubber-faced jaws should be adequate.

Tensile Testing of Laminates

The basic physics of most tensile test methods are very similar: a prismatic coupon with a straight-sided gage section is gripped at the ends and loaded in uniaxial tension. The principal differences between these tensile test coupons are the coupon cross section and the load-introduction method. The cross section of the coupon may be rectangular, round, or tubular; it may be straight-sided for the entire length (a "straight-sided" coupon) or width- or diameter-tapered from the ends into the gage section (often called "dogbone" or "bow-tie" specimens). Straight-sided coupons may use tabbed load application points. This section briefly discusses the most common tensile test methods that have been standardized for fiber-reinforced composite materials. Reference 3 includes a more detailed dis-

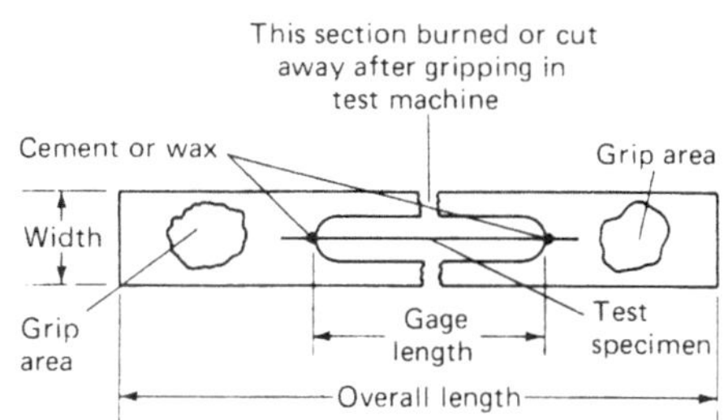

Fig. 2 Schematic showing typical specimen-mounting method for determining single-filament tensile strength

cussion and briefly reviews several nonstandard methods as well.

By changing the coupon configuration, many of the tensile test methods are able to evaluate different material configurations, including unidirectional laminates, woven materials, and general laminates. However, some coupon/material configuration combinations are less sensitive to specimen preparation and testing variations than others. Perhaps the most dramatic example of this is the unidirectional coupon. Fiber versus load axis misalignment in a 0° unidirectional coupon, which can occur due to either specimen preparation or testing problems or both, can reduce strength as much as 30% due to an initial 1° misalignment. Furthermore, bonded end tabs intended to *minimize* load-introduction problems in high-strength unidirectional materials can actually *cause* premature coupon failure (even in nonunidirectional coupons) if not applied and used properly. Because of these and similar issues, tensile testing is subject to a great deal of "art" in order to obtain legitimate data. Alternatives to problematic tests, such as the unidirectional tensile test, are often available, and careful attention must be paid to the test specification for recommendations. Reference 1 is also an excellent resource for test optimization suggestions.

In-Plane Tensile Test Methods

Straight-sided coupon tensile tests include:

- ASTM D 3039/D 3039M, "Standard Test Method for Tensile Properties of Polymer-Matrix Composites"
- ISO 527, "Plastics—Determination of Tensile Properties"
- SACMA SRM 4, "Tensile Properties of Oriented Fiber-Resin Composites"
- SACMA SRM 9, "Tensile Properties of Oriented Cross-Plied Fiber-Resin Composites"

ASTM D 3039/D 3039M, originally released in 1971 and updated several times since then, is the original standard test method for straight-sided rectangular coupons (Fig. 3). It is still the most commonly used in-plane tension method. ISO 527 parts 4 and 5 and the two SACMA tensile test methods, SRM 4 and SRM 9, are substantially based on ASTM D 3039 and as a result, are quite similar. Care should be taken, however, not to substitute one method for another, because subtle differences between them do exist. In general, the ASTM standard offers better control of testing details that may cause variability, as discussed subsequently. For this reason, it is the preferred method.

In each of the previous test methods, a tensile stress is applied to the specimen through a mechanical shear interface at the ends of the coupon, normally by either wedge or hydraulic grips. The material response is measured in the gage section of the coupon by either strain gages or extensometers, subsequently determining the elastic material properties.

If used, end tabs are intended to distribute the load from the grips into the specimen with a minimum of stress concentration. A schematic example of an appropriate failure mode of a multidirectional laminate using a tabbed tensile coupon is shown in Fig. 4. Because the straight-sided specimen provides no geometric stress-concentrated region, such as would be found in a specimen with a reduced-width gage section, failure often occurs at or near the ends of the tabs or grips. While this failure mode is not necessarily invalid, care must be taken when evaluating the data to guard against unrealistically low strengths resulting from poorly performing tabs or overly aggressive gripping.

Design of end tabs remains somewhat of an art, and an improperly designed tab interface will produce low coupon strengths. For this reason, a standard tab design has not been mandated by ASTM, although unbeveled 90° tabs are preferred. Recent comparisons confirm that the success of a tab design is more dependent on the use of a sufficiently ductile adhesive than on the tab angle. An unbeveled tab applied with a ductile adhesive will outperform a tapered tab that has been applied with an insufficiently ductile adhesive. Therefore, adhesive selection is most critical to bonded tab use. Furthermore, the use of a softer tab material is usually preferred when testing high-modulus materials (such as fiber-glass tabs on a graphite-reinforced specimen).

The simplest way to avoid bonded tab problems is to not use them. Many laminates (mostly

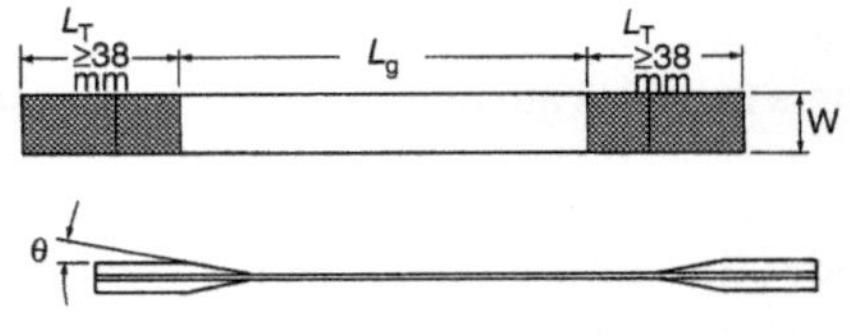

Fig. 3 Specimen for tensile testing of composites as defined in ASTM D 3039. L_g = gage length; L_T = tab length; θ = tab bevel angle; W = width. Note: the gage length is commonly 125 to 150 mm (5 to 6 in.).

nonunidirectional) can be successfully tested without tabs, or with friction rather than bonded tabs. Flame-sprayed unserrated grips have also been successfully used in tensile testing without tabs.

Other important factors that affect tension testing results include control of specimen preparation, specimen design tolerances, control of conditioning and moisture content variability, control of test machine-induced misalignment and bending, consistent measurement of thickness, appropriate selection of transducers and calibration of instrumentation, documentation and description of failure modes, definition of elastic property calculation details, and data reporting guidelines. These factors are described in detail by ASTM D 3039/D 3039M.

Limitations of the straight-sided coupon tensile methods are described subsequently.

Bonded Tabs. The stress field near the termination of a bonded tab is significantly three-dimensional, and critical stresses tend to peak at this location. Much research has been done on minimizing peak stresses, but it is impossible to make general recommendations that are appropriate for all materials and configurations. Furthermore, improperly designed tabs can significantly degrade results. As a result, tabless or tabbed configurations that use unbonded tabs are becoming more popular, when the resulting failure mode is appropriate.

Specimen Design. There are, particularly within ASTM D 3039, a number of coupon design options included in the standard, which are needed to cover the wide range of materials systems and lay-up configurations within the scope of the test method. Great care should be taken to ensure that an appropriate geometry is chosen for the material being tested.

Specimen Preparation. Specimen preparation plays a crucial role in test results. While this is true for most composite mechanical tests, it is particularly important for unidirectional tests, and unidirectional tensile tests are no exception. Fiber alignment, control of coupon taper, and specimen machining (while maintaining alignment) are the most critical steps of specimen preparation. For very low strain-to-failure materials systems or test configurations, like the 90° unidirectional test, flatness is also particularly important. Edge machining techniques (avoiding machining-induced damage) and edge surface finishes are also particularly critical to strength results from the 90° unidirectional test.

Unidirectional Testing. All the elements that make tensile testing subject to error are exacerbated in the unidirectional case, particularly in the 0° direction. This has led to the increased use of a much less sensitive $[90/0]_{ns}$-type laminate coupon (also known as the "crossply" coupon) from which unidirectional properties can be easily derived (Ref 4). Properly tested crossply coupons often produce results equivalent to the best attainable unidirectional data. While unidirectional testing is still performed, and in certain cases may be preferred or required, a straight-sided, tabless, $[90/0]_{ns}$-type coupon is now generally believed to be the lowest cost, most reliable configuration for lamina tensile testing of unidirectional materials. This straight-sided tabless configuration also works equally well for nonunidirectional material forms and for other general laminates. Another advantage is that, unlike with 0° unidirectional specimens, $[90/0]_{ns}$-type coupon failures do not usually mask indicators of improper testing/specimen preparation practices.

Width tapered coupon tensile tests are standardized in ASTM D 638, "Standard Test Method for Tensile Properties of Plastics." The test, developed for and limited to use with plastics, uses a flat, width-tapered tensile coupon with a straight-sided gage section. Several geometries are allowed, depending on the material being tested. Figure 5 shows a schematic of one general configuration. Despite its heritage, this coupon has also been evaluated and applied to composite materials. The coupon taper is accomplished by a large cylindrical radius between the wide gripping area at each end and the narrower gage section, resulting in a shape that justifies the nickname of the "dogbone" coupon. The taper makes the specimen particularly unsuited for testing of 0° unidirectional materials, because only about half of the gripped fibers are continuous throughout the gage section. This

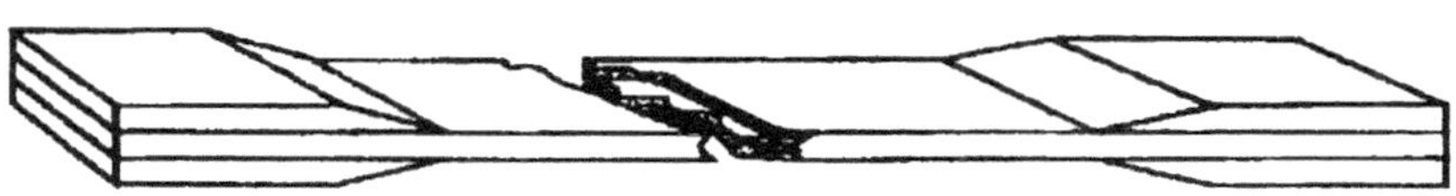

Fig. 4 Typical tension failure of multidirectional laminate using a tabbed coupon

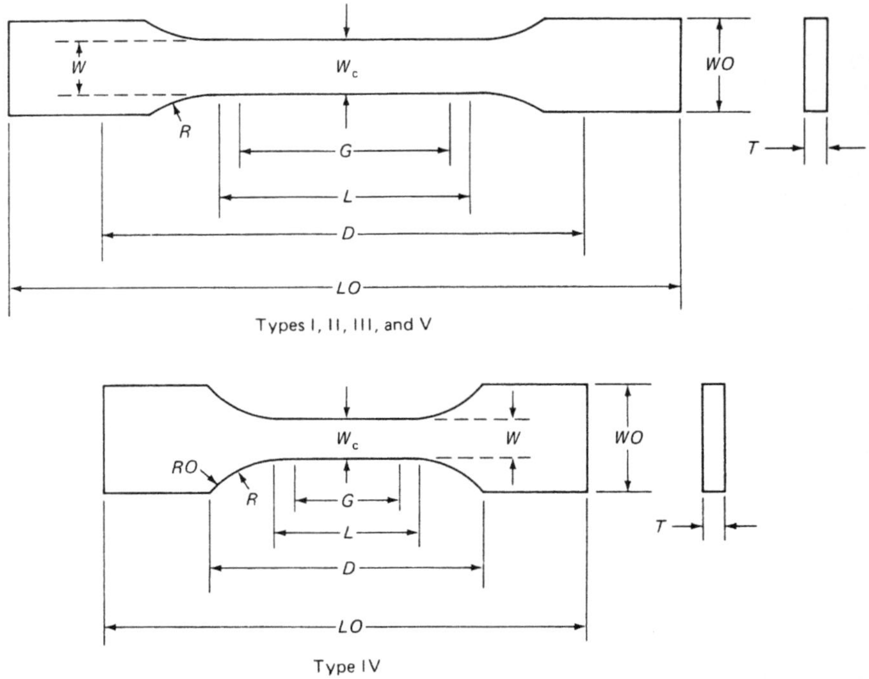

Fig. 5 Schematic of typical ASTM D 638 test specimen geometry. *W*, width; W_c, width at center; *WO*, width overall; *T*, thickness; *R*, radius at fillet; *RO*, outer radius; *G*, gage length; *L*, length; *LO*, length overall; *D*, distance between grips

usually results in failure by splitting at the radius, due to inability of the matrix to shear the load from terminated fibers into the gage section.

While the ASTM D 638 coupon configuration has been successfully used for fabric-reinforced composites and with general nonunidirectional laminates, some materials systems remain sensitive to the stress concentration at the radius. For its intended use with plastics, the coupon is molded to shape. Likewise, discontinuous fiber composites can be molded to the required geometry. To ensure valid results, care must be taken that the molding flow does not create preferentially oriented fibers. For laminated materials the coupon must be machined, ground, or routed to shape. The coupon also has the drawback of having a relatively small gage volume and is poorly suited for characterization of coarse weaves with repeating units larger than the gage width of 6.4 to 13 mm (0.25 to 0.50 in.). The standardized procedure, due to the intended scope, does not adequately cover the testing parameters required for advanced composites.

Limitations of the ASTM D 638 method are described in the following paragraphs.

Standardization. While the ASTM D 638 test is standardized, it was not developed for advanced composites and is primarily applicable to relatively low-modulus, unreinforced materi-

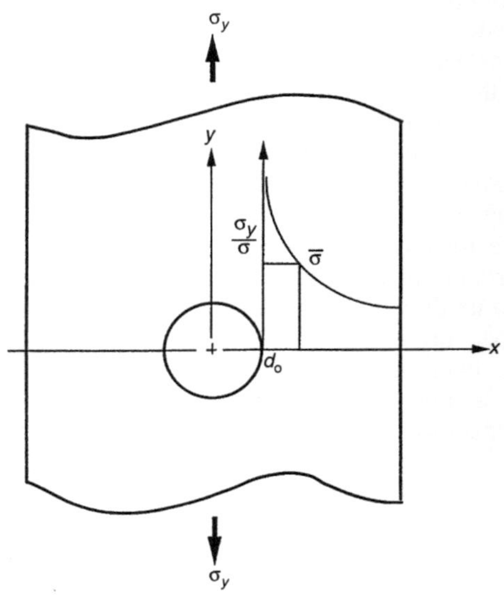

Fig. 6 Stress concentration adjacent to a hole in a composite laminate subjected to uniaxial loading

als, or low-reinforcement volume materials incorporating randomly oriented fibers.

Specimen Preparation. Special care is required to machine the taper into a laminated coupon.

Stress State. The radius transition region can dominate the failure mode and result in reduced strength results. The width-tapered coupon is not suitable for unidirectional laminates, and is limited to fabrics or nonunidirectional laminates when gage section failures can be attained.

Limited Gage Section Volume. The limited gage width makes it unsuitable for coarse fabrics.

The sandwich beam test is standardized as ASTM C 393, "Standard Test Method for Flexural Properties of Flat Sandwich Constructions." While primarily intended as a flexural test for sandwich core shear evaluation, the scope also allows use for determination of facing tensile strength. While this use is not well documented within the test method, it has been used for tensile testing of composite materials, particularly for 90° properties of unidirectional materials, or for fiber-dominated testing in extreme nonambient environments. This test specimen is claimed by some to be less susceptible to handling and specimen preparation damage than D 3039-type 90° specimens, resulting in higher strengths and less test-induced variation.

In order to assure failure in the tensile facesheet, the compression facesheet is often manufactured from the same material, but at twice the thickness as the tensile facesheet.

Limitations of the ASTM C 393 method are described subsequently.

Cost. Specimen fabrication is relatively expensive.

Stress State. The effect on the stress state of the sandwich core has not been studied in tension and could be a concern.

Standardization. While this test technically is standardized, its practical application and limitations are not well studied or documented.

Environmental Conditioning. Conditioning is problematic because of the difficulty of assuring tensile facesheet moisture equilibrium due to the moisture protection offered by the compression facesheet and the core. The extended conditioning times required also often cause adhesive breakdown prior to testing.

Out-of-Plane Tensile Test Methods

ASTM D 6415, "Standard Test Method for Measuring the Curved Beam Strength of a Fiber-Reinforced Polymer-Matrix Composite," is currently the only published standard for out-of-plane tensile testing specifically relating to composites, though modifications to ASTM C 297, C 633 and D 2095 are also often employed. These methods are not discussed here, and the reader is referred to Ref 3 and the test standards for more information.

Open Hole Tensile Test

Cutouts and holes are requirements in many structural applications. The effect of cutouts in composite laminates is greater than the effect caused by the reduction in load-carrying area alone. Stress concentrations are produced in the laminate adjacent to cutout boundaries that substantially reduce load-carrying capacity. Stress concentrations are a function of laminate anisotropy and cutout geometry. Sharp notches produce higher stress concentration factors than circular cutouts. However, the notch sensitivity of laminates is significantly influenced by laminate stacking sequence and a host of microstructural materials characteristics like matrix toughness, matrix stiffness, and fiber to matrix adhesion. High stress concentrations produce complex damage zones, which in turn redistribute the stress and increase the energy required to produce failure significantly above that predicted from the stress concentration factor alone. It has been shown that larger notches produce lower strengths, because the stress concentrations involve a larger volume, increasing the probability of failure due to a critical flaw (Ref 5). The stress distribution illustrated in Fig. 6 is the basis for the point stress criterion for notched strength prediction (Ref 6), which states that failure occurs when the stress at some characteristic distance d_0 reaches the unnotched tensile strength of the composite.

The test method for open hole tension uses a circular cutout in a test specimen (Fig. 7). The method is now standardized as test method ASTM D 5766 "Standard Test Method for Open Hole Tensile Strength of Polymer Matrix Composite Laminates." It employs a 305 mm long by 38 mm wide (12 in. by 1.5 in.) specimen containing a 6.35 mm (0.25 in.) hole. Quasi-isotropic laminate configurations are specified to be $(+45/0/-45/90°)_{2s}$ for tape or $(\pm 45/(0/90°))_{2s}$ for fabric prepregs. While other laminate configurations and geometries are possible, it is recommended that the width-to-hole diameter ratio of 6 be maintained.

The specimen should be machined to the specification shown in Fig. 7. Tolerance on the hole location relative to the specimen centerline is critical, since eccentricity can significantly decrease strengths. Specimens can be tabbed or untabbed, although untabbed specimens reduce cost. If ultimate strain and modulus are desired, specimens may be instrumented with a strain gage located on the specimen centerline 25 mm (1 in.) from the hole center ASTM D 5766, however, covers only notched strength and does not contain provisions for strain measurement.

The test is performed as a uniaxial tensile test following ASTM D 3039. The specimen is loaded until tension failure occurs through the notch. If failure occurs outside the notch, the test result should be discarded, since the failure was caused by a flaw in the material. If failures consistently fall outside the notch area, the naturally occurring flaws in the material are larger than the notch (this is possible with some sheet molding compounds). Then, the specimen design must be scaled to reflect the material inhomogeneity level. At least five specimens should be tested per test condition.

The notched strength σ_N is calculated as the tensile strength of the laminate based on the far-field stress:

$$\sigma_N = P/bd \qquad \text{(Eq 4)}$$

where P is the maximum load, b is the specimen width, and d is the specimen thickness. If the specimen is instrumented, the modulus is determined as:

$$E_x = \frac{P_3 - P_1}{0.002\ bd} \qquad \text{(Eq 5)}$$

where P_1 and P_3 are the loads at 1000 and 3000 microstrain, respectively. The strain at failure is determined from the stress-strain curve.

Notched strength data is typically used for materials screening and for determining design allowables. For design, it is necessary to generate empirical data based on the material, the laminate configuration, and the hole sizes required. In lieu of generating empirical data for every conceivable material, laminate, and hole size combination, it is possible to use the point stress criterion (PSC) analysis to interpolate notched strength over a range of hole diameters by testing a series of three different notch sizes for the material and laminate construction of interest (Ref 7, 8).

Tensile Testing of Metal-Matrix Composites

Tensile testing of metal-matrix composites is based on ASTM Standard D 3552 (Ref 9). In addition to a straight-sided coupon similar to the ASTM D 3039 specimen for polymeric-matrix composites, two tapered specimen configurations, flat and round, are available in conjunction with this test method. Flat panels are produced by such techniques as diffusion bonding, whereas composites fabricated by various liquid infiltration and other methods used for producing massive materials are better suited to circular-cross-section shapes. The flat specimen configuration is shown in Fig. 8. The circular-cross-section specimen is of limited use and will not be discussed here. A complete description of this specimen can be found in ASTM D 3552.

For 0° flat specimens, tabs are bonded to the grip section to cushion the end region from filament damage. Straight-sided coupons have a gage length of 50.8 mm (2 in.) or 76.2 mm (3 in.) and a width of 9.525 mm (0.375 in.) or 12.7 mm (0.5 in.), respectively. The recommended tab length, L_T, is 25.4 mm (1 in.). Tapered specimens have a gage length, L_G, of 25.4 mm (1 in.)

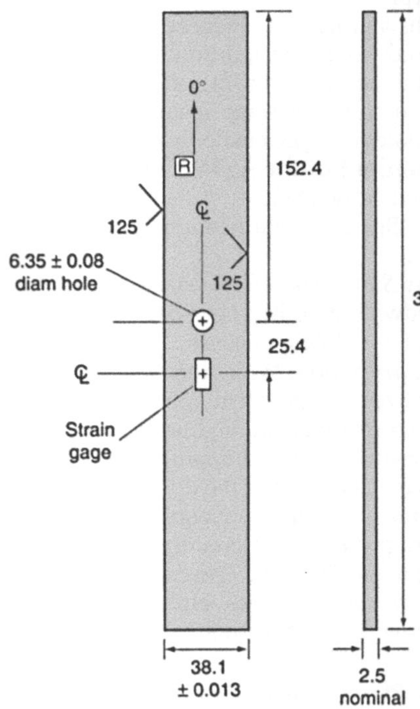

Fig. 7 Open hole tensile test specimen geometry. All dimensions are in millimeters.

and a gage-section width, W_G, of either 6.35 mm (0.25 in.) or 9.525 mm (0.375 in.). The shoulder and tab lengths, L_1 and L_T, respectively, should be 25.4 mm (1 in.). The radius of curvature of the shoulder, R, should be a minimum of 25.4 mm (1 in.). For tensile testing of materials in limited supply, a 25.4 mm (1 in.) gage section may be utilized in conjunction with a 6.35 mm (0.25 in.) gage width. The tab region may be reduced to 19.05 mm (0.75 in.) and the radius of the shoulder reduced to 12.7 mm (0.5 in.). It should be noted that with 0° tapered specimens, failure may tend to initiate at or near the fillet radius. If this occurs, a straight-sided specimen should be substituted.

Because 90° unidirectional composites tend to have low strength, larger widths are necessary to obtained reproducible data. In this case, a straight-sided coupon with a gage length of 25.4 mm (1 in.) and a width of 12.7 mm (0.5 in.) is recommended. The tab length remains at 25.4 mm (1 in.). If availability of material dictates a smaller specimen, the gage section may be reduced to 12.7 mm (0.5 in.).

As in the case of polymeric-matrix specimens, strain measurements can be obtained by utilizing an extensometer or strain gages. If Poisson's ratio is to be determined, strain must be measured in both the longitudinal and transverse directions. Gages should not measure less than 3 mm (0.1181 in.) in the longitudinal direction and not less than 1.5 mm (0.0591 in.) in the transverse direction. For specimens with short (12.7 mm, or 0.5 in.) gage sections, extensometers are not recommended.

Self-aligning wedge-type or lateral-pressure-type grips with serrated or knurled surfaces are required by ASTM Standard D 3552. Gripping pressure should be sufficient to prevent specimen slippage without damaging the end tabs. Emery cloth or a similar material can be used to distribute the pressure more uniformly if the serrations are too coarse.

Mechanical properties of metal-matrix composites are very sensitive to specimen preparation. Special care should be taken in machining or trimming. For some types of metal-matrix composites, conventional machining methods are appropriate. In other cases, grinding or electrical discharge machining (EDM) should be used. Damaging vibrations must be minimized during machining, and in the EDM method the specimen must be mounted in such a manner as to ensure good electrical contact and thus prevent extraneous arcing and resulting specimen damage.

Data Reduction

Calculations of strength, Young's modulus, and Poisson's ratio are the same for both polymeric-matrix and metal-matrix composites. Tensile strength in the load direction is determined by dividing the maximum load by the cross-sectional area of the gage section:

$$S_L = \frac{P}{hW_G} \qquad \text{(Eq 6)}$$

where S_L is ultimate tensile strength in the load direction in megapascals or pounds per square inch; P is maximum load, in newtons or pounds (force); h is specimen thickness, in millimeters or inches; and W_G is the gage-section width of the specimen, in millimeters or inches.

Young's modulus in the load direction is determined from the slope of the load-strain curve in the linear region:

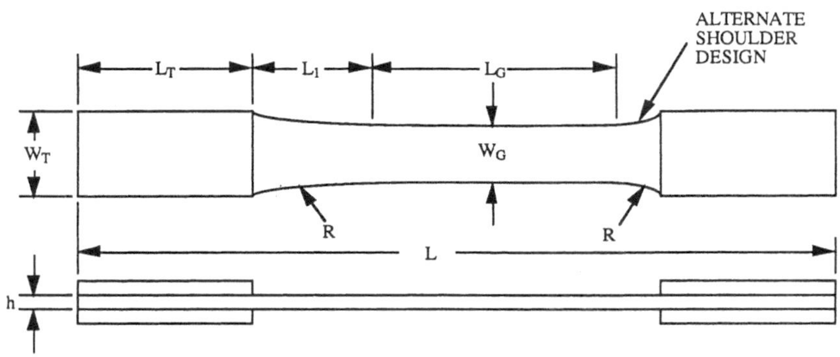

Fig. 8 Metal-matrix composite tensile specimen

$$E_L = \frac{(\Delta P/\Delta \varepsilon_L)}{hW_G} \quad \text{(Eq 7)}$$

where E_L is Young's modulus in the load direction, in megapascals or pounds per square inch; and $\Delta P/\Delta \varepsilon_L$ is the slope of the load-strain curve in the linear portion of the curve, where ε_L denotes the strain parallel to the load.

Poisson's ratio can be calculated from the relationship

$$\nu_{LT} = -\frac{\Delta \varepsilon_T}{\Delta \varepsilon_L} \quad \text{(Eq 8)}$$

where ν_{LT} is Poisson's ratio relative to the load direction; and $\Delta \varepsilon_T/\Delta \varepsilon_L$ is the slope of the strain-strain curve, where ε_T denotes the strain transverse to the load direction.

Application of Tensile Tests to Design

It is often desired to use coupon-level data for design purposes. Thus, it is appropriate to consider the merits, for design purposes, of tensile-test data generated in accordance with ASTM Standard D 3039 for polymeric-matrix composites and Standard D 3552 for metal-matrix composites. Because both of these test methods involve straight-sided specimens, one must be careful that failures do not consistently occur near the end tabs. Even for the tapered metal-matrix specimens, consistent failure near the fillets are of concern.

In addition to these obvious pitfalls, one has to be concerned with the over-all failure processes that occur in laminates. In particular cases for which the initial failure mode is delamination due to free edges, one must carefully assess whether such a failure process represents how the material will behave in the structure or whether the data is an artifact of the test method. In fact, failure modes produced at the coupon level should always be evaluated as to their applicability to behavior in a structure. This is particularly true for multidirectional fiber-reinforced composites.

Other considerations include the influence of "first ply failure" on design. In particular, matrix cracking (first ply failure) may occur far below ultimate failure in a multidirectional laminate. The effect of first ply failure on the usefulness of the laminate in the structure is an important design consideration that is not of concern to the experimentalist performing tensile tests. It may be important, however, for the experimentalist to determine first ply failure. This is usually done by observing a plateau in the stress-strain curve. For fiber-dominated laminates, such as $[0°/90°]_s$, observance of a plateau may require monitoring of the transverse stress-strain curve, because matrix failure in the 90° plies will not have an influence on the longitudinal stress-strain curve.

ACKNOWLEDGMENTS

This chapter was adapted from:

- J.M. Whitney, Tensile Testing of Fiber-Reinforced Composites, *Tensile Testing*, 1st ed., P. Han, Ed., ASM International, 1992, p 183–200
- D. Wilson and L.A. Carlsson, Mechanical Testing of Fiber-Reinforced Composites, *Mechanical Testing and Evaluation*, Vol 8, *ASM Handbook*, ASM International, 2000, p 905–932
- S. Bugaj, Constituent Materials Testing, *Composites*, Vol 21, ASM Handbook, ASM International, 2001, p 749–758
- J. Moylan, Lamina and Laminate Mechanical Testing, *Composites*, Vol 21, *ASM Handbook*, ASM International, 2001, p 766–777

REFERENCES

1. "Standard Test Method for Tensile Strength and Young's Modulus for High-Modulus Single-Filament Materials," D 3379, *Annual Book of ASTM Standards*, ASTM International
2. "Standard Test Methods for Properties of Continuous Filament Carbon and Graphite Fiber Tows," D 4018, *Annual Book of ASTM Standards*, ASTM International Testing and Materials
3. *Composite Materials*, Vol 1, Chapter 6, MIL-HDBK-17-1E, Department of Defense Handbook
4. Use of Crossply Laminate Testing to Derive Lamina Strengths in the Fiber Direction, *Composite Materials*, Vol 1, Chapter 6, MIL-HDBK-17-1E, Department of Defense Handbook
5. J.M. Ogonowski, Analytical Study of Finite

Geometry Plates with Stress Concentrations, *AIAA Paper 80-0778*, American Institute of Aeronautics and Astronautics, New York, 1980, p 694

6. J.M. Whitney and R.J. Nuismer, Stress Fracture Criteria for Laminated Composites Containing Stress Concentrations, *J. Compos. Mater.*, Vol 8, 1974, p 253
7. L.A. Carlsson and R.B. Pipes, *Experimental Characterization of Advanced Composite Materials*, 2nd ed., Technomic, Lancaster, 1987
8. R.B. Pipes, R.C. Wetherhold, and J.W. Gillespie, Jr., Notched Strength of Composite Materials, *J. Compos. Mater.*, Vol 13, 1979, p 148
9. "Test Method for Tensile Properties of Fiber Reinforced Metal Matrix Composites," D 3552, *Annual Book of ASTM Standards*, ASTM International

CHAPTER 12

Tensile Testing of Components

THE MECHANICAL EVALUATION of components requires an engineer to use many sources of information. It requires an understanding of service conditions, design, and manufacturing variables. While there are many types of component tests for a multitude of products, this chapter focuses on three examples of engineering components that undergo significant loading in tension: threaded fasteners and bolted joints; adhesive joints; and welded joints. For some components, tensile loading is not the primary concern. For example, rolling contact fatigue is the most important consideration for rolling-element bearings. Gears, in addition to rolling contact fatigue tests, are tested for resistance to wear, bending fatigue, and impact. Pressure vessels, piping, and tubing are tested for their creep and fracture resistance.

An overview of mechanical properties for component design can be found in Ref 1. Properties and design for static (tensile and compressive) loads, dynamic (impact and fracture toughness) loads, and cyclic (fatigue) loads are addressed.

Testing of Threaded Fasteners and Bolted Joints

Fastener engineering and the mechanical testing of threaded fasteners and bolted joints is an important specialty within the field of mechanical engineering. With the wide variety of fasteners and bolted joints available for use, no one set of tests can be specified to cover all applications. Fasteners are routinely tested for hardness, tensile strength, and torsional strength, as well as corrosion and hydrogen embrittlement. Before describing the standardized tensile test for externally threaded fasteners, some brief background information is provided to help the reader understand the relationships between torque, angle-of-turn, tension, and friction.

Torque, Angle, Tension, and Friction

A proper amount of tension, or clamping force, must be developed to ensure that a bolted assembly will function in a safe and reliable manner. The most common attempt to indirectly estimate fastener tension is to take torque measurements either dynamically as the fastener is tightened or with a breakaway audit after the fact. The torque that is required to produce the desired tension in a fastener is dependent on several factors, with frictional characteristics being the most important. Angle-of-turn measurements combined with torque measurements can help overcome the unknown friction-induced variability in the torque-tension relationship.

Tension. The tension that is created in a threaded fastener when it is tightened represents the clamping force that holds the assembly together. Once the assembly is brought together, the fastener responds like a tension spring, and the assembly acts like a compression spring. The interaction between the fastener and the assembly is illustrated in Fig. 1. As the fastener is turned and load is applied, the fastener is stretched, and the parts are compressed. This compression results in an elastic joint in which the fastener is normally the more flexible member, and the assembly is the more rigid member.

The amount of clamping force that the fastener must provide to hold the assembly together must be sufficient to both maintain preloading and prevent slipping of the parts or opening of the joint when the service loads are applied. The factors that primarily establish the preload requirement are the stiffness of the materials in the joint and the loads that are placed on the assembly.

Fastener tension can be measured using different devices, such as strain-gaged bolts or fastener force washers, or by using special techniques, such as ultrasonic bolt measurement. Although these devices and methods are useful in research and engineering efforts, they are often impractical or costly for evaluating fastener tension in production quality-control efforts.

Torque. The most common way to estimate clamping force is to observe the amount of torque applied to the fastener, either as the fastener is tightened or with a breakaway audit of the tightened fastener. This procedure assumes that the relationship between torque and tension is known, such that, for example, the nut factor, or K, from the simple equation $T = KDF$ (where T is torque, D is diameter, and F is clamping force) is established and known to have acceptable variability. The truth of the matter is that if torque alone is measured, it can never be known with certainty whether the desired tension has been achieved. Thus, unfortunately, it must be concluded that torque is a highly unreliable, totally inaccurate measurement for evaluation of the preload on a threaded fastener. However, for many noncritical fasteners, where safety or the functional performance of an assembly is not compromised, it may be acceptable to specify and monitor torque alone. The most common measurement tools are hand torque wrenches that are used for installation and torque audit measurements and rotary torque sensors that are used to measure installation torque dynamically.

In order for tension to be developed, the torque applied to a fastener must overcome friction under the head of the fastener and in the threads, and the fastener or nut must turn. Because the friction may absorb as much as 90 to 95% of the energy applied to the fastener, as little as 5 to 10% of the energy is left for generating fastener tension as shown in Fig. 2. If the amount of friction varies greatly, wide variations in clamping force are produced, which can mean loose or broken bolts leading to assembly failures. To ensure proper assembly of critical fasteners, more than torque must be measured.

Angle. The amount of fastener tension can be correlated to fastener rotation once the parts of an assembly are drawn firmly together. The clamping force that is developed in this zone of the assembly process, called the elastic tightening region, has been proven to be proportional to the angle-of-turn. This proportional relationship is based on the helix of the threads and is not influenced by the frictional characteristics of the joint once sufficient clamping force has been produced to firmly align the components such that a linear torque-angle signature slope is attained. More detailed information on the relationship between torque and angle-of-turn can be obtained by *torque-angle signature analysis* described in Ref 2.

Friction Measurements. Whereas fastener engineering analysis of threaded fasteners must consider material strength, surface finishes, plating, and coatings to ensure reliable performance, for predictable and repeatable assemblies it is also necessary to understand, measure, and control the frictional characteristics in both the thread and underhead regions. This is particu-

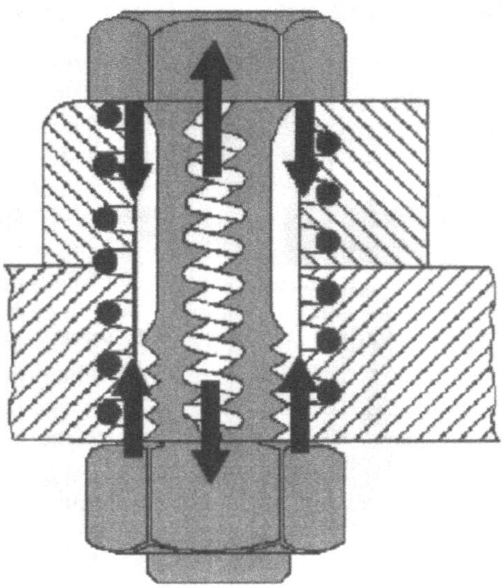

Fig. 1 Spring effect of fastener and assembly under load

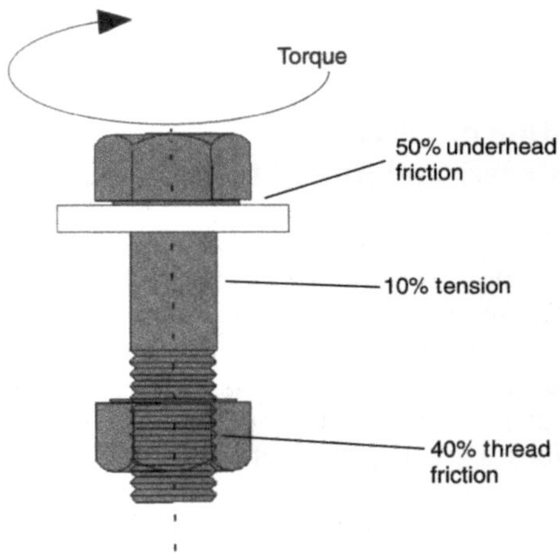

Fig. 2 Typical distribution of energy from torque applied to a bolted assembly

larly true when developing fastener-locking devices such as locknuts, serrated underheads, special thread forms, and thread-locking adhesives and friction patches. Achieving a specific clamp force during installation is always the desired result, and the roles of thread friction and underhead friction must be analyzed and understood to ensure joint integrity.

To determine both thread friction and underhead friction, measurements are taken using a torque-tension research head, as shown in Fig. 3. This device is a special load cell designed to simultaneously measure both thread torque and clamp load. When used with torque sensors that measure the input torque, it is possible to determine the underhead friction torque and the thread friction torque. With this measurement equipment, the fastener can then be tested to establish and maintain standards for friction performance.

Fig. 3 Torque-tension research head, 800 kN capacity

For example, in the test plot illustrated in Fig. 4, a locknut is initially driven onto a bolt. The thread friction torque is equal to the input torque until contact with the underhead-bearing surface is made. Once contact is made with the underhead area, the underhead friction torque is measured as the difference between the total input torque and the thread torque. As clamp force is developed, the pitch torque is calculated and subtracted from the thread torque to compute the thread-friction torque. Note that for prevailing torque locknuts, the elastic origin is located at the prevailing torque level as shown in Fig. 4, not at the zero torque level used for fasteners without prevailing torque characteristics.

Considerations in Testing. There are a number of factors that can affect the tension created in a bolt when torque is applied. Depending on the fastener and joint configuration, direct measurement of tension is not always practical or even possible by any means. Fortunately, torque and angle measurements can be taken for most bolted joints and then analyzed to assist in determination of important characteristics and properties related to strength and reliability.

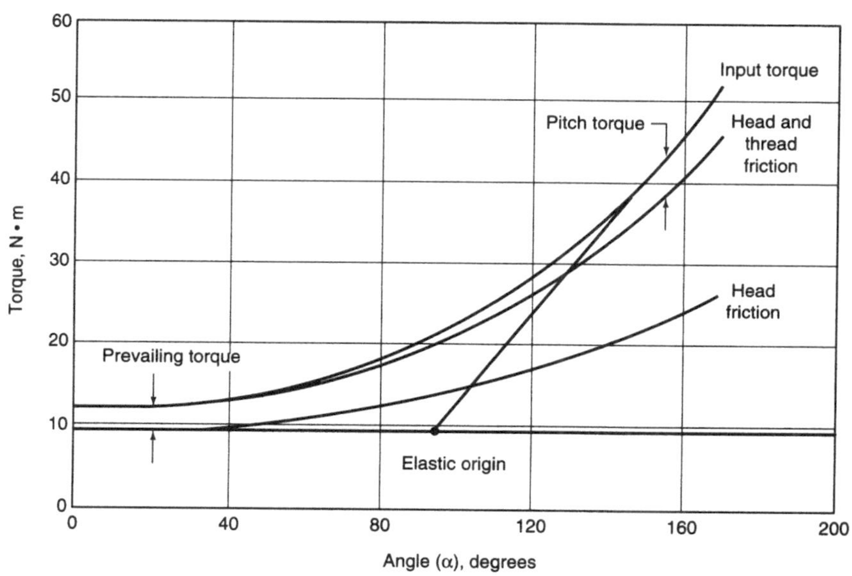

Fig. 4 Determining friction forces for prevailing torque locknut

When tightening a threaded fastener, it is almost always important to know both how much torque is applied and how far the fastener is turned. Similarly, it is always important to fully understand how friction affects the relationship of torque, angle, and tension.

To ensure that critical joints are tightened properly, it must be kept in mind that it is the control of tension that is most important, not the control of torque. This fact must always be considered when choosing and setting up tools, when monitoring production, and when performing quality control audits. The fastener-tightening process is dependent upon the energy transfer from the tightening tool into the fastener and bolted joint. The integrated area under the torque-angle signature curve is a measure of the energy absorbed by the assembly.

Standard Test Methods for Determining Materials Properties of Fasteners

The materials properties of the fastener must be known before a more detailed analysis of the bolted joint is possible. Many standards exist for the testing of fasteners. ASTM F 606M (Ref 3), a specification developed through the procedures of ASTM for metric fasteners, is considered to be one of the most complete. The corresponding standard for English threaded fasteners is ASTM F 606. More complete descriptions of the methods can be found in the standard. The text following in this section is a summary of the basic test methods according to ASTM F 606M.

The test methods described in ASTM F 606M establish procedures for conducting mechanical tests to determine the materials properties of externally and internally threaded fasteners. For externally threaded fasteners, the following test methods are described:

- Product hardness
- Proof load by length measurement, yield strength, or uniform hardness
- Axial tension testing of full-sized products
- Wedge tension testing of full-sized products
- Tension testing of machined test specimens
- Total extension at fracture testing

Product Hardness

The hardness of fasteners and studs can be determined on the ends, wrench flats, or unthreaded shanks after removal of any oxide, decarburization, plating, or other coating material. Rockwell or Vickers hardness standards may be used at the option of the manufacturer. Hardness is determined at midradius of a transverse section of the product taken at a distance of one diameter from the point end of the product. The reported hardness is the average of four hardness readings located at 90° to one another. Acceptable alternative methods of determining hardness for bolts are either at midradius, one diameter from the end, or on the side of the head of a hex-head or square-head product of all property classes after adequate preparation to remove any decarburization. As explained subsequently, uniform hardness measurement is one method for determining the proof load.

Tensile Tests

Fasteners and studs should be tested at full-size and to a minimum ultimate load in kilonewtons (kN) or stress in megapascals (MPa). Such testing includes proof-load tests (by length measurement, yield strength, or uniform hardness), axial tensile tests, wedge tensile tests, and total extension-at-fracture tests.

Proof-Load Tests. The basic proof-load test consists of stressing the product with a specified load that the product must withstand without any measurable permanent set and evaluating the fastener in terms of any change in length. Alternative tests to determine the ability of a fastener to pass the proof-load test are the yield-strength test and the uniform hardness test. Although any of the alternative test methods described may be used, the proof-load test is the arbitration method used in case of any dispute.

Method 1, Length Measurement. The overall length of the specimen is measured at its true centerline with an instrument capable of measuring changes in length of 0.0025 mm with an accuracy of 0.0025 mm in any 0.025 mm range. Measuring the length between conical centers on the centerline of the fastener or stud with mating centers on the measuring anvils is preferred. The head or body of the fastener or stud should be marked so that it can be placed in the same position for all measurements.

The product is assembled in the fixture of the tension-testing machine so that six complete threads are exposed between the grips. Tests for heavy hex structural bolts are based on four threads. This is obtained by freely running the nut or fixture to the thread runout of the specimen and then unscrewing the specimen six full turns. For continuous threaded fasteners, at least six full threads should be exposed. The fastener should be loaded axially to the proof load specified in the product specification. The speed of testing, as determined with a free-running cross head, should not exceed 3 mm/min, and the proof load should be maintained for a period of 10 s before releasing the load. Upon release of this load, the length of the fastener or stud should be measured again to determine permanent elongation. A tolerance (for measurement error only) of ±0.013 mm is allowed between the measurements made before loading and that made after loading.

Variables, such as straightness, thread alignment, or measurement error, could result in apparent elongation of the product when the specified proof load is initially applied. In such cases, the product may be retested using a 3% greater load and is considered acceptable if there is no difference in the length measurement after this loading within a 0.013 mm measurement tolerance as outlined.

Method 2, Yield Strength. The product is assembled in the testing equipment as described for method 1. As the load is applied, the total elongation of the product or any part of it that includes the exposed threads should be measured and recorded to produce a load-elongation diagram. The load or stress at an offset equal to 0.2% of the length of fastener occupied by six full threads is determined, as shown in Fig. 5.

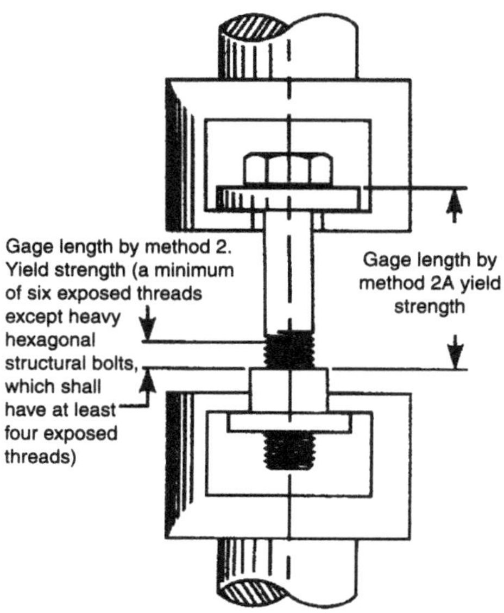

Fig. 5 Tensile testing of full-size fastener (typical set-up). Source: Ref 3

Method 2A, Yield Strength for Austenitic Stainless Steel and Nonferrous Materials. The product is assembled in the testing equipment as described in method 1. As the load is applied, the total elongation of the product should be measured and recorded in order to produce a load-elongation diagram. The load or stress at an offset equal to 0.2% strain should be determined based on the length of the bolt between the holders as shown in Fig. 5, which will be subject to elongation under load by using the yield-strength method described in the section "Tensile Testing of Machined Test Specimens."

Method 3, Uniform Hardness. The fasteners are tested for hardness as described previously, and in addition, the hardness is determined in the core. The difference between the midradius and core hardness should be not more than three points on a Rockwell C scale, and both readings must be within product specification.

Short Fasteners and Studs. Fasteners with lengths less than those shown in Table 1, or that do not have sufficient threads for proper engagement, are deemed too short for tensile testing. Acceptance is then based on a hardness test. If tests other than product hardness are required, their requirements are referenced in the product specification.

Axial Tensile Testing of Full-Sized Products. Fasteners are tested in a holder with a load axially applied between the head and a nut or in a suitable fixture as shown in Fig. 5. Sufficient thread engagement must exist to develop the full strength of the product. The nut or fixture should be assembled on the product, leaving six complete fastener threads exposed between the grips. Studs are tested by assembling one end of the threaded fixture to the thread runout. If the stud has unlike threads, the end with the finer pitch thread, or with the larger minor diameter, is used. The other end of the stud is assembled in the threaded fixture, leaving six complete

Table 1 Required minimum length of fasteners for tensile testing

Nominal product diam (*D*), mm	Min length, mm
5	12
6	14
8	20
10	25
12	30
14	35
16	40
20	45
Over 20	3*D*

Source: Ref 3

threads exposed between the grips. For continuous studs, at least six complete threads are exposed between the fixture ends.

The maximum speed of the free-running cross head should not exceed 25 mm/min. When reporting the tensile strength of the product, the thread stress area is calculated as follows:

$$A_s = 0.7854(D - 0.9382P)^2 \quad \text{(Eq 1)}$$

where A_s is the thread stress area, mm^2; D is the nominal diameter of the fasteners or stud, mm; and P is thread pitch, mm.

The product should support a load prior to fracture not less than the minimum tensile strength specified in the product specification for its size, property class, and thread series. In addition, failure should occur in the body or in the threaded section with no fracture at the juncture of the body and head.

Wedge Tensile Testing. The wedge tensile strength of a hex- or square-head fastener, socket-head cap screw, or stud is the tensile load that the product is capable of sustaining when stressed with a wedge under the head. The purpose of this test is to obtain the tensile strength and to demonstrate the head quality and ductility of the product.

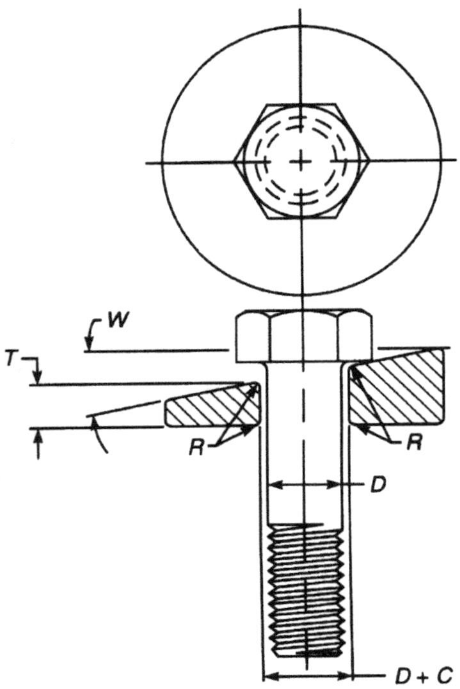

Fig. 6 Wedge-test details for fasteners. *D*, diameter of bolt; *C*, clearance of wedge hole; *R*, radius; *T*, thickness of wedge at short side hole; *W*, wedge angle

Wedge Tensile Testing of Fasteners. The ultimate load of the fastener is determined as described previously under "Axial Tensile Testing of Full-Sized Products," except to place a wedge under the fastener head. When both wedge and proof-load testing are required by the product specification, the proof-load-tested fastener for wedge testing should be used. The wedge must have a minimum hardness of 45 HRC for fasteners having an ultimate tensile strength of 1035 MPa or less, and a minimum of 55 HRC for fasteners having a tensile strength in excess of 1035 MPa. Additionally, the wedge should have the following:

- A thickness of one-half the nominal fastener diameter (measured at the thin side of the hole as shown in Fig. 6)
- A minimum outside dimension such that at no time during the test will any corner loading of the head of the product occur adjacent to the wedge
- An included angle as shown in Table 2 for the product type being tested

The hole in the wedge should have a clearance over the nominal size of the fastener and have its edges top and bottom rounded as specified in Table 3.

The fastener is then tensile tested to failure. The fastener must support a load prior to fracture

Table 2 Wedge angles for tensile testing of fasteners

	Degrees	
Nominal product diam, mm	Fasteners(a)	Studs and flange fasteners
5–24	10	6
Over 24	6	4

(a) For heat-treated fasteners that are threaded one diam or closer to the underside of the head, a wedge angle of 6° for sizes 5 to 24 mm and 4° for sizes over 24 mm should be used. Source: Ref 3

Table 3 Requirements for wedge-hole clearance and radius for tensile testing of fasteners

Nominal product diam, mm	Nominal clearance in hole, mm	Nominal radius on corners of hole, mm
To 6	0.50	0.70
Over 6–12	0.80	0.80
Over 12–20	1.60	1.30
Over 20–36	3.20	1.60
Over 36	3.20	3.20

Source: Ref 3

not less than the minimum tensile strength specified in the product specification for the applicable size, property class, and thread series. In addition, the fracture should occur in the body or threaded portion with no fractures at the junction of the head and the shank.

Wedge-Tensile Testing of Studs. When both wedge-tension and proof-load testing are required, one end of the same stud previously used for proof-load testing is assembled in a threaded fixture to the thread runout. For studs having unlike threads, the end with the finer-pitch thread or with the larger minor diameter is used. The other end of the stud should be assembled in a threaded wedge to the runout and then unscrewed six full turns, leaving six complete threads exposed between the grips as shown in Fig. 7. For continuous threaded studs, at least six full threads are exposed between the fixture ends. The angle of the wedge for the stud size and property class is as specified in Table 2.

The stud should be assembled in the testing machine and tensile tested to failure, as described previously under "Axial Tensile Testing of Full-Sized Products." The minimum hardness of the threaded wedge is 45 HRC for products having an ultimate tensile strength of less than 1035 MPa and 55 HRC for product lines having an ultimate tensile strength in excess of 1035 MPa. The length of the threaded section of the wedge must be equal to at least the diameter of the stud. To facilitate removal of the broken stud, the wedge can be counterbored. The thickness of the wedge at the thin side of the hole is equal to the diameter of the stud plus the depth of the counterbore. The thread in the wedge should have class 4H6H tolerance, except when testing studs having an interference fit thread, in which case the wedge will have to be threaded to provide a finger-free fit. The supporting fixture should have a hole clearance over the nominal size of the stud, and the top and bottom edges should be rounded or chamfered to the same limits specified for the hardened wedge in Table 3.

The stud must support a load prior to fracture of not less than the minimum tensile strength specified in the product specification for its size, property class, and thread series.

Tensile Testing of Machined Test Specimens. Where fasteners and studs cannot be tested at full-size, tests are conducted using test specimens machined from the fastener or stud. Fasteners and studs should have their shanks machined to the dimensions shown in Fig. 8. The reduction of the shank diameter of heat-treated fasteners and studs with nominal diameters larger than 16 mm should not exceed 25% of the original diameter of the product. Alternatively, fasteners 16 mm in diameter or larger may have their shanks machined to a test specimen with the axis of the specimen located midway between the axis and outside surface of the fastener as shown in Fig. 9. In either case, machined test specimens should exhibit tensile strength, yield strength (or yield point), elongation, and reduction of area equal to or greater than the values of these properties specified for the product size in the applicable product specification when tested in accordance with this section.

Tensile Properties: Yield Point. Yield point is the first stress in a material, less than the maximum obtainable stress, at which an increase in strain occurs without an increase in stress. Yield

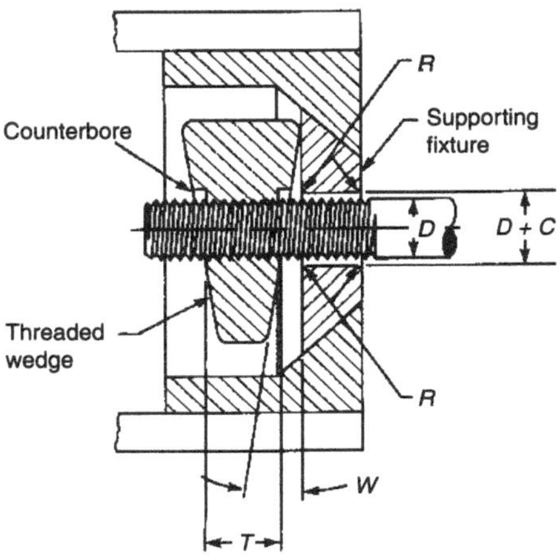

Fig. 7 Wedge-test details for studs. *D*, diameter of stud; *C*, clearance of wedge hole; *R*, radius; *T*, thickness of wedge at short side hole; *W*, wedge angle

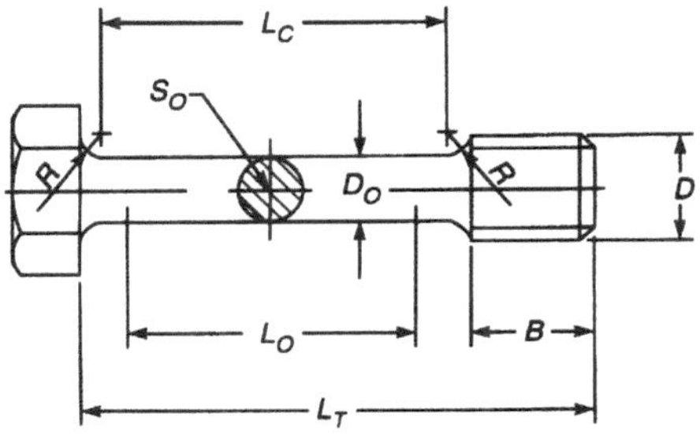

Fig. 8 Tensile-test specimen with turned-down shank. Source: Ref 3

point is intended for application only for materials that may exhibit the unique characteristic of showing an increase in strain without an increase in stress. A sharp knee or discontinuity characterizes the stress-strain diagram. The yield point can be determined by one of the following methods:

- *Drop-of-the-beam or halt-of-the-pointer method:* In this method, an increasing load is applied to the specimen at a uniform rate. When a lever and poise machine is used, the beam is kept in balance by running out the poise at an approximately steady rate. When the yield point of the material is reached, the increase of the load will stop, but the poise should be run a small amount beyond the balance position, and the beam of the machine will drop for a brief interval of time. When a machine equipped with a load-indicating dial is used, there is a halt or hesitation of the load-indicating pointer, which corresponds to the drop of the beam. The load is recorded at the drop of the beam or the halt of the pointer. This point is the yield point of the fastener or stud.
- *Autographic diagram method:* When a sharp-kneed stress-strain diagram is obtained by an autographic device, the yield point is taken as either the stress corresponding to the top of the knee, as shown in Fig. 10, or as the stress at which the curve drops, as shown in Fig. 11.
- *Total extension-under-load method:* When testing material for yield point, the test specimens may not exhibit the well-defined disproportionate deformation that characterizes a yield point as measured by the previous methods. In these cases, the following method can be used to determine a value equivalent to the yield point in its practical

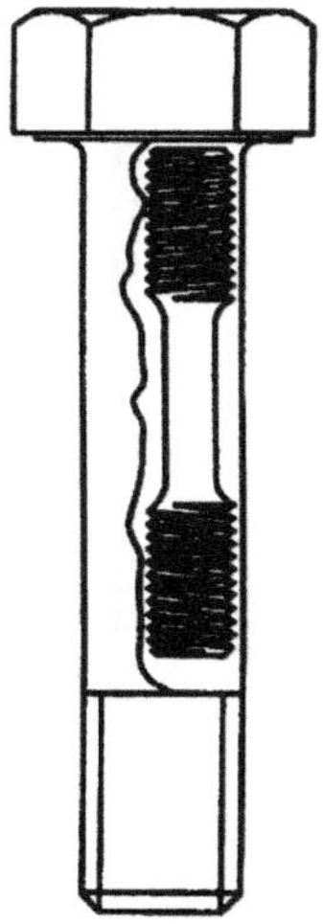

Fig. 9 Location of standard tensile-test specimen when turned from large sized fastener. Source: Ref 3

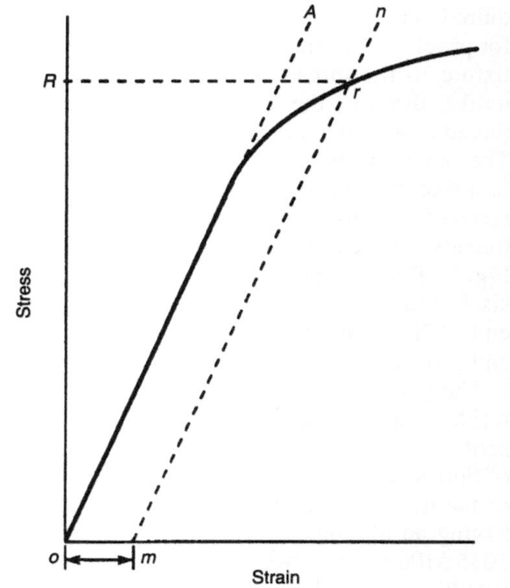

Fig. 10 Stress-strain diagram for determination of yield strength by the offset method. *o-m* is the specified offset. To determine offset yield strength, draw line *m-n* parallel to the line *o-A*. From the intersection point *r*, draw a horizontal line to determine the offset yield strength, *R*.

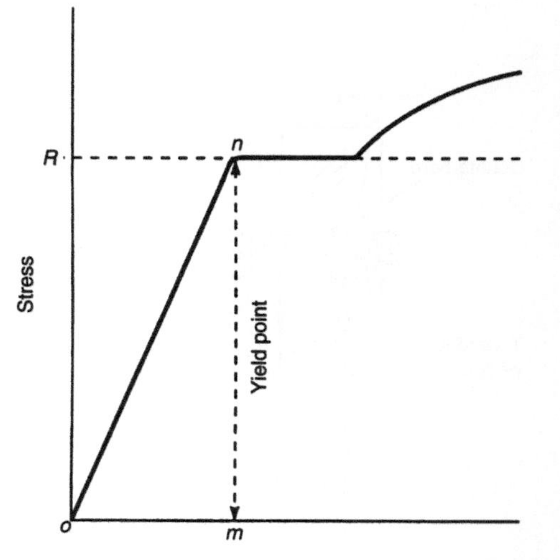

Fig. 11 Stress-strain diagram showing yield point corresponding with top of knee. *o-m*, offset to yield point. Source: Ref 3

significance that may be recorded as the yield point. A class C or better extensometer is attached to the specimen. When the load producing a specified extension is reached, the stress corresponding to the load as the yield point is recorded and the extensometer removed (Fig. 12).

Yield Strength. Yield strength is the stress at which a material exhibits a specified limiting deviation from the proportionality of stress to strain. The deviation is expressed in terms of strain, percentage of offset, total extension under load, and so on. Yield strength may be determined by the offset method or the extension-under-load method.

To determine the yield strength by the offset method, it is necessary to secure data (autographic or numerical) from which a stress-strain diagram may be drawn. Then, on the stress-strain diagram layoff, *o-m*, as shown in Fig. 10, equal to the specified value of the offset, *m-n* should be drawn parallel to *o-A* and thus locate *r*. The yield-strength load, *R*, is the load corresponding to the highest point of the stress-strain curve before or at the intersection of *m-n* and *r*. In reporting values of yield strength obtained by this method, the specified value of the offset used should be stated in parenthesis after the term yield strength, thus:

Yield strength (0.2% offset) = 360 MPa

In using this method, a minimum extensometer magnification of 250 to 1× is required. A class B1 extensometer meets this requirement. (Extensometer system classification is discussed in Chapter 4, "Testing Equipment and Strain Sensors.")

The extension-under-load method is used to determine the acceptance or rejection of materials whose stress-strain characteristics are well known from previous tests of similar materials in which stress-strain diagrams are plotted. For these tests, the total strain corresponding to the stress at which the specified offset occurs should be known as within satisfactory limits. The stress on the specimen, when total strength is reached, is the value of the yield strength. The total strain can be obtained satisfactorily by the use of a class B1 extensometer. The extension under load (mm/mm of gage length) can be determined as follows:

$$YS/E = r \qquad \text{(Eq 2)}$$

where *YS* is the specified yield strength, MPa; *E* is the modulus of elasticity, MPa; and *r* is the limiting plastic strain, mm/mm.

Tensile strength is calculated by dividing the maximum load the specimen sustains during a tensile test by the original cross-sectional area of the specimen.

Elongation. The ends of the fractured specimen are fitted together carefully and the distance between the gage marks measured to the nearest 0.25 mm for gage lengths of 50 mm or under, and to the nearest 0.5 mm of the gage length for gage lengths over 50 mm. A percentage scale reading to 0.5% of the gage length may be used. The elongation is the increase in length of the gage length, expressed as a percentage of the original gage length. In reporting elongation values, both the percentage increase and the original gage length should be given.

If any part of the fracture takes place outside the middle half of the gage length or in a punched or scribed mark with the reduced section, the elongation value obtained may not be representative of the material. If the elongation so measured meets the minimum requirements specified, no further testing is indicated, but if the elongation is less than the minimum requirements, the test should be discarded and performed again.

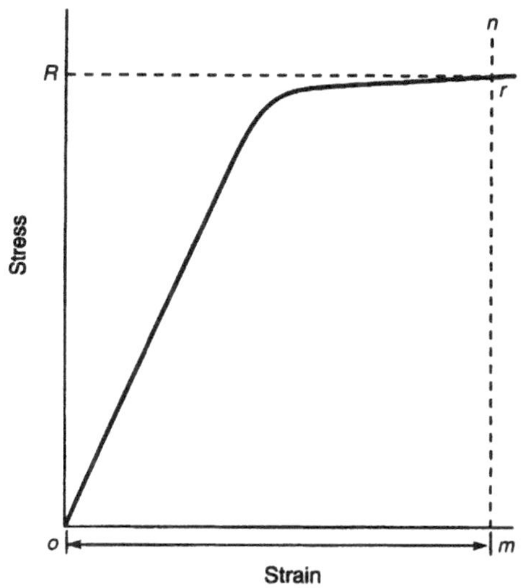

Fig. 12 Stress-strain diagram showing yield point or yield strength by extension-under-load method. *o-m*, specified extension under load. Line *m-n* is vertical, and the intersection point, *r*, determines yield strength value, *R*. Source: Ref 3

Reduction of Area. The ends of the fractured specimen are fitted together and the mean diameter or the width and thickness at the smallest cross section measured to the same accuracy as the original dimensions. The difference between the area thus found and the area of the original cross section expressed as a percentage of the original area is the reduction in area.

Total Extension at Fracture Test. The test to determine extension at fracture, A_L, is carried out on stainless steel and nonferrous products in the finished condition with the length equal to or in excess of those minimums listed in Table 1. The products to be tested are measured for total length, L_1, described as follows and shown in Fig. 13.

Both ends of the fastener or stud are marked using a permanent marking substance, such as bluing, so that the measuring reference points for determining total lengths, L_1 and L_2, are established. An open-end caliper and steel rule or other device capable of measuring to within 0.25 mm are used to determine the total length of the product (Fig. 13).

The product under test is screwed into the threaded adapter to a depth of one diameter (Fig. 5) and load applied axially until the product fractures. The maximum speed of the free-running cross head should not exceed 25 mm/min.

After the product has been fractured, the two broken pieces are fitted closely together, and the overall length, L_2, is measured. The total extension at fracture, A_L, is then calculated as follows:

$$A_L = L_2 - L_1 \quad \text{(Eq 3)}$$

The value obtained should equal or exceed the minimum values shown in the applicable specification for the product and material type.

Testing of Adhesive Joints

Adhesive bonding is a materials joining process in which an adhesive (usually a thermosetting or thermoplastic resin) is placed between the faying surfaces of the parts or bodies called adherends. The adhesive then solidifies or hardens by physical or chemical property changes to produce a bonded joint with useful strength between the adherends. The strength of the adhesive joint is determined by the following tests (Ref 4):

- Peel tests
- Lap shear tests
- Tensile tests
- Fracture mechanics tests

Tensile Tests. Most adhesive joints are designed to avoid (or at least reduce) direct tensile forces across the bond line. However, for many joints where the primary loading is shear, failure may be initiated by the induced secondary tensile stresses and the adhesive joint's tensile strength is of interest. Accordingly, the third most common type of adhesive joint strength test is the tensile test (the lap shear test geometry is the most popular followed by the peel test).

The geometries of several tensile tests for which there are specific ASTM test procedures are shown in Fig. 14 (Ref 5). Some of these test geometries seem relatively simple; however, it has been demonstrated that the stresses along the bond line have a rather complex dependence on geometric factors and adhesive and adherent properties (adhesive thickness and its variation across the bonded surface, modulus, Poisson's ratio, and so on) (Ref 6).

It is almost always difficult to load tensile adhesion specimens in an axisymmetric manner, even if the sample itself is axisymmetric. Non-

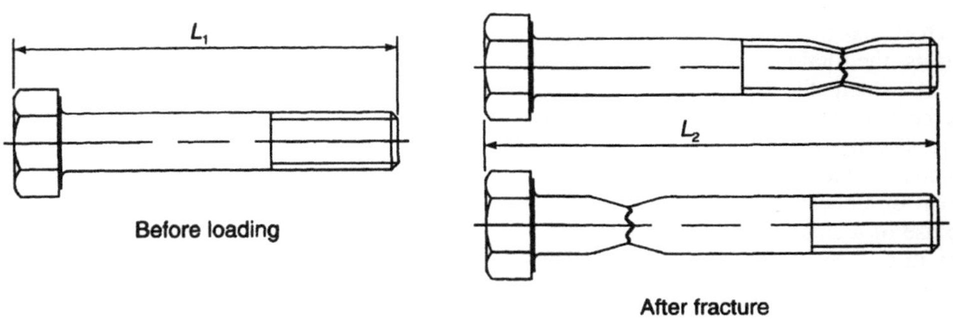

Fig. 13 Determination of total extension at fracture (A_L) for a screw product. Source: Ref 3

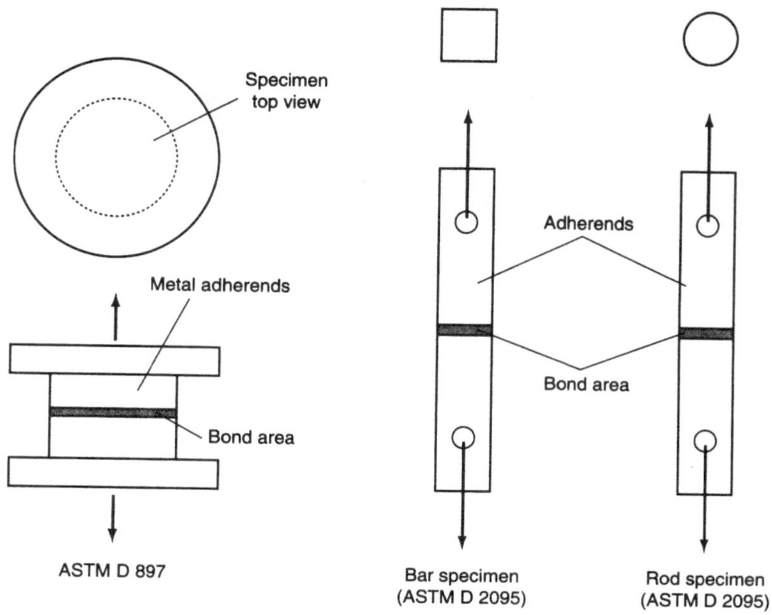

Fig. 14 Typical specimen geometries for testing the tensile strength of adhesive joints. Source: Ref 5

axisymmetric loads have been shown to reduce the bond failure load capability and to cause large scatter in the resulting failure data. Superficially, the geometry for standard tensile adhesion tests is deceptively simple. The result of the tensile adhesion test, as normally reported by experimentalists, is simply the failure load divided by the cross-sectional area of the adhesive (Ref 7). Such average stress at failure can be very misleading. Because of the differences in mechanical properties of the adhesive and adherend, the stresses may become singular at the bond edges when analyzed using linear elastic analysis (Ref 6, 8). Even if the edge singularity is neglected, the stress field in the adhesive is very complex and nonuniform, with maximum values differing markedly from the average value (Ref 6, 8).

Some sense of the complex nature of the stresses can be obtained by visualizing a butt joint of a low modulus polymer (e.g., a rubber) between two steel cylinders. As these are pulled apart, the rubber elongates much more readily than the steel. Poisson's effect will cause a tendency for the rubber to contract laterally. However, if it is tightly bound to the metal, it is restrained from contracting, and shear stresses are induced at the bond line. Reference 9 provides the results of a finite element analysis that demonstrates how these stresses vary across the sample. As noted, for an elastic analysis, both the shear and tensile stresses are singular (tending to infinity) at the outer periphery.

For the tensile specimen configurations considered to this point, the applied loading is intended to be axisymmetric. There is another class of specimen in which the dominant stress is deliberately tensile but in which the loading is obviously "off center." At least four ASTM standards describe so-called cleavage specimens and tests. The reader familiar with cohesive fracture mechanics will see a similarity between the test specimen in ASTM D 1062 ("Standard Test Method for Cleavage Strength of Metal-to-Metal Adhesive Bonds") as shown in Fig. 15, and the compact tensile specimen commonly used in fracture mechanics testing. ASTM D 1062 specifies reporting the test results as force

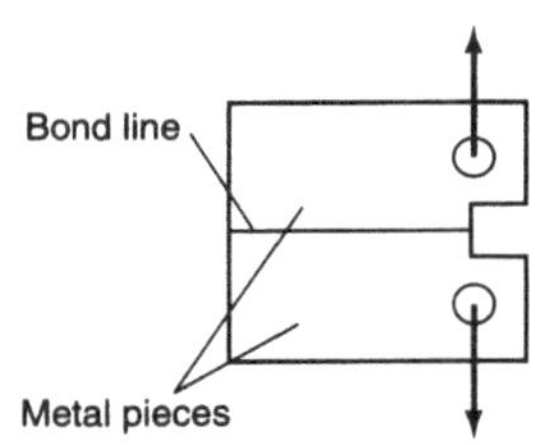

Fig. 15 Specimen for testing the cleavage strength of metal-to-metal adhesive bonds. Source: ASTM D 1062

required, per unit width, to initiate failure in the specimen, while in fracture mechanics, the results are given as G_c with units of J/m^2, which might be interpreted as the energy required to create a unit surface. A knowledgeable and enterprising reader may want to adapt the D 1062 specimen for obtaining fracture mechanics parameters. ASTM D 3807, "Standard Test Method for Strength Properties of Adhesives in Cleavage Peel by Tension Loading," uses a different geometry to measure the cleavage strength. In this case, two 25.4 mm (1 in.) wide by 6.35 mm (0.25 in.) thick plastic strips 177 mm (7 in.) long are bonded over a length of 76 mm (3 in.) on one end, leaving the other ends free and separated by the thickness of the adhesive. Approximately 25 mm (1 in.) from the end of each of these free segments, a "gripping wire" is attached as shown in Fig. 16. During testing, these wires are clamped in the jaws of a universal testing machine and the sample pulled to failure. The results are reported as load per unit width (kg/m or lb/in.).

ASTM D 5041 ("Standard Test Method for Fracture Strength in Cleavage of Adhesives in Bonded Joints") also makes use of a sample composed of two thin sheets bonded together over part of their length. In this case, forcing a wedge (45° angle) between the unbonded portion of the sheets facilitates the separation. The results are typically given as "failure initiation energy" or "failure propagation energy" (i.e., areas under the load deformation curve).

This latter test is similar to another test, formalized as ASTM D 3762 ("Standard Test Method for Adhesive-Bonded Surface Durability of Aluminum—Wedge Test") that has been found very useful for studying time-environmental effects on adhesive bonds. This test is called by various names, but is commonly referred to as the "Boeing Wedge Test" (Ref 10, 11). The test has been used by personnel at this and other aerospace companies to screen various adhesives, surface treatment, and so on for long-term loading at high temperatures and humidities. For testing, two long, slender strips of candidate structural materials are first treated with the prescribed surface treatment(s) and bonded over part of their length with a candidate adhesive (Fig. 17). As in the test described in the previous paragraph, the free ends are forced apart by a wedge. The amount of separation by the wedge (determined by wedge thickness and depth of insertion) determines the value of the stresses in the adhesive. These stresses can, of course, be adjusted and the values calculated from mechanics of material concepts. When the wedge is in place, the sample is placed in an environmental chamber. At periodic time intervals, the length of the crack is measured, and a plot of crack length versus time is constructed. The more satisfactory adhesives and/or surface treatments are those for which the crack is arrested or grows very slowly. While the environmental chamber typically contains hot, humid air, there is no reason why other environmental agents cannot be studied by the same method, including immersion in liquids.

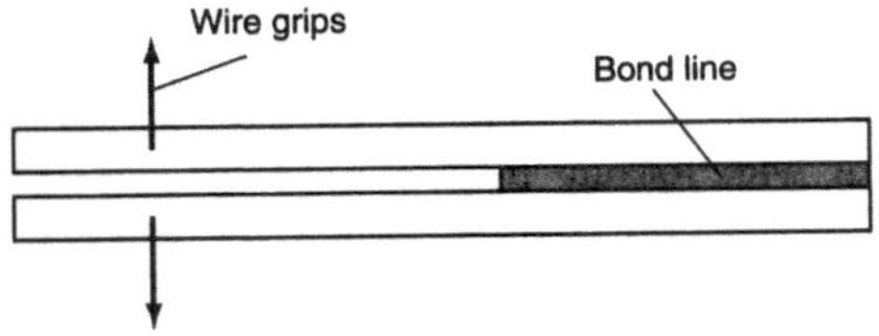

Fig. 16 Specimen for testing cleavage peel (by tension loading). Source: ASTM D 3807

Testing of Welded Joints

Testing for mechanical properties of strength and ductility for welded joints is somewhat more

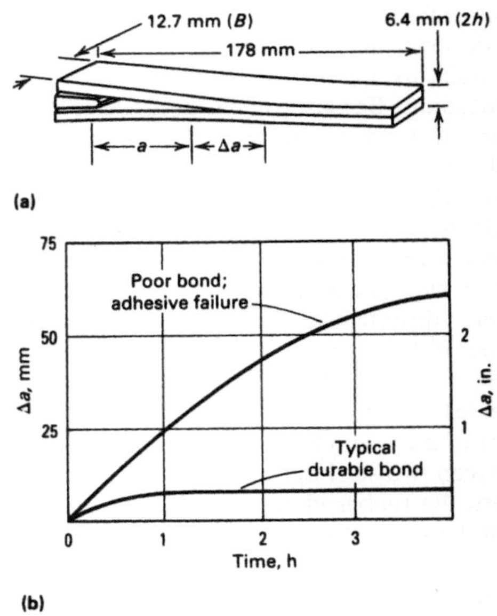

Fig. 17 Boeing wedge test (ASTM D 3762). (a) Test specimen. (b) Typical crack propagation behavior at 49 °C (120 °F) and 100% relative humidity. *a*, distance from load point to initial crack tip; Δ*a*, growth during exposure. Source: Ref 4

complicated than it is for base metal, because these properties vary across the weld metal, the adjacent heat-affected zone (HAZ), and the base metal. Several different tests may be used or combined to assess the strength of the overall welded joints. Tensile testing is widely used to measure the strength and ductility of the weld metal alone. Tensile testing of welds in place, with weld metal, HAZ, and base metal, allows an overall strength to be determined but usually cannot provide the strengths of the individual parts of the weldment.

Tensile tests of welds can also measure elastic modulus. However, except in rare cases of dissimilar metal joining, the elastic modulus is not sensitive to the differences between weld, HAZ, and base metal. So, measurement during weld tensile tests is not usually required. Also, most tensile testing procedures for weld joints cannot be relied upon to provide accurate values of elastic modulus. The specific procedures for testing of elastic modulus distributed by ASTM should be used if required (Ref 12).

Testing of Weld Material. Deposited weld metal can be tested for the mechanical properties of strength and ductility using the same test methods used for base metals (Ref 13, 14). However, a sufficient volume of deposited weld metal is required to remove a test specimen made entirely of weld metal. Often, arc welds are long only in one direction (the longitudinal direction), while the through-thickness and cross-weld directions are much smaller. This encourages all-weld-metal tensile test specimens to be removed with the long direction of the specimen corresponding to the longitudinal direction of the weld. Such longitudinal tensile test specimens are standard for all-weld-metal tests.

All-weld-metal tests are most commonly done on specimens with round cross section. The diameter of the specimen may need to be reduced from that used for base metal so that the specimen can be taken entirely from weld metal. Rectangular cross-section specimens also are used occasionally.

Ultimate tensile strength, yield strength (usually based either on yield point or a specified offset), elongation, and reduction of area are all commonly recorded.

While the specimen surface should be smooth, without deep machining marks, imperfections within the gage length due to welding should not be removed. This requirement may increase the variability of results within a group of similar specimens.

If the data required are for a class of weld material such as an electrode lot, the material can be taken from specimens that reduce the possibility of dilution of base metal into the weld, such as a built-up weld pad. If the data required are for a particular weldment, the geometry as well as the welding process and procedure should model those of the weldment as closely as possible. Some modifications of the weldment may be allowed, such as increasing the root opening by 6 mm (¼ in.) or buttering the groove faces with the weld metal to be tested. The surface of the tested section, in the gage length, is recommended to be 3 mm (⅛ in.) or more from the fusion line.

Testing of Welds in Place. When the weld metal extends over only part of the tested gage length, tensile tests can be performed similar to those performed on the round and rectangular specimen tests of weld metal. The nonuniformity of deformation or stresses of the weld, HAZ, and base metal combination limits the information normally recorded.

For transverse tests, ultimate strength and the location of fracture are the only commonly recorded parameters, because strength, elongation, and reduction in area will all be affected by the constraint of the adjacent differing materials. If the weld is undermatched, the yield strength tends to be higher than it is for an all-weld-metal specimen, while the elongation over the gage length and reduction in area are smaller. If the weld yield strength exceeds that of the base material, that is, it is overmatched, the failure tends to occur not in the adjacent HAZ, but in the base material closer to the end of the gage length, because of the constraint provided by the high-strength weld metal.

Local strain measurements, such as those made by strain gages, can add useful information to the results of transverse testing. The local strain information can be correlated to the load and displacement information to allow local strengths to be determined.

For longitudinal tests, the strain will be nearly uniform across the weld metal, HAZ, and base metal. Differences in response to the applied strain may result in stresses varying across the cross section. Only ultimate strength is commonly measured.

Testing standards may need to be varied for some specific geometries. For instance, girth welded tubes of less than 75 mm (3 in.) diameter are commonly tested in the form of tubes with central plugs at the grips. The weld is placed at the center of the gage length between the grips. The additional constraint induced by the hoop direction continuity tends to increase the measured strengths and decrease the measured ductilities for tube welds tested in this manner compared to a similar joint between flat sheets.

ACKNOWLEDGMENTS

This chapter was adapted from:

- R.S. Shoberg, Mechanical Testing of Threaded Fasteners and Bolted Joints, *Mechanical Testing and Evaluation,* Vol 8, *ASM Handbook,* ASM International, 2000, p 811–835
- K.L. Devries and P. Borgmeier, Testing of Adhesive Joints, *Mechanical Testing and Evaluation,* Vol 8, *ASM Handbook,* ASM International, 2000, p 836–844
- W. Mohr, Mechanical Testing of Welded Joints, *Mechanical Testing and Evaluation,* Vol 8, *ASM Handbook,* ASM International, 2000, p 845–852

REFERENCES

1. H.E. Fairman, Overview of Mechanical Properties for Component Design, *Mechanical Testing and Evaluation,* Vol 8, *ASM Handbook,* ASM International, 2000, p 789–797
2. R.S. Shoberg, Mechanical Testing of Threaded Fasteners and Bolted Joints, *Mechanical Testing and Evaluation,* Vol 8, *ASM Handbook,* ASM International, 2000, p 811–835
3. "Standard Test Method for Determining the Mechanical Properties of Externally and Internally Threaded Fasteners, Washers, and Rivets (Metric)," ASTM F 606M, Annual Book of ASTM Standards, ASTM
4. K.L. Devries and P. Borgmeier, Testing of Adhesive Joints, *Mechanical Testing and Evaluation,* Vol 8, *ASM Handbook,* ASM International, 2000, p 836–844
5. Adhesives, *Annual Book of ASTM Standards,* Vol 15.06, ASTM (updated annually)
6. G.P. Anderson and K.L. DeVries, Analysis of Standard Bond-Strength Tests, *Treatise on Adhesion and Adhesives,* Vol 6, R.L. Patrick, K.L. DeVries, and G.P. Andersen, Ed., Marcel Dekker, 1989
7. J.K. Strozier, K.J. Ninow, K.L. DeVries, and G.P. Anderson, *Adhes. Sci. Rev.,* Vol 1, 1987, p 121
8. G.P. Anderson, D.H. Brinton, K.J. Ninow, and K.L. DeVries, A Fracture Mechanics Approach to Predicting Bond Strength, *Advances in Adhesively Bonded Joints, Proceedings of a Conference at the Winter Annual Meeting of ASME,* 27 Nov–2 Dec 1988 (Chicago), S. Mall, K.M. Liechti, and J.K. Vinson, Eds., ASME, 1988, p 98–101
9. G.P. Anderson and K.L. DeVries, Predicting Strength of Adhesive Joints from Test Results, *Int. J. Fract.,* Vol 39, 1989, p 191–200
10. V.L. Hein and F. Erodogan, Stress Singularities in a Two-Material Wedge, *Int. J. Fract.,* Vol 7, 1971, p 317
11. J.A. Marceau, Y. Moji, and J.C. McMillan, A Wedge Test for Evaluating Adhesive Bonded Surface Durability, *21st SAMPE Symposium,* Vol 21, 6–8 April 1976
12. "Standard Test Method for Young's Modulus, Tangent Modulus, and Chord Modulus," E 111, *Annual Book of ASTM Standards,* Vol 3.01, ASTM, 1999
13. "Standard Methods of Tension Testing of Metallic Materials," E 8, *Annual Book of ASTM Standards,* Vol 3.01, ASTM, 1999
14. "Standard Methods of Tension Testing Wrought and Cast Aluminum, and Magnesium Alloy Products," B 557, *Annual Book of ASTM Standards,* Vol 3.01, ASTM, 1999

CHAPTER 13

Hot Tensile Testing

HIGH-TEMPERATURE MECHANICAL PROPERTIES of metals are determined by three basic methods:

- Short-term tests at elevated temperatures
- Long-term tests of creep deformation at elevated temperatures
- Short-term and long-term tests following long-term exposure to elevated temperatures

This chapter focuses on *short-term* tensile testing at high temperatures. Such tests are commonly referred to as hot tensile, or hot tension, tests. The basic methods and specimens for these tests are similar to room-temperature testing, although the specimen heating, test setup, and material behavior at higher temperatures do introduce some additional complexities and special issues for hot tensile testing.

Emphasis in this chapter has been put on one of the most important reasons for conducting hot tensile tests—the determination of the hot working characteristics of metallic materials. The proper hot-working temperature and deformation rate must be established to produce high-quality wrought products of complicated geometries. It is also important that product yield losses (from either grinding to remove surface cracks or excessive cropping to remove end splits) be held to a minimum, while avoiding the formation of internal cavities (pores). Severe cracking is ordinarily the result of high surface tensile stresses introduced when hot working is conducted either above or below the temperature range of satisfactory ductility. Similarly, cavitation is associated with *internal* tensile stresses, which, for a given material, depend on the temperature, the deformation rate, and workpiece/die geometry.

The first and most important step in specifying appropriate hot-working practice is to determine suitable hot-working conditions. In particular, the tensile ductility (e.g., fracture strain), the flow stress, and cavity formation conditions should be established as a function of temperature and strain rate. A curve of ductility versus temperature or strain rate shows what degree of deformation the material can tolerate without failure. On the other hand, a plot of the flow stress versus temperature, along with workpiece size and strain rate, indicates the force levels required of the hot-working equipment. Last, a curve of cavity volume fraction versus strain, strain rate, and temperature shows what processing parameters should be selected in order to produce high-quality products.

Although commercial metalworking operations cannot be analyzed in terms of a simple stress state, workpiece failures are caused by localized tensile stresses in most instances (Ref 1–4). In rolling of plate, for example, edge cracking is caused by tensile stresses that form at bulged (unrestrained) edges (Ref 1, 3, 4). The geometry of these unrestrained surfaces affects the magnitude of tensile stresses at these locations. Moreover, tensile stresses are also created on the unrestrained surface of a round billet being deformed with open dies. Therefore, to obtain a practical understanding of how well a material will hot work during primary processing, it is essential to know how it will respond to tensile loading at the strain rates to be imposed by the specific hot-working operation.

The ideal hot-workability test is one in which the metal is deformed uniformly, without instability, at constant true strain rate under well-controlled temperature conditions with continuous measurement of stress, strain, and temperature during deformation followed by instantaneous quenching to room temperature. Two types of hot tensile tests are discussed in this chapter: the Gleeble test and the conventional isothermal hot-tensile test. The major advantage of the hot tensile test is that its stress/strain state simulates the conditions that promote cracking in most industrial metalworking operations. However, even though the tensile test is simple in nature, it may provide misleading information if not properly designed. Specifically, parameters such as the specimen geometry, tension machine characteristics, and strain rate and temperature control all influence the results of the tension test. Therefore, the tensile test should be designed and conducted carefully, and testing procedures should be well documented when data are reported.

Equipment and Testing Procedures

The apparatus used to conduct hot tensile tests comprises a mechanical loading system and equipment for sample heating. A variety of equipment types are used for applying forces (loads) to test specimens. These types range from very simple devices to complex systems that are controlled by a digital computer. The most common test configurations utilize universal testing machines, which have the capability to test material in tension, compression, or bending. The word *universal* refers to the variety of stress states that can be applied by the machine, in contrast to other conventional test machines that may be limited to either tensile loading or compressive loading, but not both. Universal test machines or tension-only test frames may apply loads by a gear (screw)-driven mechanism or hydraulic mechanisms, as discussed in more detail in the section "Frame-Furnace Tensile-Testing Equipment" in this chapter.

The heating method used for hot tensile testing varies with the application. The most common heating techniques are direct-resistance heating (in the case of Gleeble systems) and indirect-resistance or induction heating with conventional load frames. In some cases, universal test machines may include a special chamber for testing in either vacuum or controlled atmosphere. Specimen (testpiece) temperatures typically are monitored and controlled by thermocouples, which may be attached on the specimen surface or located very close to the specimen. In some cases, temperature is measured by optical or infrared pyrometers. Accurate measurement and control is very critical for obtaining reliable data. To this end, the use of closed-loop temperature controllers is indispensable.

The occurrence of deformation heating may also be an important consideration, especially at high strain rates, because it can significantly raise the specimen temperature.

Gleeble Testing Equipment

The Gleeble system (Ref 5) has been used since the 1950s to investigate the hot tensile behavior of materials and thus to generate important information for the selection of hot-working parameters. A Gleeble unit is a high-strain-rate, high-temperature testing machine where a solid, buttonhead specimen is held horizontally by water-cooled grips, through which electric power is introduced to resistance heat the test specimen (Fig. 1). Specimen temperature is monitored by a thermocouple welded to the specimen surface at the middle of its length. The thermocouple, with a function generator, controls the heat fed into the specimen according to a programmed cycle. Therefore, a specimen can be tested under

(a) (b)

Fig. 1 Gleeble test unit used for hot-tension and hot-compression testing. (a) Specimen in grips showing attached thermocouple wires and linear variable differential transformer (LVDT) for measuring strain. (b) Close-up of a test specimen. Courtesy of Duffers Scientific, Inc.

time and temperature conditions that simulate hot-working sequences.

Contemporary Gleeble systems (e.g., see www.gleeble.com) are fully integrated servohydraulic setups that are capable of applying as much as 90 kN (10 tons) of force in tension at displacement rates up to 2000 mm/s (80 in./s). Different load cells allow static-load measurement to be tailored to the specific application. Control modes that are available include displacement, force, true stress, true strain, engineering stress, and engineering strain.

The direct-resistance heating system of the Gleeble machine can heat specimens at rates of up to 10,000 °C/s (18,000 °F/s). Grips with high thermal conductivity (e.g., copper) hold the specimen, thus making the system capable of high cooling rates as well. Thermocouples or pyrometers provide signals for accurate feedback control of specimen temperature. Because of the unique high-speed heating method, Gleeble systems typically can run hot-tension tests several times faster than conventional systems based on indirect-resistance (furnace) heating methods.

A digital-control system provides all the signals necessary to control thermal and mechanical test variables simultaneously through the digital closed-loop thermal and mechanical servo systems. The Gleeble machine can be operated totally by computer, by manual control, or by any combination of computer and manual control needed to provide maximum versatility in materials testing.

Sample Design. A typical specimen configuration used in Gleeble testing is shown in Fig. 2. This solid buttonhead specimen, with an overall length of 88.9 mm (3.5 in.), has an unreduced test-specimen diameter of 6.35 mm (0.25 in.). The length of the sample between the grips at the beginning of the test is also an important consideration. Generally, this length is 25.4 mm (1 in.). Shorter lengths produce a narrow hot zone and restrict hot deformation to a smaller, constrained region; consequently, the apparent reduction of area is diminished. On the other hand, a long sample length generally produces higher apparent ductility/elongation values. For example, Smith, et al. (Ref 6) have shown that a grip separation of 36.8 mm (1.45 in.) produces a hot zone about 12.7 mm (0.5 in.) long. When specimen diameter is increased, as is necessary in testing of extremely coarse-grain materials, the grip separation should also be increased proportionately to maintain a constant ratio of hot-zone length to specimen diameter.

Test Procedures. It is essential that hot tensile tests be conducted at accurately controlled temperatures because of the usually strong dependence of tensile ductility on this process variable. To this end, temperature is monitored by a thermocouple percussion welded to the specimen surface. Using a function generator, heat input to the specimen is controlled according to a predetermined programmed cycle chosen by the investigator. However, the temperature measured from this thermocouple junction does not coincide exactly with the specimen temperature because (a) heat is conducted away from the junction by the thermocouple wires, and (b) the junction resides above the specimen surface and radiates heat at a rate higher than that of the specimen itself. Consequently, the thermocouple junction is slightly colder than the test specimen. Furthermore, specimen temperature is highest midway between the grips and decreases toward the grips. In general, the specimen will fracture in the hottest plane perpendicular to the specimen axis. Therefore, it is important to place the thermocouple junction midway between the grips in order that the hottest zone of the specimen, which will be the zone of fracture, is monitored. The longitudinal thermal gradient does not present a serious problem because the specimen deforms in the localized region where the temperature is monitored. Consequently, the measured values of reduction of area and ultimate tensile strength represent the zone where the thermocouple is attached.

Strain rate is another important variable in the hot tensile test. However, strain rate varies during hot tensile testing under constant-crosshead-speed conditions and must be taken into account when interpreting test data. An analysis of the strain-rate variation during the hot-tension test and how it correlates to the strain rates in actual metalworking operations is presented later in this chapter.

The load may be applied at any desired time in the thermal cycle. Temperature, load, and

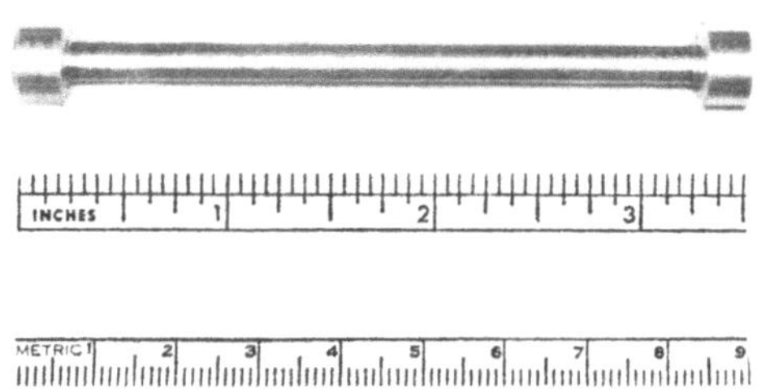

Fig. 2 Typical specimen used for Gleeble testing

crosshead displacement are measured versus time and captured by the data acquisition system. From these measurements, standard mechanical properties such as yield and ultimate tensile strength can be determined. The reduction of area at failure is also readily established from tested samples.

If hot-working practices are to be determined for an alloy for which little or no hot-working information is available, the preliminary test procedure usually comprises the measurement of data "on heating." In such tests, samples are heated directly to the test temperature, held for 1 to 10 min, and then pulled to fracture at a strain rate approximating the rate calculated for the metalworking operation of interest. The reduction of area for each specimen is plotted as a function of test temperature; the resulting "on-heating" curve will indicate the most suitable temperature range to be evaluated to determine the optimal preheat* temperature. This temperature, as indicated from the plot in Fig. 3, lies between the peak-ductility (PDT) and zero-ductility (ZDT) temperatures.

To confirm the appropriate selection of preheat temperature, specimen blanks should be heat treated at the proposed preheat temperature for a time period equal to that of a furnace soak commensurate with the intended workpiece size and hot-working operation. These specimens should be water quenched to eliminate any structural changes that could result from slow cooling. Subsequently, the specimens should be tested by heating to the proposed furnace temperature, holding at this preheat temperature for a moderate period of time (1 to 10 min) to redissolve any phases that may have precipitated, cooling to various temperatures at intervals of 25 or 50 °C (45 or 90 °F) below the preheat temperature, holding for a few seconds at the desired test temperature, and finally pulling in tension to fracture at the calculated strain rate. These "on-cooling" data demonstrate how the material will behave after being preheated at a higher temperature. Testing "on cooling" is necessary because the relatively short hold times during testing "on heating" may not develop a grain size representative of that hold temperature and may be insufficient to dissolve or precipitate a phase that may occur during an actual furnace soak prior to hot working. Also, most industrial hot metalworking operations are conducted as workpiece temperature is decreasing. The "on-cooling" data will indicate how closely the ZDT can be approached before hot ductility is seriously or permanently impaired. In addition, if deformation heating (Ref 7) during "onheating" tests has resulted in a marked underestimation

*In the context of this chapter, preheat temperature is the temperature at which the test specimen or workpiece is held prior to deformation at *lower* temperatures. In actual metalworking operations, preheat temperature usually refers to the actual furnace temperature.

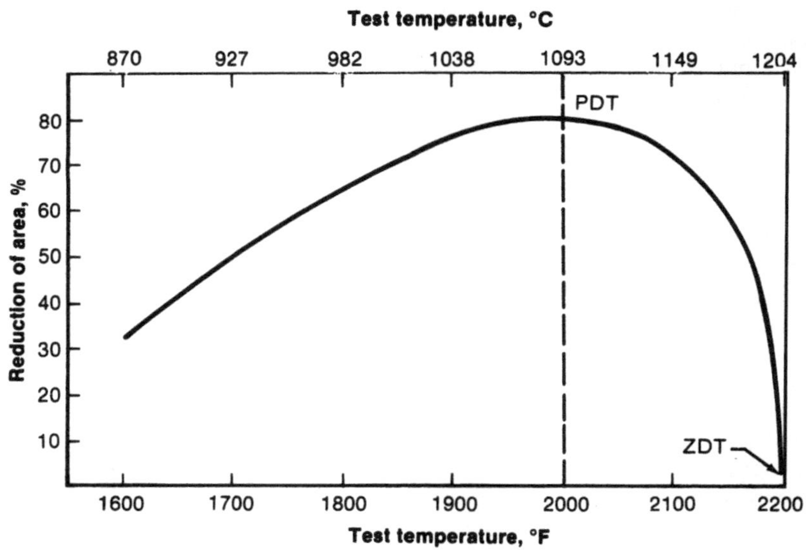

Fig. 3 Hypothetical "on-heating" Gleeble curve of specimen reduction of area as a function of test temperature

of the maximum preheat temperature, this will be revealed and can be rectified by examination of "on-cooling" data.

Frame-Furnace Tensile-Testing Equipment

Universal testing machines and tensile-test frames can be used for hot tensile tests by attaching a heating system to the machine frame. The frame may impart loading by either a screw-driven mechanism or servohydraulic actuator. Screw-driven (or gear-driven) machines are typically electromechanical devices that use a large actuator screw threaded through a moving crosshead. The screws can turn in either direction, and their rotation moves a crosshead that applies a load to the specimen. A simple balance system is used to measure the magnitude of the force applied.

Loads may also be applied using the pressure of oil pumped into a hydraulic piston. In this case, the oil pressure provides a simple means of measuring the force applied. Closed-loop servohydraulic testing machines form the basis for the most advanced test systems in use today. Integrated electronic circuitry has increased the sophistication of these systems. Also, digital computer control and monitoring of such test systems have steadily developed since their introduction around 1965. Servohydraulic test machines offer a wider range of crosshead speeds of force ranges with the ability to provide economically forces of 4450 kN (10^6 lbf) or more. Screw-driven machines are limited in their ability to provide high forces due to problems associated with low machine stiffness and large and expensive loading screws, which become increasingly more difficult to produce as the force rating goes up.

For either a screw-driven or servohydraulic machine, the hot tensile test system is a load frame with a heating system attached. A typical servohydraulic universal testing machine with a high-temperature chamber is shown in Fig. 4. The system is the same as that used at room temperature, except for the high-temperature capabilities, including the furnace, cooling system, grips, and extensometer. In this system, the grips are inside the chamber but partly protected by refractory from heating elements. Heating elements are positioned around a tensile specimen. Thermocouple and extensometer edges touch the specimen. The grip design and the specimen geometry depend on the specific features of the frame and the heating unit as well as the testing conditions. Temperature is measured by thermocouples attached on or located very near to the specimen. In some cases, a pyrometer can also be used.

The most common methods of heating include induction heating and indirect-resistance heating in chamber. Typical examples are shown in Fig. 5. Induction heating (Fig. 5a) usually allows faster heating rates than indirect heating does, but accurate temperature control requires extra care. Induction-heating systems can reach testing temperatures within seconds. Induction heating heats up the outer layer of the specimen first. Furnaces with a lower frequency have better penetration capability. Coupling the heating coil and the specimen also plays an important role in heating efficiency. The interior of the specimen is heated through conduction. With the rapid heating rate, the temperature is often overshot and nonuniform heating often occurs.

Indirect-resistance heating may provide better temperature control/monitoring than induction heating can. Indirect-resistance heating can be combined readily with specially designed chambers for testing either in vacuum or in a controlled atmosphere (e.g., argon, nitrogen, etc.). Vacuum furnaces are expensive and have high maintenance costs. The furnace has to be mounted on the machine permanently, making it inconvenient if another type of heating device is to be used. The heating element is expensive and oxidizes easily. The furnace can only be opened at relatively lower temperatures to avoid oxidation. Quenching has to be performed with an inert gas, such as helium.

Environmental chambers (Fig. 5b), which are less expensive than vacuum chambers, have a circulation system to maintain uniform temperature inside the furnace. Inert gas can flow through the chamber to keep the specimen from oxidizing. Temperature inside the chamber can be kept to close tolerance (e.g., about ±1 °C, or ±2 °F). However, the maximum temperature of an environmental chamber is usually 550 °C (1000 °F), while that of a vacuum furnace can be as high as 2500 °C (4500 °F). The chamber can either be mounted on the machine or rolled in and out on a cart. Split-furnace designs (Fig. 5c) are also cost effective and easy to use. When not in use, it can be swung to the side. The split furnace shown in Fig. 5(c) has only one heating zone. More sophisticated split furnaces have three heating zones for better temperature control. Heating rate is also programmable. When furnace heating is used, it is a common practice to use a low heating rate. In addition, the specimen is typically "soaked" at the test temperature for about 10 to 30 min prior to the application of the load.

The mechanical and thermal control systems are similar to those described in the previous section on the Gleeble testing apparatus. The main advantage of hot tensile test machines is that the test specimen is heated uniformly along its entire gage length, and hence other useful materials properties such as total tensile elongation, plastic anisotropy parameter, cavity formation, and so forth can be determined in addition to yield/ultimate tensile strength and reduction of area. On the other hand, the overall time needed to conduct a single test may be longer than in the Gleeble test method.

Specimen Geometry. In tensile testing, elongation values are influenced by gage length. It is thus necessary to state the gage length over which elongation values are measured. When the ductilities of different materials (or of a single material tested under different conditions) are compared in terms of total elongation, the specimen gage length also should be adjusted in proportion to the cross-sectional area. This is of great importance in the case of small elongations, because the neck strain contributes a significant portion to the total strain. On the other hand, the neck strain represents only a small portion of the elongation in the case of superplastic deformation. Unwin's equation (Ref 8) also shows the rationale for a fixed ratio of gage length with cross-sectional area, expressed as a

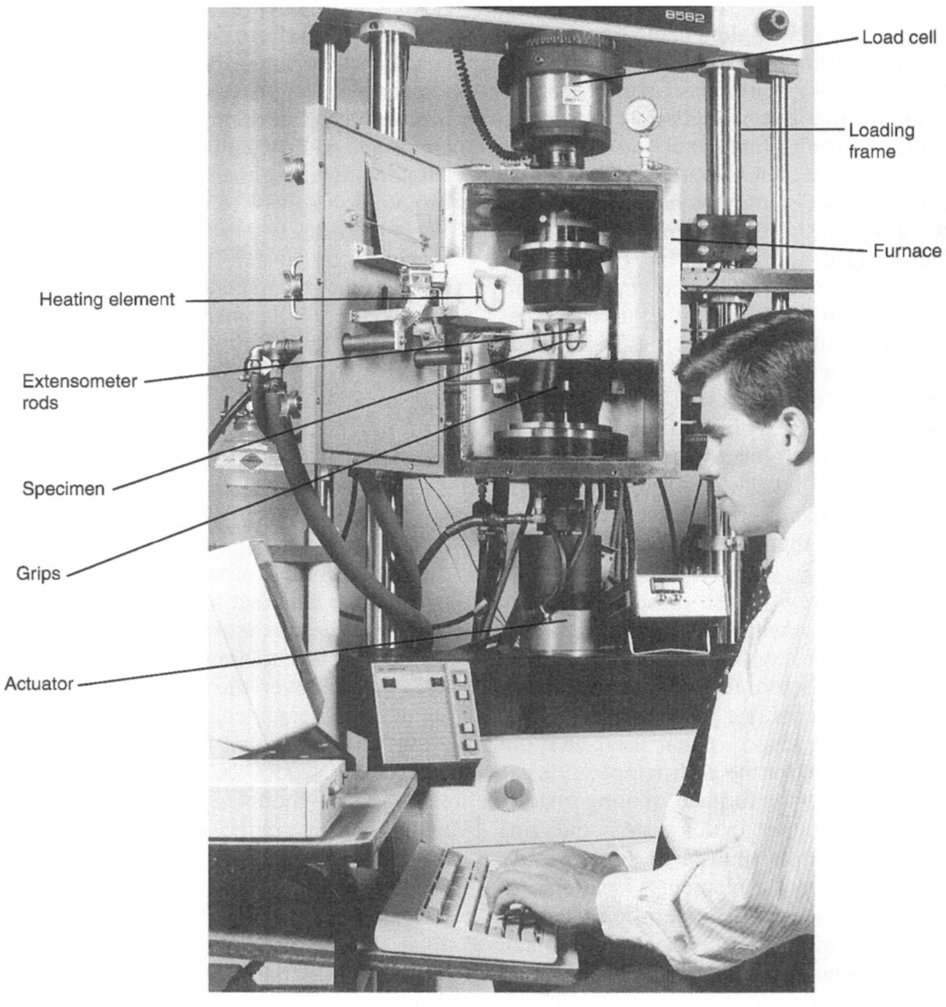

Fig. 4 Typical servohydraulic universal testing machine with a chamber and instrumentation for high-temperature testing

fixed ratio of gage length to diameter (for round bars) or gage length to the square root of the cross-sectional area (sheet specimens). This reinforces the importance of stating the gage length used in measuring elongation values. Usually, the length-to-diameter ratio is between 4 and 6.

In addition to the gage length, the specimen shoulder geometry and hence the gripping system are also important design considerations. It is desirable that specimen deformation takes place only within its gage length; the shoulder should remain undeformed. This is not always the case, as can be seen from Fig. 6. The macrographs of Fig. 6(a) (Ref 9) show the initial and deformed condition of a specimen in which measurable deformation has occurred within the grip area. On the other hand, the specimen with a different shoulder geometry (Fig. 6b) deformed essentially only along its gage length (Ref 10). The shoulder-deformation problem is not insurmountable. In this regard, analyses and techniques, such as those developed by Friedman and Ghosh (Ref 9), should be applied in order to eliminate the effect of shoulder deformation from measured hot-tensile data.

Hot Ductility and Strength Data from the Gleeble Test

The reduction of area (RA) and strength are the key parameters measured in hot tensile tests conducted with a Gleeble machine (Ref 1, 11, 12). Because RA is a very structure-sensitive property, it can be used to detect small ductility variations in materials of low to moderate ductility, such as specialty steels and superalloys. However, it should be recognized that RA will not effectively reveal small variations in materials of extremely high ductility (Ref 2). Yield

Fig. 5 Typical examples of heating methods for load-frame tensile testing. (a) Induction heating. (b) Environmental chamber. (c) Split-furnace setup

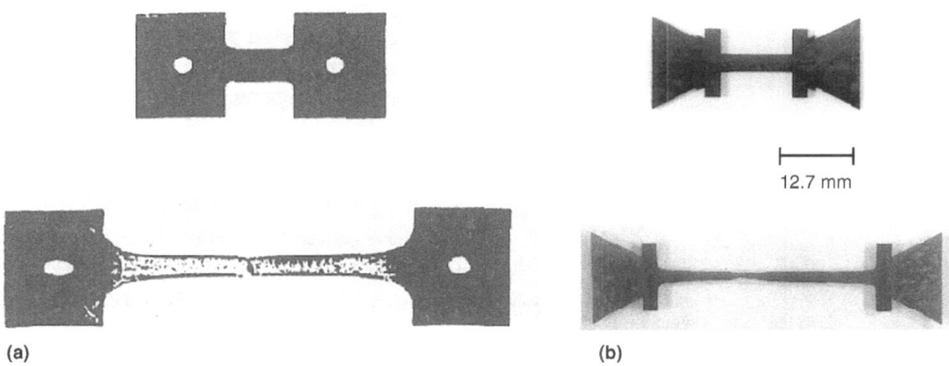

Fig. 6 Initial specimen geometry and deformed specimen for cases in which (a) shoulder deformation occurred (Ref 9) or (b) the shoulder remained undeformed (Ref 10)

and tensile strength can be used to select required load capacity of production processing equipment.

Ductility Ratings

Experience has indicated that the qualitative ratings given in Table 1 for hot ductility as a function of Gleeble reduction-of-area data can be used to predict hot workability, select hot-working temperature ranges, and establish hot-reduction parameters. "Normal reductions"* may be taken on superalloys when the reduction of area exceeds 50%, but lighter reductions are necessary when ductility falls below this level. Thus, in this rating system, the *minimum* hot-working temperature is designated by the temperature at which the reduction of area falls below approximately 30 to 40%. The *maximum* hot-working temperature is determined from "on-cooling" data. The objective is to determine which preheat temperature provides the highest ductility over the broadest temperature range without risking permanent structural damage by overheating.

An alloy with hot-tensile ductility rated as marginal or poor may be hot worked, but smaller reductions and fewer passes per heating are required, perhaps in combination with insulating coatings and/or coverings. In extreme instances, it may be necessary to minimize development of tensile strains by employing special dies for deforming under a strain state that more nearly approaches hydrostatic compression (e.g., extrusion).

*"Normal reductions" as used in this chapter depend on both the alloy system being hot worked and the equipment being used. For example, normal reductions for low-carbon steels would be much greater than those for superalloy systems.

It should be emphasized that the hot tensile test reflects the inherent hot ductility of a material, that is, its natural ability to deform under deformation conditions. If a workpiece possesses defects or flaws, it may crack due to localized stress concentration in spite of good inherent hot ductility.

Figure 3 illustrates how hot tensile data are used to select a hot-working temperature. The safe, maximum hot-working temperature lies between the PDT and the ZDT. In this hypothetical curve of "on-heating" data, the PDT is 1095 °C (2000 °F) and the ZDT is 1200 °C (2200 °F). "On-cooling" data should be determined using preheat temperatures between the PDT and the ZDT. For example, 1095, 1150, 1175, and 1200 °C (2000, 2100, 2150, and 2200 °F) would be good preheat temperatures for "on-cooling" studies. Typical "on-cooling" results are depicted in Fig. 7. A 1200 °C (2200 °F) preheat temperature results in marginal or poor hot workability over the possible working range, whereas an 1175 °C (2150 °F) preheat temperature results in acceptable hot workability over a relatively narrow temperature range. Both 1150 and 1095 °C (2100 and 2000 °F) preheats result in good hot ductility over a relatively narrow temperature range. The 1150 °C preheat temperature is preferred over the 1095 °C preheat temperature because it provides good hot ductility over a broader temperature range.

Hot workability usually is enhanced by greater amounts of prior hot deformation. This occurs because second phases and segregation-prone elements are distributed more uniformly and the grain structure is refined. Deformation at high and intermediate temperatures during commercial hot-working operations often refines the grain structure by dynamic (or static)

Table 1 Qualitative hot-workability ratings for specialty steels and superalloys

Hot-tension reduction of area(a), %	Expected alloy behavior under normal hot reductions in open die	Remarks regarding alloy hot-working practice
<30	Poor hot workability. Abundant cracks	Preferably not rolled or open-die forged. Extrusion may be feasible. Rolling or forging should be attempted only with light reductions, low strain rates, and an insulating coating.
30–40	Marginal hot workability. Numerous cracks	This ductility range usually signals the minimum hot-working temperature. Rolled or press forged with light reductions and lower-than-usual strain rates.
40–50	Acceptable hot workability. Few cracks	Rolled or press forged with moderate reductions and strain rates
50–60	Good hot workability. Very few cracks	Rolled or press forged with normal reductions and strain rates
60–70	Excellent hot workability. Occasional cracks	Rolled or press forged with heavier reductions and strain rates
>70	Superior hot workability. Rare cracks. Ductile ruptures can occur if strength is too low.	Rolled or press forged with heavier reductions and higher strain rates than normal provided that alloy strength is sufficiently high to prevent ductile ruptures.

(a) Ratings apply for Gleeble tensile testing of 6.25 mm (0.250 in.) diam specimens with 25.4 mm (1 in.) head separation.

recrystallization, thereby augmenting subsequent hot ductility at lower temperatures. Because a specimen tested "on cooling" to the low-temperature end of the hot-working range has not been deformed at a temperature where grains dynamically recrystallize, the grain structure is unrefined. Thus, ductility values will tend to be somewhat lower than those experienced in an actual metalworking operation in which deformation at higher temperatures has refined the structure. The fact that the low-temperature end of the "on-cooling" ductility range is lower than the values that would result in plant metalworking operations is not sufficient to alter the practical translation of the results. This feature serves as a safety factor for establishing the minimum hot-working temperature.

Although some alloys will recover hot ductility when cooled from temperatures in the vicinity of the ZDT, it is nonetheless wise to avoid hot-working preheat temperatures approaching the ZDT in plant practice because interior regions of the workpiece may not cool sufficiently to allow recovery of ductility, thereby causing center bursting. Because industrial furnaces do not control closer than approximately ±14 °C (±25 °F), the recommended furnace temperature ordinarily should be at least 14 °C (25 °F) lower than the maximum temperature indicated by testing "on cooling."

Strength Data

In the hot-working temperature range, strength generally decreases with increasing temperature. However, the strength data plotted in Fig. 8 demonstrate that deformation resistance does not vary with preheat temperature to the same degree as does ductility. Furthermore, strength measured "on heating" is usually greater than that measured "on cooling."

To calculate the force required to deform a metal in an industrial hot-working operation, accurate measurement of *flow stress* is desirable. Ultimate tensile stress measured in the hot tensile test is only slightly greater than flow stress at the high-temperature end of the hot-working range because work hardening is negligible. However, the difference between ultimate tensile stress and flow stress increases as temperature decreases because restoration processes cease. Furthermore, the Gleeble tensile test does not accurately determine flow stress because the strain rate is not constant. Nonetheless, the test still provides useful, comparative information concerning how the strength of an alloy varies as a function of temperature within a given strain-rate range. For example, by analyzing strength values for common alloys in relation to the load-bearing capacity of a given mill, it may be possible to use test data for a new or unfa-

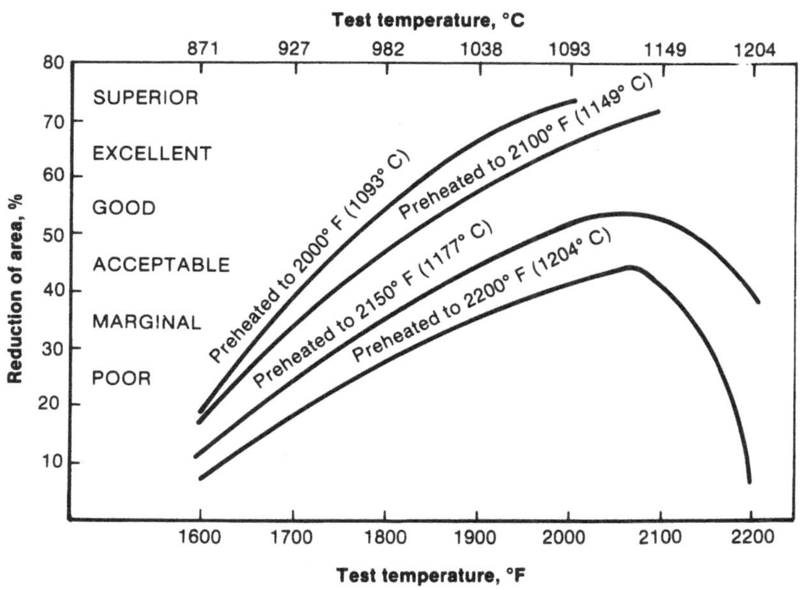

Fig. 7 Typical "on-cooling" Gleeble curves of specimen reduction of area as a function of test and preheat temperatures with typical hot-workability ratings indicated

miliar alloy in judging whether the equipment is capable of forming the new or unfamiliar alloy.

Hot Tensile Data for Commercial Alloys

For illustrative and comparative purposes, Gleeble hot ductility and strength curves for some commercial alloys are presented in Fig. 9. The nominal compositions of these materials are given in the table accompanying Fig. 9.

The hot tensile strengths for the cobalt- and nickel-base superalloys over the hot-working temperature range are substantially higher than those for the high-speed tool steel and the high-strength alloy, which are iron-base materials. Furthermore, the ductility data reveal that René 41 has the narrowest hot-working temperature range (ΔT) of 140 °C (250 °F) of the three superalloys. Hot working of this alloy below 1010 °C (1850 °F) will lead to severe cracking. This characteristic, coupled with its high deformation resistance, makes this alloy relatively difficult to hot work. On the other hand, HS 188 has high deformation resistance, but it has high ductility over a broad hot-working temperature range from 1190 °C (2175 °F) to below 900 °C (1650 °F). Therefore, the permissible reduction per draft may be relatively small if the hot-working equipment is not capable of high loads, but HS 188 can be hot worked over a broader temperature range than René 41. However, if the equipment has high load capacity, then heavier reductions can be taken on HS 188 than on René 41. From the hot-working curves established for the iron-base, high-strength alloy AF 1410, the low deformation resistance coupled with high ductility over a broad temperature range indicate that this material has extremely good hot workability. Mill experience has verified this. The curves shown for M42, the high-speed tool steel, reveal that it is intermediate in hot workability between the superalloys and AF 1410; this conclusion has also been verified by mill experience.

Figures 10 and 11 illustrate the variation of hot tensile ductility values at various temperatures. These results were correlated to the fracture surfaces and structures of the test specimens. For the high-strength iron-base alloy AF 1410, the "on-heating" curve in Fig. 10 shows that the ZDT is never reached at practical upper-limit hot-working temperatures. At the highest temperature tested (1230 °C, or 2250 °F) and at the PDT (1120 °C, or 2050 °F), where hot-tensile ductility is extremely high, the fracture appearance is ductile and dynamic recrystallization occurs, leading to an equiaxed grain structure. At the higher temperature, a coarser grain structure results from grain growth, which accounts for the drop in ductility. At the opposite end of the hot-working temperature range (842 °C, or 1548 °F), the elongated grain structure reveals that dynamic recrystallization does not occur, and the fracture surfaces indicate a less ductile fracture mode. The correlation among fracture appearance, microstructure, and hot tensile ductility was even more evident for a developmental solid-solution-strengthened cobalt-base superalloy (Fig. 11). At the PDT (1150 °C, or 2100 °F), dynamic recrystallization occurs, fracture

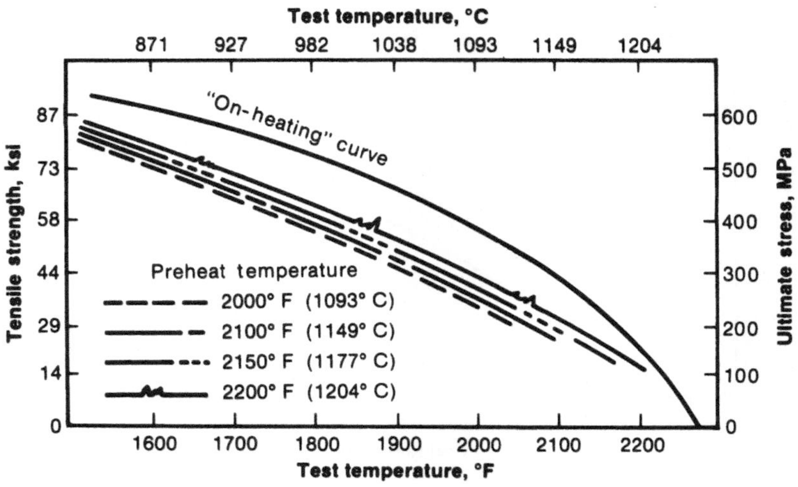

Fig. 8 Typical "on-heating" deformation-resistance data obtained in Gleeble testing

was primarily transgranular, and the fracture appearance was ductile. At the ZDT (1200 °C, or 2200 °F), both static recrystallization and grain growth were obvious, but incipient melting was not evident in the microstructure. Microstructural evidence of incipient melting at the ZDT is observed for some alloys, but not for others (Ref 13).

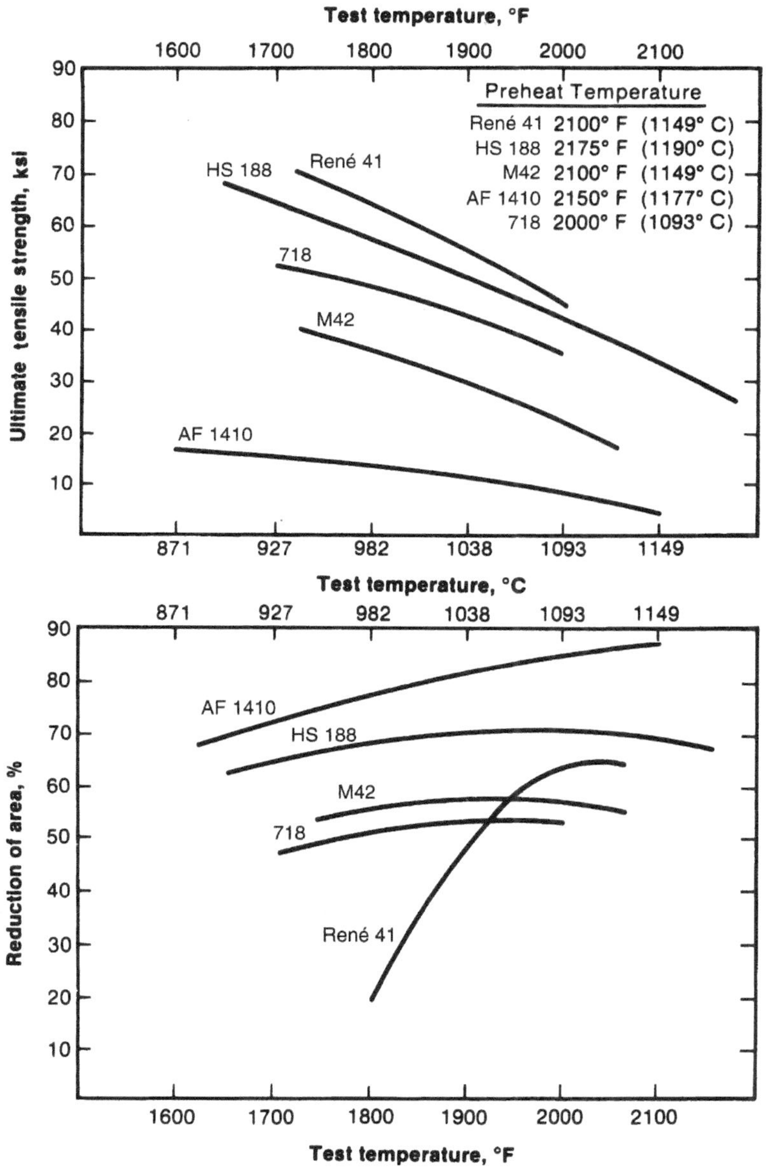

Material	Alloy type	Nominal composition, wt%
René 41	Nickel-base superalloy	0.09 C, 19 Cr, 10 Mo, 11 Co, 3 Ti, 1.5 Al, 1.35 Fe, bal Ni
Alloy 718	Nickel-base superalloy	0.05 C, 18 Cr, 3 Mo, 5 Nb, 1 Ti, 0.5 Al, 19 Fe, bal Ni
HS 188	Cobalt-base superalloy	0.08 C, 22 Cr, 22 Ni, 14 W, 1.5 Fe, 0.05 La, bal Co
M42	High-speed tool steel	1.10 C, 3.8 Cr, 9.5 Mo, 8 Co, 1.5 W, 1.2 V, bal Fe
AF 1410	High-strength steel	0.16 C, 2 Cr, 1 Mo, 14 Co, 10 Ni, bal Fe

Fig. 9 Typical "on-cooling" Gleeble curves of strength and ductility as functions of test temperature for several commercial alloys.

An example of the sensitivity of the hot tensile Gleeble test is shown in Fig. 12 for iron/nickel-base superalloy Alloy 901 (Ref 14). A small amount of lanthanum added to one heat (top curve) was sufficient to reduce the analyzed sulfur content to the 1 to 5 ppm range. This resulted in a small improvement in the hot tensile ductility according to Gleeble hot-tensile data.

Isothermal Hot Tensile Test Data

From the *isothermal* hot tensile test, information can be obtained about a number of material parameters that are important with regard to metalworking process design. These include plastic-flow (stress-strain) behavior, plastic anisotropy, tensile ductility, and their variation with the test temperature and the strain rate.

Stress-Strain Curves

Engineering stress-strain curves from isothermal hot tensile tests are constructed from load-elongation measurements. The engineering, or nominal, stress is equal to the average axial stress and is obtained by dividing the instantaneous load by the original cross-sectional area

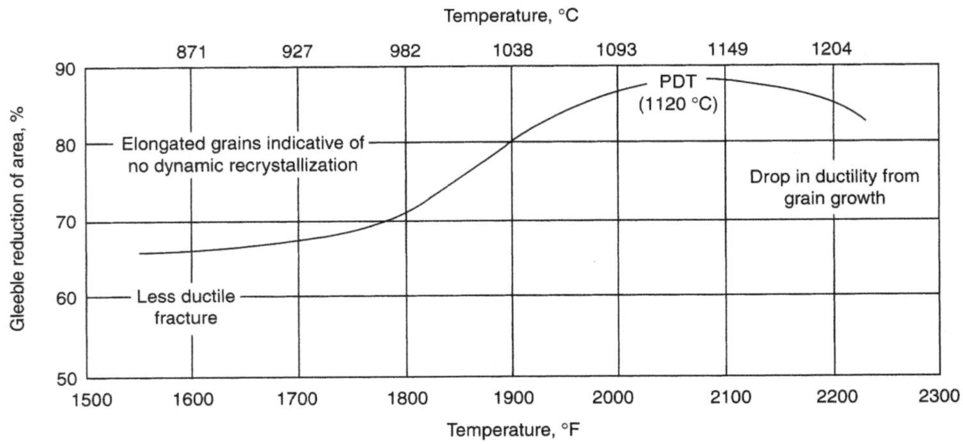

Fig. 10 Typical Gleeble curve of reduction of area versus test temperature for an aircraft structural steel (AF 1410). At the PDT, dynamic recrystallization occurs leading to an equiaxed grain structure. Fracture appearance is ductile.

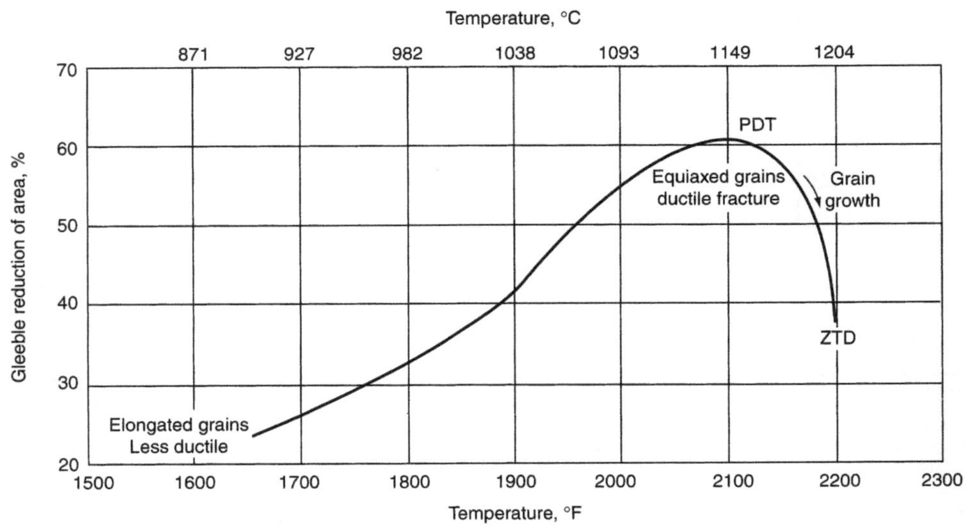

Fig. 11 Typical Gleeble curve of reduction area versus test temperature for a cobalt-base superalloy.

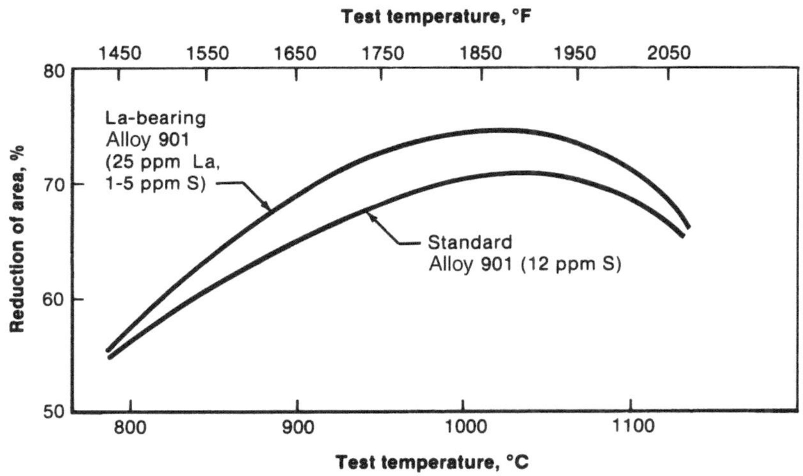

Fig. 12 Gleeble ductility curves for lanthanum-bearing and standard Alloy 901 tested on cooling from 1120 °C (2050 °F). Note that the lanthanum-bearing heat displays slightly higher ductility. Specimens represent transverse orientation on a nominal 25 cm square billet. Specimen blanks were heat treated at 1095 °C (2005 °F) for 2 h and then water quenched prior to machining. Specimens were heated to 1120 °C (2050 °F), held for 5 min, cooled to test temperature and held for 10 s before being tested at a nominal strain rate of 20 s^{-1} (crosshead speed 5 cm/s; jaw spacing, 2.5 cm). Source: Ref 14

of the specimen. Similarly, the engineering, or nominal, strain represents the average axial strain and is obtained by dividing the elongation of the gage length of the specimen by its original length. Hence, the form of the engineering stress-strain curve is exactly the same as that of the load-elongation curve. Examples of engineering stress-strain curves obtained from hot-tension testing of an orthorhombic titanium aluminide alloy (Ref 15) at 980 °C (1800 °F) and a range of nominal (initial) strain rates are shown in Fig. 13. The curves exhibit a stress maximum at strains less than 10%, a regime of quasi-stable flow during which a diffuse neck develops and the load drops gradually, and, lastly, a period of rapid load drop during which the flow is highly localized (usually in the center of the specimen gage length) and failure occurs. The engineering stress-strain curve does not give a true indication of the deformation characteristics of a metal because it is based entirely on the original dimensions of the specimen. These dimensions change continuously during the test. Such changes are very significant when testing is performed at elevated temperatures.

The true stress and true strain are based on actual (instantaneous) cross-sectional area and length measurements at any instant of deformation. The true-stress/true-strain curve is also known as the *flow curve* since it represents the basic plastic-flow characteristic of the material under the particular (temperature-strain rate) testing conditions. Any point on the flow curve can be considered the yield stress for a metal strained in tension by the amount shown on the curve. An example of the variation of the true stress versus true strain for Al-8090 alloy deformed under superplastic conditions (T = 520 °C, $\dot{\varepsilon} = 7.8 \times 10^{-4}\ s^{-1}$) is shown in Fig. 14 (Ref 16). Under these conditions, it is apparent that the flow stress is almost independent of strain. For ideal superplastic materials, the flow stress is independent of strain. A nearly constant, or steady-state, flow stress is also frequently observed at hot-working temperatures in materials that undergo dynamic recovery. In these cases, steady-state flow is achieved at strains of the order of 0.2, at which the rate of strain hardening

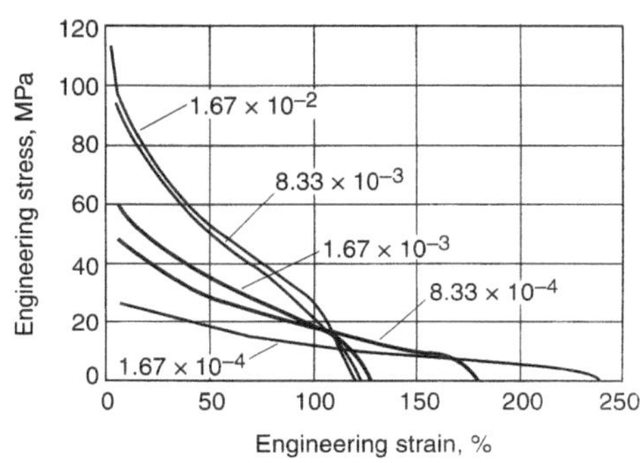

Fig. 13 Engineering stress-strain curves for an orthorhombic titanium alloy (Ti-21Al-22Nb) tested at 980 °C (1795 °F) and a range of initial strain rates (s^{-1}). Source: Ref 15

due to dislocation multiplication is exactly balanced by the rate of dislocation annihilation by dynamic recovery.

The variation of true stress with true strain can also give insight into microstructural changes that occur during hot deformation. For example, for superplastic materials, an increase in the flow stress with strain is normally indicative of strain-enhanced grain growth. A decrease in flow stress, particularly at high strains, can often imply the development of cavitation damage (see the section "Cavitation During Hot Tensile Testing" in this chapter) or the occurrence of dynamic recrystallization. As an example, true-stress/true-strain curves for a γ-TiAl submicrocrystalline alloy deformed at temperatures between 600 and 900 °C (1110 and 1650 °F) and a nominal (initial) strain rate of $8.3 \times 10^{-4}\ s^{-1}$ are shown in Fig. 15 (Ref 17). These curves reveal that deformation at low temperatures, at which nonsuperplastic conditions prevail, is characterized by an increase of flow stress with strain due to the strain hardening. At higher temperatures, the effect of strain on the flow stress decreases until it becomes negligible at the highest test temperature, thus indicating the occurrence of superplastic flow.

Material Coefficients from Isothermal Hot Tensile Tests

A number of material coefficients can be obtained from isothermal hot tensile tests. These include measures of strain and strain-rate hardening and plastic anisotropy. The strain-hardening exponent (usually denoted by the symbol n) describes the change of flow stress (with an effective stress, $\bar{\sigma}$) with respect to the effective strain, $\bar{\varepsilon}$, such that:

$$n = \frac{\partial \ln \bar{\sigma}}{\partial \ln \bar{\varepsilon}} \quad \text{(Eq 1a)}$$

For a uniaxial tensile test, and prior to the development of a neck, the distinction of effective stress and strain is not necessary because they are equal to the axial stress σ and strain ε, so that the expression is simply:

$$n = \frac{\partial \ln \sigma}{\partial \ln \varepsilon} \quad \text{(Eq 1b)}$$

The strain-hardening exponent may have values from $n = 0$ for a perfectly plastic solid to $n = 1$ for an elastic solid; negative values of n may also be found for materials that undergo flow softening due to changes in microstructure or crystallographic texture during deformation. According to Eq 1, if the constitutive equation for stress-strain behavior is of the form $\sigma = K\varepsilon^n$, then a logarithmic plot of true stress versus true strain results in a straight line with a slope equal

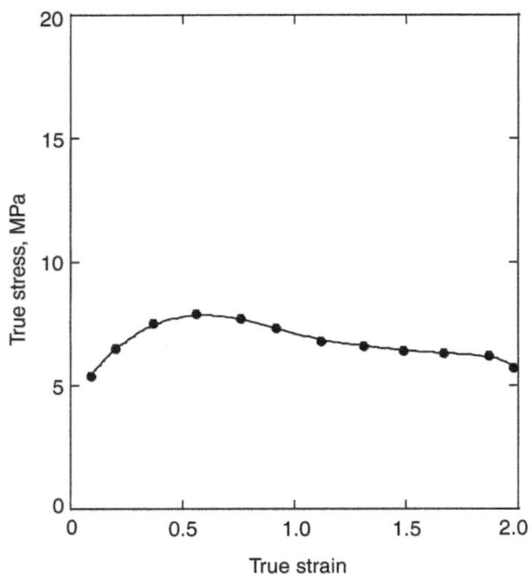

Fig. 14 True-stress/true-strain data for an Al-8090 alloy deformed in tension at 520 °C (970 °F) and a true strain rate of $7.8 \times 10^{-4}\ s^{-1}$. Source: Ref 16

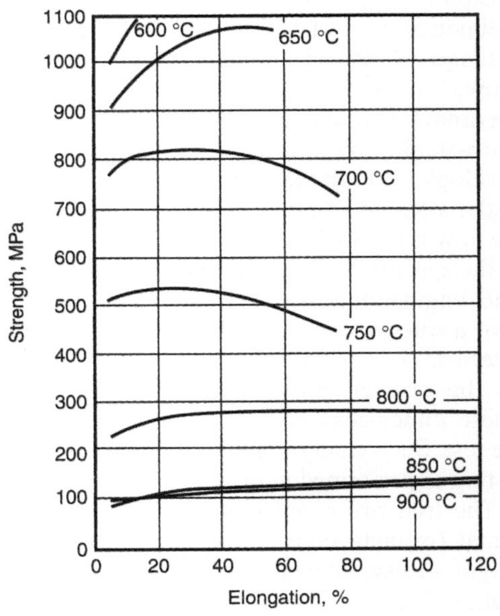

Fig. 15 True-stress/true-strain curves obtained from tensile testing of submicrocrystalline TiAl samples. Source: Ref 17

to n. However, this is not always found to be the case and reflects the fact that this relationship is only an empirical approximation. Thus, when the plot of ln(σ) versus ln(ε) [or the plot of log(σ) versus log(ε)] results in a nonlinear value of n, then the strain-hardening exponent is often defined at a particular strain value. In general, n increases with decreasing strength level and decreasing ease of dislocation cross slip in a polycrystalline material.

The strain-rate sensitivity exponent (usually denoted by the symbol m) describes the variation of the flow stress with the strain rate. In terms of effective stress ($\bar{\sigma}$) and effective strain rate ($\dot{\bar{\varepsilon}}$), it is determined from the following relationship:

$$m = \frac{\partial \ln \bar{\sigma}}{\partial \ln \dot{\bar{\varepsilon}}} \quad \text{(Eq 2a)}$$

which is simplified for the condition of pure uniaxial tension as:

$$m = \frac{\partial \ln \sigma}{\partial \ln \dot{\varepsilon}} \quad \text{(Eq 2b)}$$

Deformation tends to be stabilized in a material with a high m value. In particular, the presence of a neck in a material subject to tensile straining leads to a locally higher strain rate and thus to an increase in the flow stress in the necked region due to *strain-rate* hardening. Such strain-rate hardening inhibits further development of the strain concentration in the neck. Thus, a high strain-rate sensitivity imparts a high resistance to necking and leads to high tensile elongation or superplasticity. Materials with values of m equal to or greater than approximately 0.3 exhibit superplasticity, assuming cavitation and fracture do not intercede. An empirical relation between tensile elongation and the m value is revealed in the data collected by Woodford (Ref 18) shown in Fig. 16. In addition, a number of theoretical analyses have been conducted to relate m and tensile failure strain, ε_f (Ref 19–21). For example, Ghosh (Ref 19) derived:

$$\varepsilon_f = -m \ln(1 - f^{1/m}) \quad \text{(Eq 3)}$$

in which f denotes the size of the initial geometric (area) defect at which flow localization occurs.

The plastic anisotropy parameter r (or R) characterizes the resistance to thinning of a *sheet* material during tensile testing and is defined as

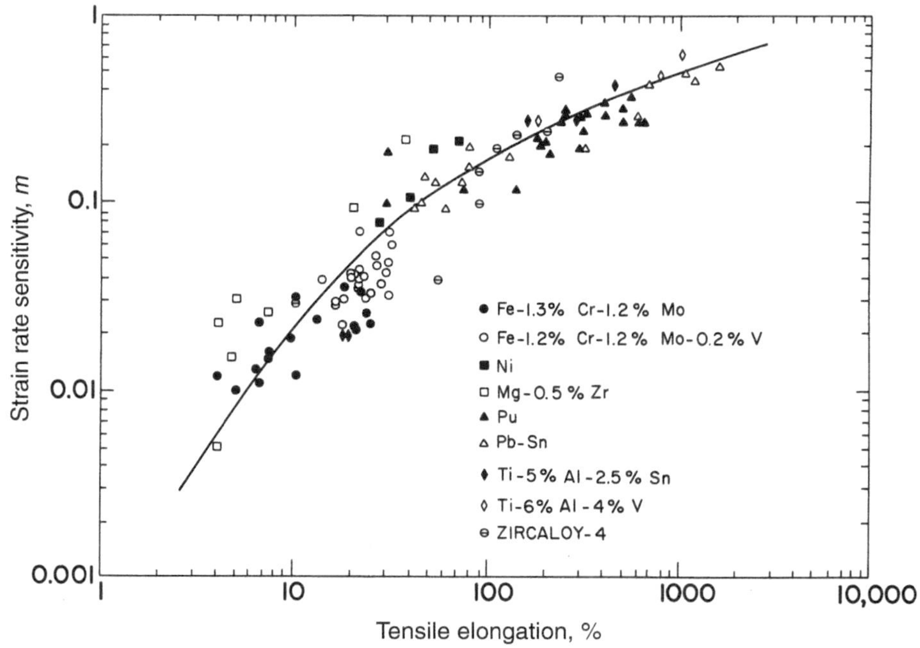

Fig. 16 Tensile elongation as a function of the strain-rate sensitivity. Source: Ref 18

the slope of a plot of width strain, ε_w, versus thickness strain, ε_t, (Ref 22), that is:

$$r = \frac{d\varepsilon_w}{d\varepsilon_t} \qquad \text{(Eq 4)}$$

A material that possesses a high r value has a high resistance to thinning and hence good formability, especially during deep-drawing operations. Materials with values of r greater than unity have higher strength in the thickness direction than in the plane of the sheet.

The plastic anisotropy parameter can be readily measured using specimens deformed in uniaxial tension. However, caution should be exercised when making such measurements to ensure that the stress state along the gage length is uniaxial. Therefore, measurements in regions near the sample shoulder and the failure site (where a stress state of hydrostatic tension may develop during necking) should be avoided. Figure 17 shows an example of such data from a Ti-21Al-22Nb sample pulled to failure in uniaxial tension at a nominal strain rate of 1.67×10^{-4} s^{-1} and at a temperature of 980 °C (1800 °F). Within experimental scatter, the r value is constant for the majority of deformation. Apparently, lower values of r at low strains (near the specimen shoulder) or very high strains (at the fracture tip) are invalid due to constraint or flow-localization effects, respectively, and hence conditions that are not uniaxial.

The r value of a sheet material may be sensitive to the testing conditions and in particular to the strain rate and temperature. This is a result of variation of the mechanism that controls deformation (e.g., slip, grain-boundary sliding, etc.) with test conditions. For the orthorhombic titanium aluminide material discussed previously (Ti-21Al-22Nb), the normal plastic anisotropy parameter shows a very weak dependence on strain, but a noticeable variation with strain rate (Fig. 18). This trend can be attributed to the presence of mechanical and crystallographic texture and the effect of strain rate on the operative deformation mechanism.

Effect of Test Conditions on Flow Behavior

When considering the effect of test conditions on flow behavior, it must be understood that testing for the modeling of deformation processes is very different from testing for static mechanical properties at very low (quasi-static) loading rates. Testing conditions for deformation processes must cover a range of strain rates and may require high strain rates of 1000 s^{-1} or more. For tensile testing, conventional test frames are applicable for strain of rates less than 0.1 s^{-1}, while special servohydraulic frames have a range from 0.1 to 100 s^{-1} (see Chapter 15, "High Strain Rate Tensile Testing"). For strain rates from 100 to 1000 s^{-1}, the Hopkinson (Kolsky)-bar method is used. This chapter and the following discussions only consider isothermal conditions and strain rates below 0.1 s^{-1}, where inertial effects can be neglected.

Effect of Strain Rate and Temperature on Flow Stress. At hot-working temperatures, most metals exhibit a noticeable dependence of flow stress on strain rate and temperature. For instance, the variation of flow stress with strain rate for Ti-6Al-4V (with a fine equiaxed microstructure) deformed at 927 °C is shown in Fig. 19 (Ref 23). For the strain-rate range shown in Fig. 19, a sigmoidal variation of the flow stress with strain rate is observed. From these data, the

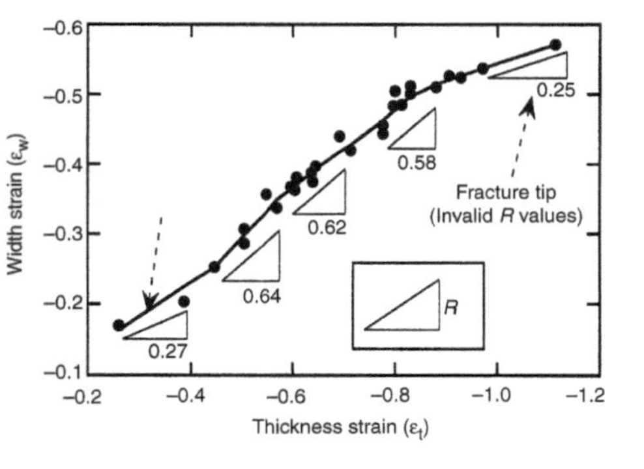

Fig. 17 Width versus thickness strain (ε_w versus ε_t) for an orthorhombic titanium aluminide specimen deformed at 980 °C (1795 °F) and a nominal strain rate of 1.67×10^{-4} s^{-1}. Source: Ref 10

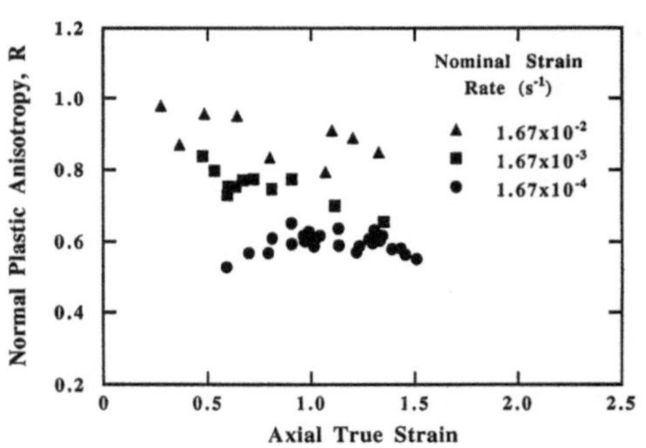

Fig. 18 Anisotropy parameter r versus the local axial true strain for various nominal strain rates. Data correspond to a Ti-21Al-22Nb alloy. Source: Ref 10

strain-rate sensitivity (m value) can be readily calculated. The result of these calculations (Fig. 20) shows that m is low at low strain rates and then increases and passes through a maximum after which it decreases again. This behavior is typical of many metals with fine-grain microstructures and reveals that superplasticity is not manifested in either the low-stress, low-strain-rate region I or the high-stress, high-strain-rate region III (refer to Fig. 20). Rather, superplasticity is found only in region II in which the stress increases rapidly with increasing strain rate. The superplastic region II is displaced to higher strain rates as temperature is increased and/or grain size is decreased. Moreover, the maximum observed values of m increase with similar changes in these parameters.

The stress-strain curve and the flow and fracture properties derived from the hot-tension test are also strongly dependent on the temperature at which the test is conducted. In both single-crystal and polycrystalline materials, the strength decreases with temperature because the critical resolved shear stress decreases sharply with an increase in temperature. On the other hand, the tensile ductility *increases* with temperature because of the increasing ease of recovery and recrystallization during deformation. However, the increase in temperature may also cause microstructural changes such as precipitation, strain aging, or grain growth that may affect this general behavior.

The flow stress dependence on temperature and strain rate is generally given by a functional form that incorporates the Zener-Hollomon parameter, $Z = \dot{\varepsilon} \exp(Q/RT)$ (Ref 24) in which Q is the apparent activation energy for plastic flow, R the universal gas constant, and T is the absolute temperature.

Effect of Crosshead Speed Control on Hot Tensile Data. The selection of constant-strain-rate versus constant-crosshead-speed control in conducting isothermal, hot tensile tests is an important consideration, especially for materials that are superplastic. When experiments are conducted under constant-crosshead-speed conditions, the specimen experiences a decreasing strain rate during the test, thus making the interpretation of results difficult, especially in the superplastic regime. A method to correct for the strain-rate variation involves continuously changing the crosshead speed during the tension test to achieve nearly constant strain rate. This approach assumes uniform deformation along the gage length and no end effects and leads to the following relation between crosshead speed $\dot{\delta}$, desired strain rate $\dot{\varepsilon}$, the initial gage length l_o, and time t:

$$\dot{\delta} = \dot{\varepsilon}\, l_o \exp(-\dot{\varepsilon} t) \qquad \text{(Eq 5)}$$

The crosshead-speed schedule embodied in Eq 5 has been used successfully for a test of Ti-6Al-4V (Ref 25). Verma et al. (Ref 26) have also shown the efficacy of this approach by conducting tensile tests at constant crosshead speed as well as constant strain rate on superplastic 5083-Al specimens. Figure 21 compares stress-strain characteristics determined under constant-cross-

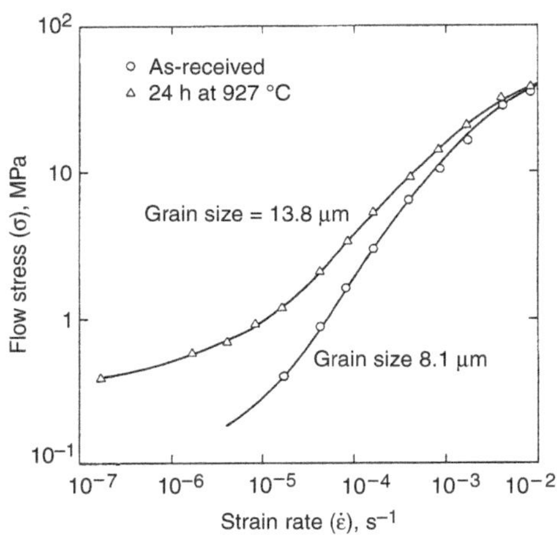

Fig. 19 Flow stress as a function of strain rate and grain size for a Ti-6Al-4V alloy deformed at 927 °C (1700 °F). The strain level was about 0.24. Source: Ref 23

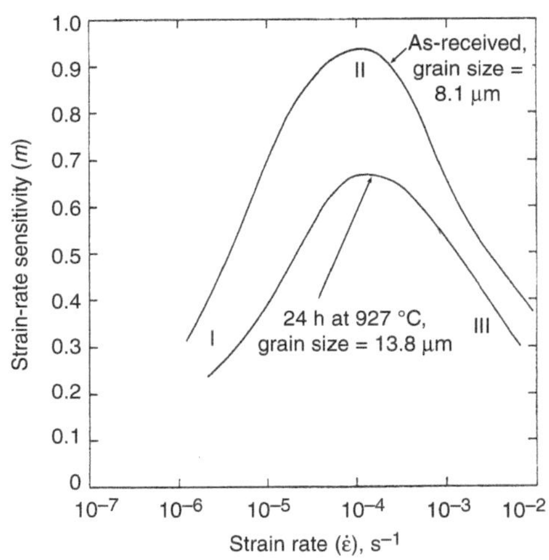

Fig. 20 Strain-rate sensitivity (m) versus strain rate ($\dot{\varepsilon}$) for the data corresponding to Fig. 19. Source: Ref 23

head-speed conditions with those from constant-strain-rate tests for two different initial strain rates. Constant-crosshead-speed tests showed consistently lower strain hardening (lower flow stresses) and larger strain to failure (higher tensile elongations) than the corresponding constant-strain-rate tests did. The above finding highlights the importance of the test control mode; in addition, this mode should be clearly stated when elongation and/or flow stress data are reported.

Effect of Gage Length on Strain Distribution. Under superplastic deformation conditions, specimen geometry (especially shoulder design) plays an important role in the determination of hot tensile characteristics. In the section "Frame-Furnace Tensile-Testing Equipment," two different specimen designs are discussed (Fig. 6). For one of these designs, deformation was limited essentially to the gage section, while the other had experienced deformation in the shoulder section. For the specimen geometry in Fig. 6(a), tensile tests indicated that significant straining can occur in the grip regions and that large strain gradients exist within the gage section of the specimen. The strain gradient (variation) along the gage length and the deformation of the grip section depend on the gage length and tensile strain rate. As can be seen in Fig. 22, the strain gradient of the smaller gage length (12.7 mm, or 0.5 in.) specimen geometry is much steeper than that of the larger one (63.5 mm, or 2.5 in.). With regard to the smaller gage length specimen, it is observed that the strain gradient becomes steeper as the strain rate increases. Furthermore, a reduction of deformation in the shoulder can be achieved by decreasing the *width* of the gage section because of the decrease in deformation load and hence stress level generated in the shoulder. However, there are constraints in gage-width reductions arising from the microstructural characteristics of a particular material; in some cases, there may be an insufficient number of grains across the specimen section.

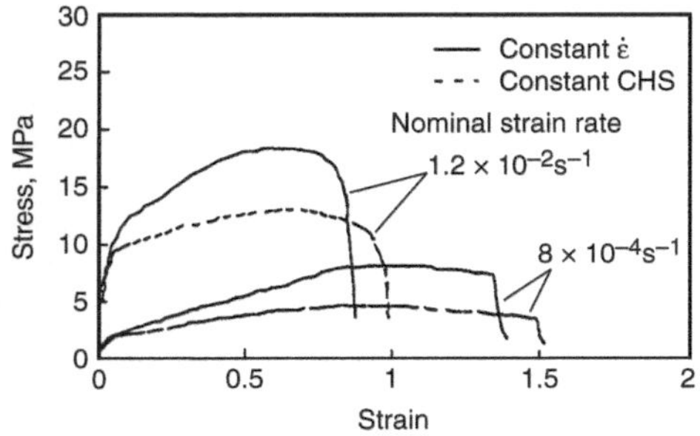

Fig. 21 Comparison of stress versus strain for constant nominal strain rate (constant crosshead speed, CHS) and constant true strain rate ($\dot{\varepsilon}$) for Al-5083 at 550 °C (1020 °F). Source: Ref 26

Modeling of the Isothermal Hot Tensile Test

The detailed interpretation of data from the isothermal hot tensile test frequently requires some form of mathematical analysis. This analysis is based on a description of the local stress state during tension testing and some form of numerical calculation. The approach is described briefly in this section.

Stress State at the Neck

Prior to necking, the stress state in the tension test is uniaxial. However, the onset of necking is accompanied by the development of a triaxial (hydrostatic*) state of stress in the neck. Because the flow stress of a material is strongly dependent on the state of stress, a correction must be introduced to convert the measured average axial stress into the effective uniaxial flow stress; that is:

$$\sigma_l^{av} = \bar{\sigma}/F_T \quad \text{(Eq 6)}$$

in which σ_l^{av} denotes the average axial stress required to sustain further deformation, $\bar{\sigma}$ is the effective (flow) stress, and F_T is the stress triaxiality factor. The magnitude of F_T (which essentially determines the magnitude of the *average* hydrostatic stress within the neck) depends on the specimen shape (round bar or sheet) and the geometry of the neck. Bridgman (Ref 27) conducted a rigorous, theoretical analysis with regard to the stress state at the neck for both round-bar and for sheet specimen geometries. For a plastically isotropic material, the following

*The term *hydrostatic stress* is defined as the mean value of the normal stresses. The term *triaxial stress* is often used to imply the presence of a hydrostatic stress. However, the term *triaxial stress* is not equivalent to *hydrostatic stress,* because the presence of a triaxial stress state could be a combination of shear stresses and/or normal stresses or only normal stresses. The term hydrostatic stress is thus preferred and more precise in describing solely normal stresses in three orthogonal directions.

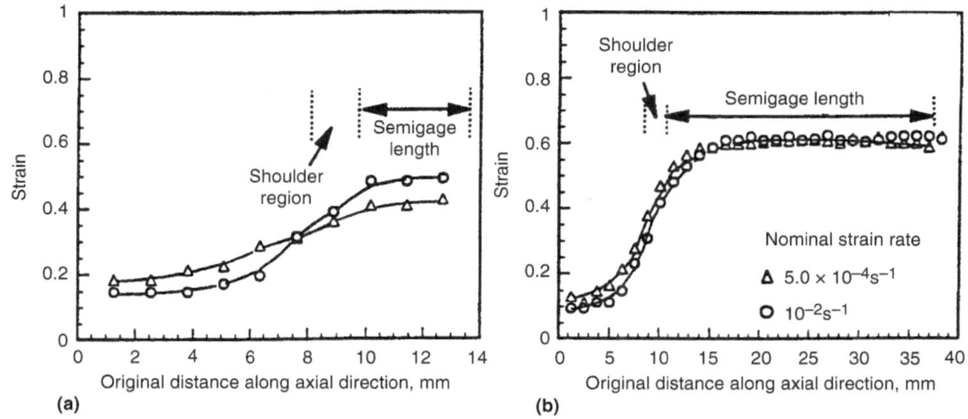

Fig. 22 Strain distribution for 12.7 mm (a) and 63.5 mm (b) gage length specimens for two different strain rates. Length strains are plotted versus original axial position along the gage length. Source: Ref 9

equations were derived for the stress triaxiality factor of round-bar (F_T^r) and sheet (F_T^s) specimens in the *symmetry plane* of the neck:

$$F_T^r = \left\{\left[1 + \left(2\frac{R}{a}\right)\right]\ln\left[1 + \left(\frac{a}{2R}\right)\right]\right\}^{-1} \quad \text{(Eq 7)}$$

$$F_T^s = \left\{\left(1 + 2\frac{R}{a}\right)^{1/2}\ln\left[1 + \frac{a}{R} + \left(\frac{2a}{R}\right)^{1/2}\left(1 + \frac{1}{2}\frac{a}{R}\right)^{1/2}\right] - 1\right\}^{-1} \quad \text{(Eq 8)}$$

in which a represents the specimen half radius or width, and R is the radius of curvature of the neck. For $a/R < 0$ (in particular, for $-2 < a/R < 0$), the stress triaxiality factor for sheet tensile specimens with a *convex* curvature is given by:

$$F_T\{[2\sqrt{Q}\ \arctan\ (1/\sqrt{Q})] - 1\}^{-1} \quad \text{(Eq 9)}$$

where $Q = -(1 + 2R/a)$.

The variation of the stress triaxiality factor for round-bar (F_T^r) and sheet (F_T^s) specimens as a function of the a/R ratio is shown in Fig. 23. For a positive a/R value (concave neck profiles), F_T is less than unity, thus promoting flow stabilization. On the other hand, for negative a/R (convex neck profiles), $F_T > 1$; thus, flow tends to be destabilized.

In a rigorous sense, the closed-form equations for F_T (Eq 7 and 8) are applicable only for the plane of symmetry at the neck. At other locations, the solution for the exact form of F_T is not available. However, as has been shown from finite-element method (FEM) analyses (Ref 28, 29), Eq 7 and 8 provide a good estimate for the stress triaxiality factor in regions away from the symmetry plane provided that the local values of a and R are inserted into the relations.

Numerical Modeling of the Hot Tensile Test

Two types of methods have been employed to model the tension behavior of materials: the FEM and the somewhat simpler finite-difference ("direct-equilibrium") method originally presented by G'sell et al. (Ref 30), Ghosh (Ref 31), and Semiatin et al. (Ref 32). Both approaches involve solutions that satisfy the axial force equilibrium equation and the appropriate bound-

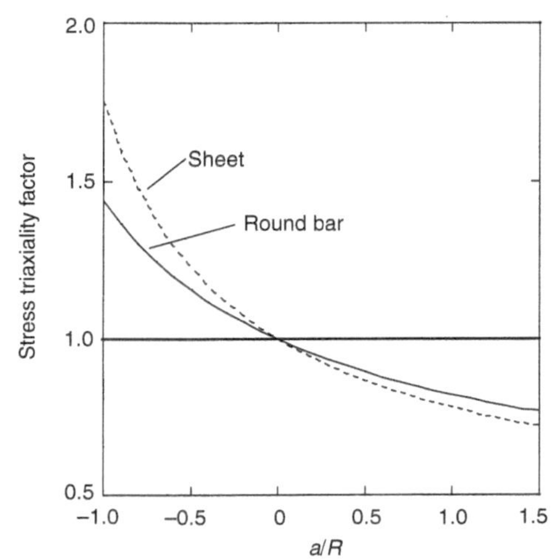

Fig. 23 Stress triaxiality factor for sheet and round-bar specimens

ary conditions. These models enable the prediction of important parameters such as neck profile, failure mode, axial-strain distribution, and ductility. A comparison of simulation results (e.g., nominal stress-strain curves, axial-strain distribution, and total elongation) obtained from FEM analyses to those of the direct-equilibrium method has shown that the latter approach gives realistic predictions (Ref 29). To this end, a brief description of this simpler method is given in the following paragraphs.

Model Formulation. The formulation of the direct-equilibrium method is based on discretization of the sample geometry, description of the material flow behavior, and development of the appropriate load-equilibrium equation. The specimen geometry (dimensions, cross-section shape, geometrical defects, etc.) is first specified. The specimen is divided along the axial direction into horizontal slices/elements (Fig. 24). For the material flow behavior, the simple engineering power-law formulation has been used in most modeling efforts, that is:

$$\bar{\sigma} = K\bar{\varepsilon}_s^n\dot{\bar{\varepsilon}}_s^m \quad \text{(Eq 10)}$$

in which $\bar{\sigma}$, $\bar{\varepsilon}_s$, and $\dot{\bar{\varepsilon}}_s$ denote the effective stress, effective strain, and effective strain rate, respectively, of the material. K, n, and m represent the strength coefficient, strain-hardening exponent, and the strain-rate-sensitivity index, respectively.

At any instant of deformation, the axial load P should be the same in each element in order to maintain force equilibrium. The load borne by each slice is equal to the product of its load-bearing cross-sectional area and axial stress; the axial stress is equal to the flow stress corrected for stress triaxiality due to necking and evaluated at a strain rate corresponding to that which the material elements experience. (In case of cavitating material, the strain rate is that of the matrix-material element, *not* the matrix-cavity continuum.) The load-equilibrium condition is thus described by:

$$\bar{\sigma}_i A_{\text{lb}}^i / F_T^i = \bar{\sigma}_j A_{\text{lb}}^j / F_T^j \quad \text{(Eq 11)}$$

or, using Eq 10:

$$\dot{\bar{\varepsilon}}_{s_i}^m \bar{\varepsilon}_{s_i}^n A_{\text{lb}}^i / F_T^i = \dot{\bar{\varepsilon}}_{s_j}^m \bar{\varepsilon}_{s_j}^n A_{\text{lb}}^j / F_T^j \quad \text{(Eq 12)}$$

in which the subscripts and/or superscripts i and j denote the corresponding parameters for elements i and j, respectively, F_T represents the stress triaxiality factor, and A_{lb} is the load-bearing area.

For the case in which the material cavitates during tension testing (see the section "Cavitation During Hot-Tensile Testing" in this chapter), Eq 11 and 12 must be modified. In particular, the presence of cavities affects the external dimensions of the specimen (because they lead to a volume increase) and hence the load-bearing area, the stress triaxiality factor, and the strain rate at which the material deforms.

As discussed by Nicolaou et al. (Ref 33), spherical and uniformly distributed cavities increase each of the three dimensions (length, width, and thickness and, for round-bar specimen geometries, diameter) of the tension specimen by the same amount. The relationship between the macroscopic area (i.e., the external area of the specimen) A_m, the load-bearing area A_{lb}, and the initial (uncavitated) area A_o^{sp} is then simply:

$$A_m = A_{\text{lb}}/(1 - C_v)^{2/3} \quad \text{(Eq 13)}$$

where C_v is the cavity volume fraction and A_{lb} is given by:

$$A_{\text{lb}} = A_o^{\text{sp}} \exp(-\bar{\varepsilon}_s) \quad \text{(Eq 14)}$$

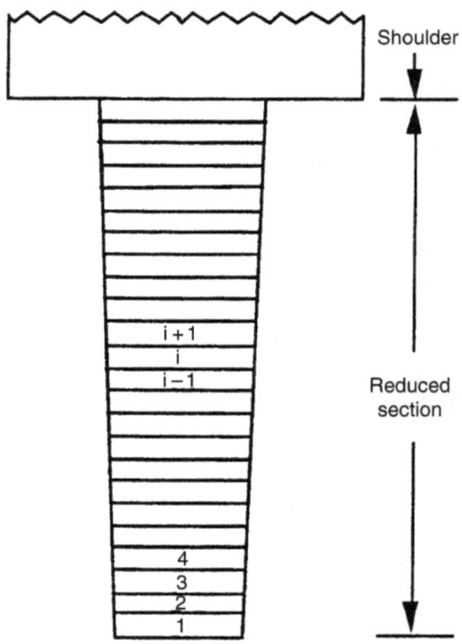

Fig. 24 Discretization of the sheet specimen for the simulations of the isothermal hot tensile test (Ref 33). The specimen geometry corresponds to the specimen shown in Fig. 6(b) (Ref 10, 15)

In addition, the length l, width w, and thickness t, for a sheet specimen, or radius r of a round-bar specimen increase according to:

$$d' = \frac{d}{(1 - C_v)^{1/3}} \quad \text{(Eq 15)}$$

in which d' denotes any of the dimensions (l', w', t', or r') for the case when cavities are present in the material, and d represents the respective dimension changes with strain alone.

The matrix strain rate $\dot{\varepsilon}_s$, can also be related to the macroscopic sample strain rate $\dot{\varepsilon}$. Using power-dissipation arguments, the relation between the two strain rates is found to be (Ref 34):

$$\dot{\varepsilon}_s = (1/\phi\rho)\dot{\varepsilon} \quad \text{(Eq 16)}$$

in which ρ is the relative density of the specimen ($\rho = 1 - C_v$) and ϕ is the stress-intensification factor, which for spherical and uniformly distributed cavities is (Ref 33):

$$\phi = 1/\rho^{2/3} \quad \text{(Eq 17)}$$

Algorithm. After having specified the specimen geometry and the material-flow relation, the equations for model formulation can be inserted into an algorithm to simulate the tensile test. At any instant of deformation, the axial variation in strain rate is calculated based on the load-equilibrium equation, which provides the ratios of the strain rates in the elements, and the boundary condition (e.g., constant crosshead speed), which provides the specific magnitudes of the strain rates. The strain rates are then used to update the macroscopic (and microscopic) strain and cavity volume fraction (for a cavitating material) in each element. The simulation steps are:

1. An increment of deformation is imposed, and a/R and F_T are calculated for each slice.
2. From the true strain and the cavity-growth rate (see the section "Cavitation During Hot Tensile Testing"), the cavity volume fraction is determined.
3. The true-strain-rate distribution is calculated for each element, using the equilibrium equation and the boundary condition.
4. From the true-strain, cavity volume fraction, and strain-rate distributions, the engineering stress and strain are calculated.
5. Steps 1 to 4 are repeated until a sharp neck is formed (localization-controlled failure) or the cavity volume fraction at the central element reaches a value of 0.3 (fracture/cavitation-controlled failure).

Example Applications. Several results illustrate the types of behavior that can be quantified using the direct-equilibrium modeling approach. The first deals with the effect of specimen taper on tensile elongation. Tensile test specimens usually have a small (≤2%) reduction in the cross-sectional area from the end to the center of the reduced section in order to control the location of failure. The predicted effect of reduced-section taper on the engineering stress-strain curves for non-strain-hardening materials is shown in Fig. 25 (Ref 29). The effect of the absence of a taper on increased elongation is quite dramatic, especially as the strain-rate sensitivity increases from $m = 0.02$ to $m = 0.15$. For materials deformed at cold-working temperatures ($m < 0.02$), tensile flow will still localize in the absence of a taper because the reduced section itself acts as the defect relative to the greater cross-sectional area of the shoulder (Ref 31, 32). In contrast to the results for samples with and without a 2% taper, the predictions for samples with a 1% versus a 2% taper show much less difference.

With appropriate modification, the direct-equilibrium modeling approach may also be used to analyze the uniaxial hot tensile testing of sheet materials that exhibit normal plastic anisotropy (Ref 35). Selected results are shown in Fig. 26. The engineering stress-strain curves exhibit a load maximum, a regime of quasi-stable flow during which the diffuse neck develops

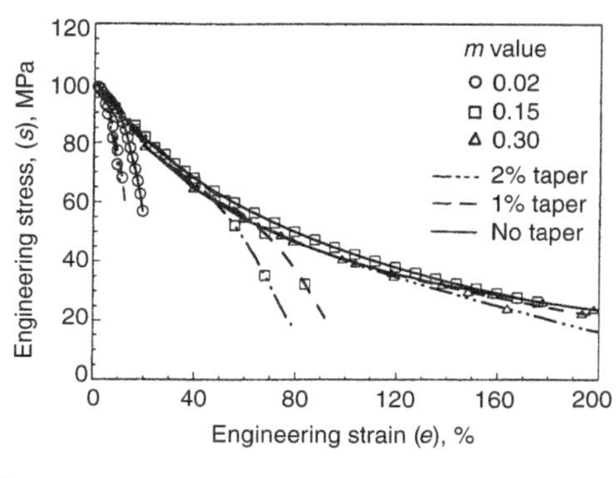

Fig. 25 Comparison of the engineering stress-strain curves for non-strain-hardening samples without or with a 1 or 2% taper predicted using the direct-equilibrium approach. Source: Ref 29

and the stress decreases gradually, and finally a period of rapid load drop during which flow is highly localized in the center of the gage length. When the m value is low, an increase in r increases the amount of quasi-stable flow; that is, it stabilizes the deformation in a manner similar to the effect of strain-rate sensitivity. In addition, the simulation results reveal that the flow-stabilizing effect of r decreases as m increases and in fact becomes negligible for conditions that approach superplastic flow (i.e., $m > 0.3$).

Cavitation during Hot Tensile Testing

A large number of metallic materials form microscopic voids (or cavities) when subjected to large strains under tensile modes of loading. This formation of microscopic cavities, which primarily occurs in the grain boundaries during high-temperature deformation, is referred to as cavitation. In some cases, cavitation may lead to premature failure at levels of deformation far less than those at which flow-localization-controlled failure would occur. For a given material, the extent of cavitation depends on the specific deformation conditions (i.e., strain rate and temperature). A wide range of materials exhibit cavitation; these materials include aluminum alloys (Fig. 27a), conventional titanium alloys (Fig. 27b), titanium aluminides, copper alloys, lead alloys, and iron alloys (Ref 36–38).

An important requirement for cavitation during flow under either hot-working or superplastic conditions is the presence of a tensile stress. On the other hand, under conditions of homogeneous compression, cavitation is not observed; in fact, cavities that may be produced under tensile flow can be removed during subsequent compressive flow. In addition, it has also been demonstrated that the superposition of a hydrostatic pressure can reduce or eliminate cavitation (Ref 39). Hot isostatic pressing can also heal the deformation damage of nucleated cavities.

Cavitation is a very important phenomenon in hot working of materials because not only may it lead to premature failure during forming, but it also may yield inferior properties in the final part. Therefore, it has been studied extensively, primarily via the tensile test.

Cavitation Mechanisms/Phenomenology

Cavitation occurs via three often-overlapping stages during tensile deformation: cavity nucleation, growth of individual cavities, and cavity coalescence. Each stage is briefly described in the following sections, while a more detailed review of ductile fracture mechanisms is in the article "Mechanisms and Appearances of Duc-

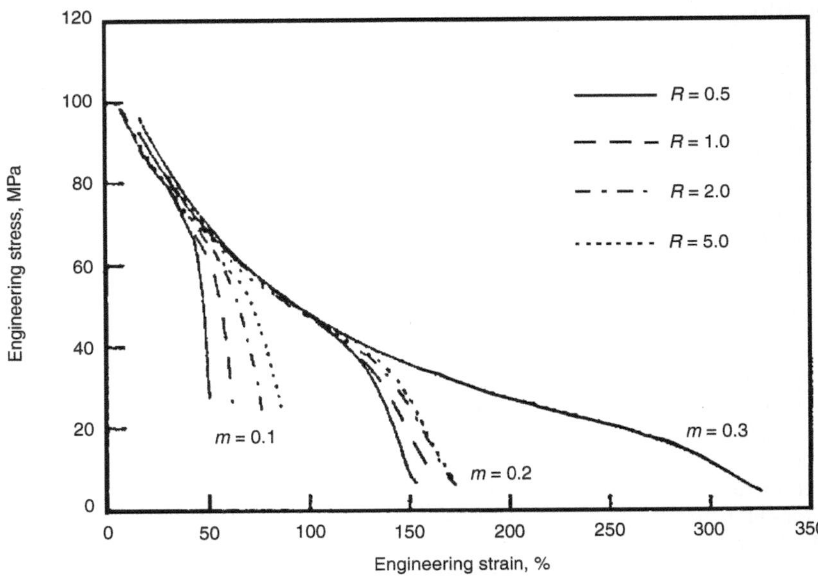

Fig. 26 Direct-equilibrium simulation predictions of engineering stress-strain curves at hot-working temperatures for various values of the strain-rate sensitivity and the normal plastic anisotropy parameter. Source: Ref 35

tile and Brittle Fracture in Metals" in *Failure Analysis and Prevention,* Volume 11 of the *ASM Handbook* (2002, p 587–626).

Cavity Nucleation. Several possible cavity-nucleation mechanisms have been established including (a) intragranular slip intersections with nondeformable second-phase particles and grain boundaries, (b) sliding of grains along grain boundaries that is not fully accommodated by diffusional transport into those regions, and (c) vacancy condensation on grain boundaries (Ref 40). A frequently used cavity-nucleation criterion based on stress equilibrium at the cavity interface is:

$$r_c = 2(\gamma + \gamma_p - \gamma_i)/\sigma \qquad \text{(Eq 18)}$$

in which r_c is the critical cavity radius above which a cavity is stable; γ, γ_p, and γ_i denote the interfacial energies of the void, the particle, and the particle-matrix interface, respectively; and σ is the applied stress. This criterion requires flow hardening, which is minimal in superplastic materials except in cases of significant grain growth, in order to nucleate cavities at less favorable sites, such as smaller particles. In addition, such surface-energy considerations require stresses for initiation and early growth that are unrealistically high. Therefore, the development of other (constrained-plasticity) approaches based on nucleation and growth from inhomogeneities/regions of high local stress triaxiality has been undertaken (Ref 41).

The cavity-nucleation rate N is defined as the number of cavities nucleated per unit area and unit strain. N may either be constant or decrease or increase with strain. However, such strain dependencies are usually not strong. Measurements have shown than N can be bracketed between 10^4 and 10^6 cavities per mm^3 per unit increment of strain (Ref 16, 41, 42).

The constrained plasticity analysis suggests that a size distribution of second-phase constituents/imperfections may lead to a variety of cavity-growth rates at the nano/submicron cavity-size level. From an operational standpoint, this effect may thus lead one to conclude that cavities nucleate continuously rather than merely become microscopically observable continuously. Irrespective of the exact mechanism, it is thought that the assumption of continuous nucleation of cavities of a certain size (e.g., 1 μm) still produces the same "mechanical" effect on failure via cavitation or flow localization as the postulated actual physical phenomenon.

Cavity-growth mechanisms can be classified into two broad categories: diffusional growth and plasticity-controlled growth (Ref 43). Diffusional growth dominates when the cavity size is very small. As cavity size increases, diffusional growth decreases very quickly, and plastic flow of the surrounding matrix becomes the dominant cavity-growth mechanism. An illustration of a cavity-growth-mechanism map (Ref 44) is shown in Fig. 28. From an engineering viewpoint, plasticity controlled growth is of greatest interest. In such cases, the growth of an isolated, noninteracting cavity is described for the case of uniaxial tension deformation by:

$$V = V_o \exp(\eta \varepsilon)$$

or

$$r = r_o \exp\left(\frac{\eta}{3}\varepsilon\right) \qquad \text{(Eq 19)}$$

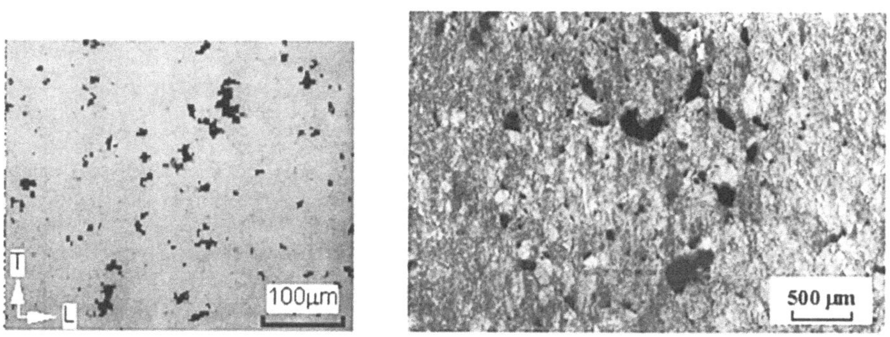

(a) (b)

Fig. 27 Examples of cavitation. (a) In aluminum (Al-7475) alloy. Courtesy of A.K. Ghosh. (b) In titanium (Ti-6Al-4V) alloy. Source: Ref 37 in which V and r are the cavity volume and radius, respectively, V_o and r_o are the volume and radius of the cavity when it becomes stable, ε denotes axial true strain, and η is the individual cavity-growth rate.

Several analyses have been conducted to correlate the cavity-growth rate η with material parameters and the deformation conditions. For example, Cocks and Ashby (Ref 45) derived the following relation between η and m for a planar array of spherical, noninteracting, grain-boundary cavities under tensile straining conditions:

$$\eta = 1.5\left(\frac{m + 1}{m}\right)\sinh\left[\frac{2}{3}\frac{(2 - m)}{(2 + m)}\right] \quad \text{(Eq 20)}$$

It should be noted that this theoretical relationship between m and the cavity-growth parameter η for an individual cavity follows the same general trend as the experimentally determined correlation between the strain-rate sensitivity and the *apparent* cavity-growth rate η_{APP}. The parameter η_{APP}, which is readily derived from experimental data (Ref 46), is defined by:

$$C_v = C_{v_o} \exp(\eta_{APP}\varepsilon) \quad \text{(Eq 21)}$$

in which C_v is the cavity volume fraction at a true strain ε, η_{APP} is the apparent cavity-growth rate, and C_{v_o} is the so-called initial cavity volume fraction.

Cavity coalescence is the interlinkage of neighboring cavities due to a microscopic flow-localization process within the material ligament between them. Coalescence occurs when the width of the material ligament reaches a critical value that depends on initial cavity spacing and the strain-rate sensitivity. Coalescence can occur along both the longitudinal and transverse directions with the latter being more important because it eventually leads to failure. According to Pilling (Ref 47), cavity coalescence may be regarded as a process that in effect increases the mean cavity-growth rate. In particular, the effect of pairwise coalescence on the average cavity-growth rate $d\bar{r}/d\varepsilon$ can be estimated from:

$$\frac{d\bar{r}}{d\varepsilon} = \frac{8C_v\Phi\eta(0.13r - 0.37(dr/d\varepsilon)_i\delta\varepsilon) + (dr/d\varepsilon)_i}{1 - 4C_v\Phi\eta\delta\varepsilon} \quad \text{(Eq 22)}$$

in which C_v is the instantaneous volume fraction of cavities, η is defined from Eq 19, $\delta\varepsilon$ ($= d\varepsilon$) is a small increment of strain, $(dr/d\varepsilon)_i$ is the rate of growth per unit strain of an isolated cavity ($= \eta r/3$ from Eq 19), and Φ is given by:

$$\Phi = (1 + \eta\delta\varepsilon/3 + (\eta\delta\varepsilon)^2/27) \quad \text{(Eq 23)}$$

The phenomenon of cavity coalescence was further investigated by Nicolaou and Semiatin (Ref 48, 49), who conducted a numerical analysis of the uniaxial tension test considering: the temporal and spatial location of the cavities inside the specimen and the temporal cavity radius. Two cases were considered: a stationary cavity array (similar to the analysis of Pilling) and continuous cavity nucleation. The analysis of the stationary cavity array led to a much simpler expression than Eq 22, that is:

$$\frac{d\bar{r}}{d\varepsilon} = \eta\bar{r}(1/3 + C_v) \quad \text{(Eq 24)}$$

This simple equation gives predictions very similar to the more complex relation of Eq 22.

With regard to the continuous cavity nucleation case, it was found that the average cavity radius was described by:

$$\frac{d\bar{r}}{d\varepsilon} = \eta\bar{r}(0.2 + C_v) \quad \text{(Eq 25)}$$

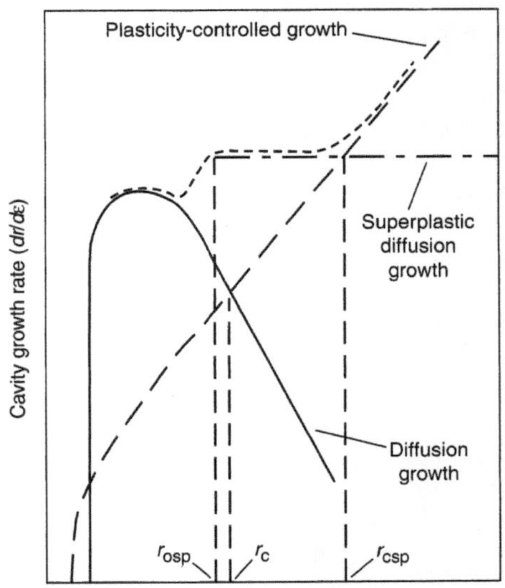

Fig. 28 Variation of the cavity-growth rate for different mechanisms. r_c, critical cavity radius; r_{osp}, cavity radius for onset of superplastic deformation; r_{csp}, critical cavity radius for superplastic deformation. Source: Ref 44

A comparison with experimental cavity size measurements (e.g., Fig. 29) revealed that actual results are bounded by cavity-growth-and-coalescence models that assume either a constant, continuous nucleation rate (lower limit) or a preexisting cavity array with no nucleation of new cavities (upper limit), that is, Eq 25 and 24, respectively.

Stress-Strain Curves

The work of Nicolaou et al. (Ref 33) also shed light on the effect of cavitation on stress-strain behavior. Engineering stress-strain (*S-e*) curves for a range of strain-rate sensitivities (*m* values) and cavity-growth rates η_{APP} were predicted using the direct-equilibrium modeling approach (Fig. 30); for all of the cases, the strain-hardening exponent *n* was 0. The cavity volume fraction (C_V) in the central element at failure is also indicated in the plots.

Examination of the engineering stress-strain curves reveals that cavitation causes a noticeable reduction in total elongation; this reduction is quantified and discussed in more detail in the next section. Figure 30 also shows that the stress-strain curves for cavitating and noncavitating samples with the same value of *m* are very close to each other, except at elongations close to failure. Surprisingly, the engineering stress at a given elongation for a cavitating material is *higher* than the corresponding stress (at the same elongation) for a noncavitating material. This intuitively unexpected result was interpreted by the examination of the effective (load-bearing) area at the same elongation of a cavitating and a noncavitating specimen. In particular, the analysis of Nicolaou et al. (Ref 33) revealed that for a given elongation the effective area of the cavitating specimen is larger than the area of a noncavitating one. Therefore, the load and hence the engineering stress required to sustain deformation is higher in the case of a cavitating material.

As shown in Fig. 30, the difference between the engineering stress-strain curves of cavitating

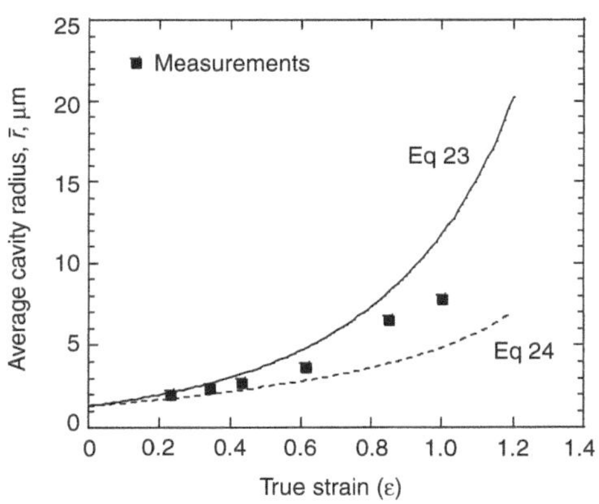

Fig. 29 Comparison of measurements and predictions of the evolution of average cavity radius with strain for an Al-7475 alloy assuming continuous nucleation (Eq 25) or a preexisting cavity array (Eq 24). Source: Ref 49

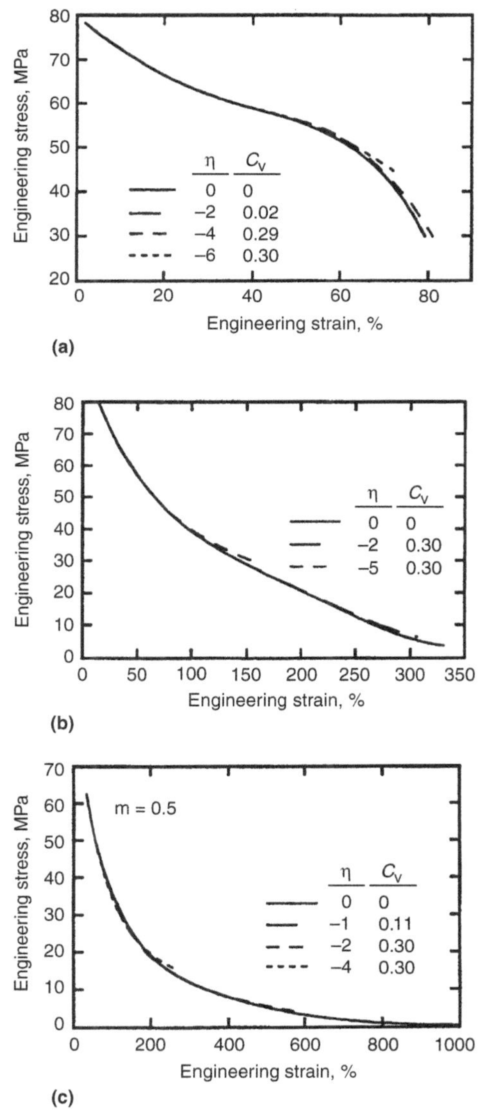

Fig. 30 Predicted engineering stress-strain curves for tensile testing of sheet samples with a 2% taper, assuming strain-hardening exponent $n = 0$, initial cavity volume fraction $C_{V_0} = 10^{-3}$, various cavity-growth rates η, and a strain-rate sensitivity exponent *m* equal to (a) 0.1, (b) 0.3, or (c) 0.5. Source: Ref 33

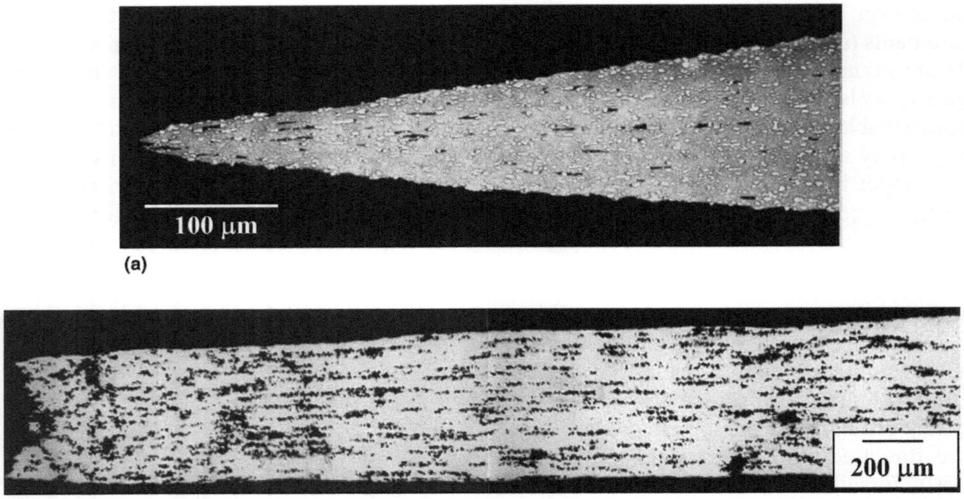

Fig. 31 Micrographs of titanium aluminide specimens that failed in tension. (a) Orthorhombic titanium aluminide that failed in tension by flow localization. Source: Ref 10. (b) Near-γ titanium aluminide that failed in tension by fracture (cavitation). Source: Ref 51

and noncavitating materials is not very large. Therefore, it can be deduced that Considere's criterion, if implemented in the usual fashion using data from a tensile test (i.e., a plot of load versus the elongation of the gage section), can be used to test whether fracture of a tensile specimen occurs due to instability, regardless of the presence of extensive internal cavities in the material and whether the volume of the material is conserved (Ref 50).

Failure Modes during Hot Tensile Testing

Cavitating hot tensile specimens may fail by either localized necking ("flow localization") or fracture/cavitation. The second mode of failure occurs without flow localization in the neck and resembles a brittle type of fracture because the fracture tip has a considerable area. Micrographs of these modes of failure are presented in Fig. 31. The localization type of failure shown in Fig. 31(a) is for an orthorhombic titanium aluminide (Ti-21Al-22Nb) deformed at 980 °C (1795 °F) and a nominal strain rate of 1.6×10^{-3} s^{-1}. On the other hand, Fig. 31(b) displays the fracture-controlled failure of a γ titanium aluminide alloy (Ref 51) deformed in tension at 1200 °C (2190 °F) and a nominal strain rate of 10^{-3} s^{-1}.

The particular mode of failure of a material tested under tension conditions can be predicted by the magnitude of the strain-rate sensitivity m and the apparent cavity-growth rate η_{APP}. The corresponding failure-mechanism map for non-strain-hardening materials is plotted in Fig. 32. For deformation under superplastic conditions ($m > 0.3$) and $\eta_{APP} > 2$, the map shows that failure is fracture/cavitation-controlled. On the other hand, flow-localization-controlled failure is seen to predominate only for small values of the cavity-growth rate. In Fig. 32, experimental observations of the failure mode of γ and orthorhombic titanium aluminides are also plotted. The solid data points correspond to fracture-controlled failures, while the open data points cor-

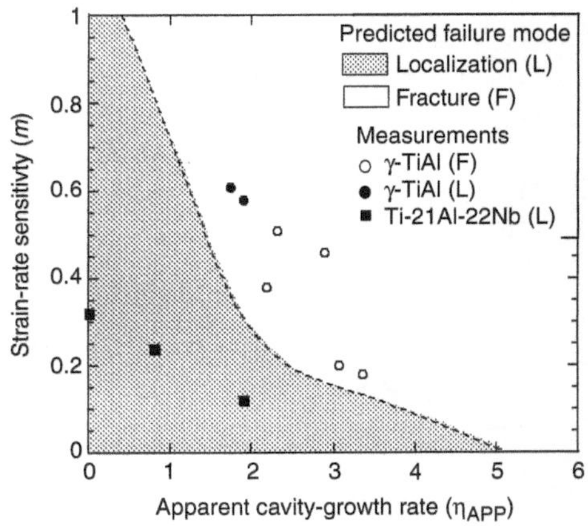

Fig. 32 Failure-mode map developed from simulations of the sheet tensile test. Experimental data points are also shown on the map.

respond to localization-controlled failures. Given the assumptions made in deriving such maps, it can be concluded that the prediction of failure mode from the magnitudes of m and η_{APP} provides good agreement with actual behavior.

Total Tensile Elongations

As shown in Fig. 33, cavitation may lead to premature failure and thus to a significant reduction in the tensile elongation compared to that measured by Woodford (Ref 18) for noncavitating metals. For a fixed value of m, the reduction in elongation for fracture-controlled failure depends on several factors, such as the cavity-nucleation rate, cavity-growth rate, cavity shape and distribution, and the cavity architecture.

Several analyses have been conducted to quantify the effect of cavitation on the tensile ductility. These include the two-slice approach* by Lian and Suery (Ref 52), micromechanical approaches by Zaki (Ref 53), and Nicolaou and Semiatin (Ref 54), as well as approaches based on the direct-equilibrium approach described in the section "Numerical Modeling of the Hot Tensile Test" in this chapter. Several results from the direct-equilibrium model serve to illustrate the efficacy of such techniques.

The results shown in Fig. 33 correspond to a non-strain-hardening material ($n = 0$) with a 2% taper and $C_{vo} = 10^{-3}$. The topmost curve in this plot depicts the total elongation as a function of m for a noncavitating sample, that is, Woodford's trend line. For such a material, the elongation is controlled, of course, by the onset of localized necking. The remaining curves in Fig. 33 for *cavitating* samples indicate the decrement in elongation due to the occurrence of fracture prior to localized necking. For low values of η_{APP} (<2) and m (<0.3), the decrement is equal to zero because failure is still necking controlled. On the other hand, the decrement is largest for large values of η_{APP} and m, for which the critical volume fraction of cavities for fracture (assumed to be 0.30) is reached much before the elongation at which necking occurs. In fact, for $\eta_{APP} > 5$, the total elongation is almost independent of m for $m > 0.3$ because fracture in these cases intercedes during largely uniform, quasi-stable flow.

*The two-slice approach assumes that the specimen comprises two regions, one of them consisting of the central plane of the specimen that contains an initial geometric or strength defect. The deformation of each region obeys the flow rule while at any instant of deformation the *load* is the same in both regions (slices).

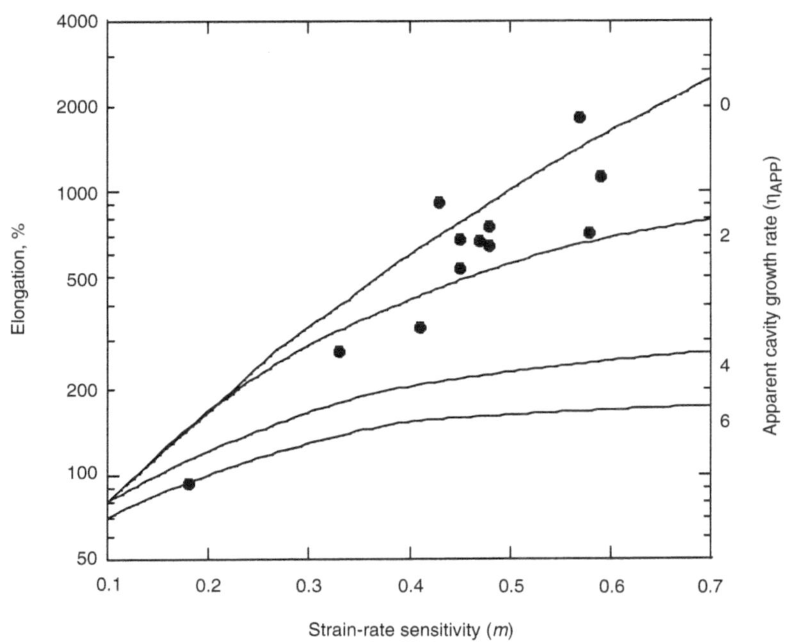

Fig. 33 Elongation as function of the strain-rate sensitivity and (apparent) cavity-growth rate predicted from direct equilibrium simulations. The individual data points represent experimental data. Source: Ref 33

Table 2 Experimental data from the literature for the deformation and failure of cavitating materials

Data point	Material	*m*	η	Tensile elongation, %
1	γ-TiAl (as received)	0.38	2.2	219
2	γ-TiAl (as received)	0.51	2.3	350
3	γ-TiAl (as received)	0.62	1.8	446, 532
4	γ-TiAl (heat treated)	0.18	3.4	93, 104
5	γ-TiAl (heat treated)	0.15	8.0	51
6	5083 Al	0.50	5.2	172
7	Zn-22Al	0.45	1.5	400
8	α/β brass	0.60	2.3	425
9	Coronze 638	0.33	4.5	275

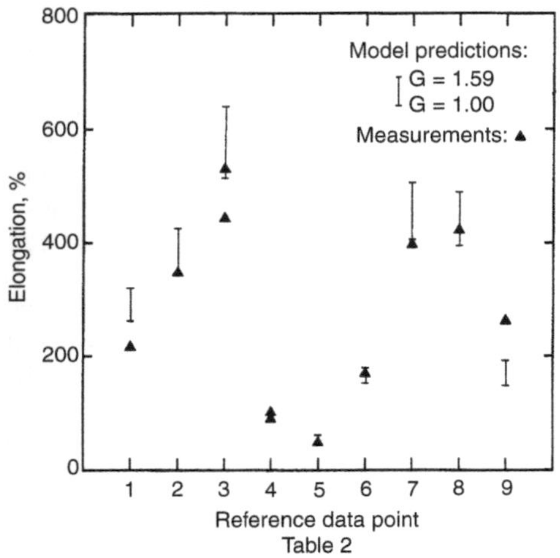

Fig. 34 Comparison of experimentally determined total elongations with (microscopic) model predictions that incorporate the cavity architecture. Source: Ref 54

In most cases, rigorous comparisons of predicted tensile elongations and experimental data (Fig. 33) cannot be made because the cavity-growth rate and the cavity-size population and shape were not measured, while other important parameters such as the specimen geometry were not reported. Therefore, a general comparison based only on the value of *m* can be made. From the results of Fig. 33, it is seen that most of the data points overlie the predicted curves.

Comparisons of reported tensile elongations data (Table 2) to predictions of a microscopic model (Ref 54) in which the cavity architecture has been taken into account through the parameter *G* are shown in Fig. 34 (*G* is a factor that describes the geometry of the ligament between two cavities as a function of the cavity architecture within the specimen). With the exception of one data point (No. 9), the major deviations of the predictions tend to be on the high side. These deviations could be a result of the neglect in the microscopic model of the macroscopic strain gradient in the diffuse neck of real tension samples. In addition, as mentioned previously, specimen geometry and deformation in the shoulder region of actual tension specimens have an effect on measured ductilities, which is difficult to quantify. Nevertheless, agreement between the measured and predicted ductilities is reasonably good.

ACKNOWLEDGMENT

This chapter was adapted from P.D. Nicolaou, R.E. Bailey, and S.L. Semiatin, Chapter 7, Hot-Tension Testing, *Handbook of Workability and Process Design,* G.E. Dieter, H.A. Kuhn, and S.L. Semiatin, Ed., ASM International, 2003, p 68–85.

REFERENCES

1. *Met. Ind.,* Vol 11, 1963, p 247–249
2. M.G. Cockroft and D.J. Latham, Ductility and the Workability of Metals, *J. Inst. Met.,* Vol 96, 1968, p 33–39
3. *J. Inst. Met.,* Vol 92, 1964, p 254–256
4. G. Cusminsky and F. Ellis, An Investigation into the Influence of Edge Shape on Cracking During Rolling, *J. Inst. Met.,* Vol 95 (No. 2), Feb 1967, p 33–37
5. E.F. Nippes, W.F. Savage, B.J. Bastian, and R.M. Curran, An Investigation of the Hot Ductility of High-Temperature Alloys, *Weld. J.,* Vol 34, April 1955, p 183s–196s
6. D.F. Smith, R.L. Bieber, and B.L. Lake, "A New Technique for Determining the Hot Workability of Ni-Base Superalloys," presented at IMD-TMS-AIME meeting, 22 Oct 1974
7. S.I. Oh, S.L. Semiatin, and J.J. Jonas, An Analysis of the Isothermal Hot Compression Test, *Metall. Trans. A,* Vol 23A (No. 3), March 1992, p 963–975

8. W.C. Unwin, *Proc. Inst. Civil Eng.*, Vol 155, 1903, p 170–185
9. P.A. Friedman and A.K. Ghosh, Microstructural Evolution and Superplastic Deformation Behavior of Fine Grain 5083Al, *Metall. Mater. Trans. A,* Vol 27A (No. 12), Dec 1996, p 3827–3839
10. P.D. Nicolaou and S.L. Semiatin, High-Temperature Deformation and Failure of an Orthorhombic Titanium Aluminide Sheet Material, *Metall. Mater. Trans. A.,* Vol 27A (No. 11), Nov 1996, p 3675–3681
11. R. Pilkington, C.W. Willoughby, and J. Barford, The High-Temperature Ductility of Some Low-Alloy Ferritic Steels, *Metal Sci. J.,* Vol 5, Jan 1971, p 1
12. D.J. Latham, *J. Iron Steel Inst.,* Vol 92, 1963–1964, p 255
13. R.E. Bailey, *Met. Eng. Quart.,* 1975, p 43–50
14. R.E. Bailey, R.R. Shiring, and R.J. Anderson, *Superalloys Metallurgy and Manufacture,* Proc. of the Third International Conference, Claitor's Publishing, Sept 1976, p 109–118
15. P.D. Nicolaou and S.L. Semiatin, An Investigation of the Effect of Texture on the High-Temperature Flow Behavior of an Orthorhombic Titanium Aluminide Alloy, *Metall. Mater. Trans. A.,* Vol 28A (No. 3A), March 1997, p 885–893
16. J. Pilling and N. Ridley, *Superplasticity in Crystalline Solids,* The Institute of Metals, London, UK, 1989
17. R.M. Imayev and V.M. Imayev, Mechanical Behaviour of TiAl Submicrocrystalline Intermetallic Compound at Elevated Temperatures, *Scr. Met. Mater.,* Vol 25 (No. 9), Sept 1991, p 2041–2046
18. D.A. Woodford, Strain-Rate Sensitivity as a Measure of Ductility, *Trans. ASM,* vol 62 (No. 1), March 1969, p 291–293
19. A.K. Ghosh and R.A. Ayres, On Reported Anomalies in Relating Strain-Rate Sensitivity (*m*) to Ductility, *Metall. Trans. A,* Vol 7A, 1976, p 1589–1591
20. J.W. Hutchinson and K.W. Neale, Influence of Strain-Rate Sensitivity on Necking Under Uniaxial Tension, *Acta Metall.,* Vol 25 (No. 8), Aug 1977, p 839–846
21. F.A. Nichols, Plastic Instabilities and Uniaxial Tensile Ductilities, *Acta Mater.,* Vol 28 (No. 6), June 1980, p 663–673
22. M.A. Meyers and K.K. Chawla, *Mechanical Metallurgy Principles and Applications,* Prentice Hall, 1984.
23. N.E. Paton and C.H. Hamilton, Microstructural Influences on Superplasticity in Ti-6Al-4V, *Metall. Trans. A,* Vol 10A, 1979, p 241–250
24. G.E. Dieter, *Mechanical Metallurgy,* 3rd ed., McGraw-Hill, 1986
25. A.K. Ghosh and C.H. Hamilton, Mechanical Behavior and Hardening Characteristics of a Superplastic Ti-6Al-4V Alloy, *Metall. Trans. A,* Vol 10A, 1979, p 699–706
26. R. Verma, P.A. Friedman, A.K. Ghosh, S. Kim, and C. Kim, Characterization of Superplastic Deformation Behavior of a Fine Grain 5083 Al Alloy Sheet, *Metall. Mater. Trans. A,* Vol 27A (No. 7), July 1996, p 1889–1898
27. P.W. Bridgman, *Studies in Large Plastic Flow and Fracture,* McGraw-Hill, 1952
28. A.S. Argon, J. Im, and A. Needleman, Distribution of Plastic Strain and Negative Pressure in Necked Steel and Copper Bars, *Metall. Trans. A,* Vol 6A, 1975, p 815–824
29. C.M. Lombard, R.L. Goetz, and S.L. Semiatin, Numerical Analysis of the Hot Tension Test, *Metall. Trans. A,* Vol 24A (No. 9), Sept 1993, p 2039–2047
30. C. G'sell, N.A. Aly-Helal, and J.J. Jonas, *J. Mater. Sci.,* Vol 18, 1983, p 1731–1742
31. A.K. Ghosh, A Numerical Analysis of the Tensile Test for Sheet Metals, *Metall. Trans. A,* Vol 8A, 1977, p 1221–1232
32. S.L. Semiatin, A.K. Ghosh, and J.J. Jonas, A "Hydrogen Partitioning" Model for Hydrogen Assisted Crack Growth, *Metall. Trans. A,* Vol 16A, 1985, p 2039–47
33. P.D. Nicolaou, S.L. Semiatin, and C.M. Lombard, Simulation of the Hot-Tension Test under Cavitating Conditions, *Metall. Mater. Trans. A.,* Vol 27A (No. 10), Oct 1996, p 3112–3119
34. S.L. Semiatin, R.E. Dutton, and S. Shamasundar, Materials Modeling for the Hot Consolidation of Metal Powders and Metal-Matrix Composites, *Processing and Fabrication of Advanced Materials IV,* T.S. Srivatsan and J.J. Moore, Ed., The Minerals, Metals, and Materials Society, 1996, p 39–52
35. P.D. Nicolaou and S.L. Semiatin, *Scr. Mater.,* Vol 36, 1997, p 83–88
36. M.M.I. Ahmed and T.G. Langdon, Exceptional Ductility in the Superplastic Pb-62 Pct Sn Eutectic, *Metall. Trans. A,* Vol 8A, 1977, p 1832–1833

37. S.L. Semiatin, V. Seetharaman, A.K. Ghosh, E.B. Shell, M.P. Simon, and P.N. Fagin, Cavitation During Hot Tension Testing of Ti-6Al-4V, *Mater. Sci. Eng. A,* Vol A256, 1998, p 92–110
38. C.C. Bampton and J.W. Edington, The Effect of Superplastic Deformation on Subsequent Service Properties of Fine-Grained 7475 Aluminum, *J. Eng. Mater. Tech.,* Vol 105 (No. 1), Jan 1983, p 55–60
39. J. Pilling and N. Ridley, Cavitation in Aluminium Alloys During Superplastic Flow, *Superplasticity in Aerospace,* H.C. Heikkenen and T.R. McNelley, The Metallurgical Society, 1988, p 183–197
40. G.H. Edward and M.F. Ashby, Intergranular Fracture During Powder-Law Creep, *Acta Metall.,* Vol 27, 1979, p 1505–1518
41. A.K. Ghosh, D.-H. Bae, and S.L. Semiatin, Initiation and Early Stages of Cavity Growth During Superplastic and Hot Deformation, *Mater. Sci. Forum,* Vol 304–306, 1999, p 609–616
42. S. Sagat and D.M.R. Taplin, Fracture of a Superplastic Ternary Brass, *Acta Metall.,* Vol 24 (No. 4), April 1976, p 307–315
43. A.H. Chokshi, The Development of Cavity Growth Maps for Superplastic Materials, *J. Mater. Sci.,* Vol 21, 1986, p 2073–2082
44. B.P. Kashyap and A.K. Mukherjee, Cavitation Behavior During High Temperature Deformation of Micrograined Superplastic Materials—A Review, *Res Mech.,* Vol 17, 1986, p 293–355
45. A.C.F. Cocks and M.F. Ashby, Creep Fracture by Coupled Power-Law Creep and Diffusion Under Multiaxial Stress, *Met. Sci.,* Vol 16, 1982, p 465–478
46. P.D. Nicolaou, S.L. Semiatin, and A.K. Ghosh, An Analysis of the Effect of Cavity Nucleation Rate and Cavity Coalescence on the Tensile Behavior of Superplastic Materials, *Metall. Mater. Trans. A,* Vol 31A, 2000, p 1425–1434
47. J. Pilling, Effect of Coalescence on Cavity Growth During Superplastic Deformation, *Mater. Sci. Technol.,* Vol 1 (No. 6), June 1985, p 461–466
48. P.D. Nicolaou and S.L. Semiatin, Modeling of Cavity Coalescence During Tensile Deformation, *Acta Mater,* Vol 47, 1999, p 3679–3686
49. P.D. Nicolaou and S.L. Semiatin, The Influence of Plastic Hardening on Surface Deformation Modes Around Vickers and Spherical Indents, *Acta Mater,* Vol 48, 2000, p 3441–3450
50. L. Weber, M. Kouzeli, C. San Marchi, and A. Mortensen, On the Use of Considere's Criterion in Tensile Testing of Materials Which Accumulate Internal Damage, *Scr. Mater.* Vol 41, 1999, p 549–551
51. C.M. Lombard, "Superplasticity in Near-Gamma Titanium Aluminides," Ph.D. Thesis, University of Michigan, Ann Arbor, MI, 2001
52. J. Lian and M. Suery, Effect of Strain Rate Sensitivity and Cavity Growth Rate on Failure of Superplastic Material, *Mater. Sci. Technol.,* Vol 2, 1986, p 1093–1098
53. M. Zaki, Micronecking and Fracture in Cavitated and Superplastic Materials, *Metall. Mater. Trans. A,* Vol 27A, 1996, p 1043–1046
54. P.D. Nicolaou and S.L. Semiatin, A Theoretical Investigation of the Effect of Material Properties and Cavity Architecture/Shape on Ductile Failure During the Hot Tension Test, *Metall. Mater. Trans. A.,* Vol 29A, 1998, p 2621–2630

CHAPTER 14

Tensile Testing at Low Temperatures

THE SUCCESSFUL USE of engineering materials at low temperatures requires that knowledge of material properties be available. Numerous applications exist where the service temperature changes or is extreme. Therefore, the engineer must be concerned with materials properties at different temperatures. Some of the typical materials properties of concern are strength, elastic modulus, ductility, fracture toughness, thermal conductivity, and thermal expansion. The lack of low temperature engineering data, as well as the use of less common engineering materials at low temperatures, results in the need for low-temperature testing.

The terms "high temperature" and "low temperature" are typically defined in terms of the homologous temperature (T/T_M), (where T is the exposure temperature, and T_M is the melting point of a material (both given on the absolute temperature scale, K). The homologous temperature is used to define the range of application temperatures in terms of the thermally activated metallurgical processes that influence mechanical behavior.

The term "low temperature" is typically defined in terms of boundaries where metallurgical processes change. One general definition of "low-temperature" is $T < 0.5\ T_M$. For many structural metals, another definition of low temperature is $T < 0.3\ T_M$, where recovery processes are not possible in metals and where the number of slip systems is restricted. For these definitions, room temperature (293 K) is almost always considered a low temperature for a metal with a few exceptions, such as metals that have melting temperatures below 200 °C (indium and mercury). In a structural engineering sense, low temperature may be one caused by extreme cold weather. A well-known example of this is the brittle fracture of ship hulls during WWII that occurred in the cold seas of the North Atlantic (Ref 1). For many applications, low temperature refers to the cryogenic temperatures associated with liquid gases. Gas liquefaction, aerospace applications, and super-conducting machinery are examples of areas in engineering that require the use of materials at very low temperatures. The term cryogenic typically refers to temperatures below 150 K. Service conditions in superconducting magnets that use liquid helium for cooling are in the 1.8 to 10 K range.

The mechanical properties of materials are usually temperature dependent. The most common way to characterize the temperature dependence of mechanical properties is to conduct tensile tests at low temperatures. Depending on the data needed, a test program can range from a full characterization of the response of a material over a temperature range, to a few specific tests at one temperature to verify a material performance. Many of the rules for conducting low temperature tests are the same as for room temperature tests. Low-temperature test procedures and equipment are detailed in this chapter. The role that temperature plays on the properties of typical engineering materials is discussed also. Important safety concerns associated with low-temperature testing are reviewed.

Mechanical Properties at Low Temperatures

In general, lowering the temperature of a solid increases its flow strength and fracture strength. The effect that lowering the temperature of a solid has on the mechanical properties of a material is summarized below for three principal groups of engineering materials: metals, ceramics, and polymers (including fiber-reinforced polymer, or FRP composites). An excellent source for an in-depth coverage of material properties at low temperatures is Ref 2.

Metals. Most metals are polycrystalline and have one of three relatively simple structures: face-centered cubic (fcc), body-centered cubic (bcc), and close-packed hexagonal (hcp). The temperature dependence of the mechanical properties of the fcc materials are quite distinct from those of the bcc materials. The properties of hcp materials are usually somewhere in between fcc and bcc materials. The general aspects of temperature-dependent mechanical behavior may be discussed using the deformation behavior maps shown in Fig. 1(a) and (b). The axes of these graphs are normalized for temperature and stress. Temperature is normalized to the melting temperature, while stress is normalized to the room temperature shear modulus, *G* (Ref 2).

The behavior characteristic of a pure, annealed fcc material is shown in Fig. 1(a). The small increase of yield strength that occurs upon cooling is characteristic of the fcc behavior. The ultimate strength, which is shown as the ductile failure line, increases much more than the yield strength on cooling. The large increase in ultimate strength coupled with the relatively small increase in yield strength in fcc materials results from ductile, rather than brittle, failure (Ref 2).

Figure 1(b) illustrates the classic bcc behavior. The large temperature dependence of the yield strength, the smaller temperature dependence of the ultimate strength, and a region where the specimen fails before any significant plastic deformation occurs should be noted (Ref 2).

The previous discussion is for pure annealed metals. Engineering alloys may behave somewhat differently, but the trends are relatively consistent. Solid solution strengthening typically increases yield and ultimate strengths of the fcc alloys while giving the yield strength an increased temperature dependence. The temperature dependence of the ultimate strength is still greater than that of the yield strength, allowing the alloy to maintain its ductile behavior. The ultimate tensile strengths of the fcc metals have stronger temperature dependence than those of bcc metals. Austenitic stainless steels have fcc structures and are used extensively at cryogenic temperatures because of their ductility, toughness, and other attractive properties. Some austenitic steels are susceptible to martensitic transformation (bcc structure) and low-temperature embrittlement. Plain carbon and low alloy steels having bcc structures are almost never used at cryogenic temperatures because of their extreme brittleness. Cases of anomalous strength behavior have been reported where a maximum strength is reached at temperatures above 0 K. These cases are unique and usually involve single crystal research materials or very soft materials, although yield strengths of commercial brass alloys are reported to be higher at 20 K than at 4 K (Ref 2).

Ceramics. Ceramics are inorganic materials held together by strong covalent or ionic bonds. The strong bonds give them the desirable properties of good thermal and electrical resistance and high strength but also make them very brittle. Graphite, glass, and alumina are ceramics

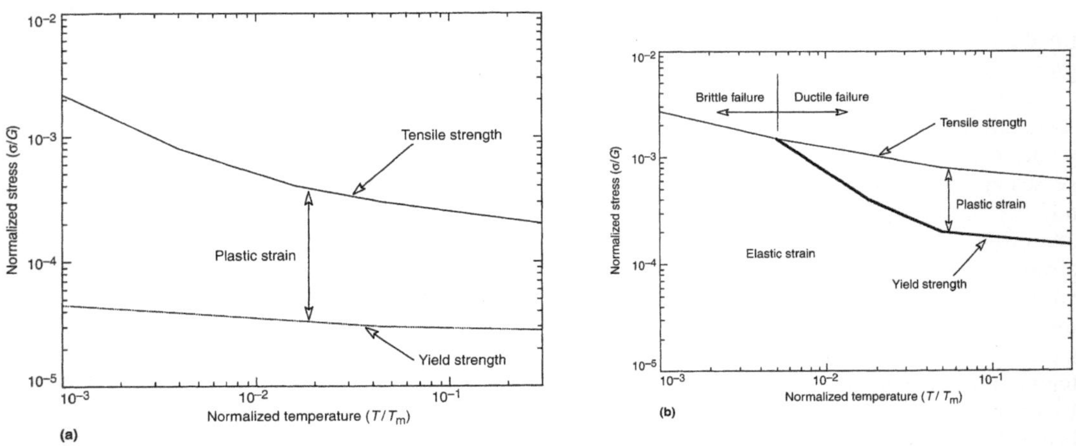

Fig. 1 Simplified deformation behavior (Ashby) maps (a) for face-centered cubic metals and (b) for body-centered cubic metals. Source: Ref 2

used at low temperature usually in the form of fibers that reinforce polymer-matrix composite materials. The high temperature (~77 K) superconducting compounds are ceramics that pose challenging problems with respect to using brittle materials at low temperatures.

Polymers and Fiber-Reinforced Polymer (FRP) Composites. Polymers are rather complex materials having many classifications and a wide range of properties. Two important properties of polymers are the melting temperature, T_m, and the glass transition temperature, T_g, both of which indicate the occurrence of a phase change. The glass transition temperature, the most important material characteristic related to the mechanical properties of polymer, is influenced by degree of polymerization. The T_g is the temperature, upon cooling, at which the amorphous or crystalline polymer changes phase to a glassy polymer. For most polymers at temperatures below T_g, the stress-strain relationship becomes linear-elastic, and brittle behavior is common. Some ductile or tough polymers exhibit plastic yielding at temperatures below T_g (Ref 3). The T_g represents the temperature below which mass molecular motion (such as chain sliding) ceases to exist, and ductility is primarily due to localized strains. Suppression of T_g helps to produce tougher polymers. The strong temperature dependence of the modulus is a distinguishing feature of polymers compared to metals or ceramics.

Fiber-reinforced polymer composites are used extensively at low temperatures because of their high strength-to-weight ratio and their thermal and electrical insulating characteristics. The FRPs tend to have excellent tensile strength that increases with decreasing temperature. Reinforcing fibers commonly used in high-performance composites for low-temperature applications are alumina, aramid, carbon, and glass. Typical product forms are high-pressure molded laminates (such as cotton/phenolics and G-10) and filament-wound or pultruded tubes, straps, and structures. Although the FRP composites have desirable tensile strength, other mechanical properties such as fatigue and interlaminar shear strength are sometimes questionable. Two good sources of properties of structural composites at low temperature are Ref 4 and 5.

Test Selection Factors: Tensile versus Compression Tests

Tensile and compression tests produce engineering data but also facilitate study of fundamental mechanical-metallurgical behavior of a material, such as deformation and fracture processes. If obtaining engineering data is the objective and the materials application is at low temperature, the designer must be sure that mechanical properties are stable at the desired temperatures. One important factor related to low-temperature testing is that the low temperature may cause unstable brittle fracture behavior that tensile or compression tests may fail to reveal. The cooling of materials, especially bcc metals and polymers, can cause the materials to undergo a ductile-to-brittle transition. This behavior is not unique to steel but has its counterpart in many other materials. Brittle fracture occurs in the presence of a triaxial stress state to which a simple tensile or compression test will not subject the material. Brittle fracture is caused by high tensile stress, while ductile behavior is related to shear stress. A metal that flows at low stress and fractures at high stress will always be ductile. If, however, the same material is retreated so that its yield strength approaches its fracture strength, its behavior may become altered, and brittleness may ensue (Ref 1). If the materials application is at low temperature, the designer must be sure that mechanical properties are stable, because the possibility of brittle fracture requires modification of the design approach.

If the material in question is a new material or a material for which little or no low-temperature data exist, screening tests that can assess susceptibility to brittle fracture are advisable. Two such screening tests are Charpy impact tests and notch tensile tests. Conducting Charpy or notch tensile tests at various temperatures can detect a ductile-to-brittle transition over a temperature range. Ultimately, if the fracture toughness of the material is an issue, fracture toughness testing should be performed.

The intended service condition for the material should influence the test temperature and the decision to perform tensile tests. It is good practice to determine the properties while simulating the service conditions. Of course, life is not always this simple, and actual service conditions may not be easily achieved with an axial stress test at a given temperature.

Tensile testing is the most common test of mechanical properties and is usually easier than other test methods, such as compression testing, to conduct properly at any temperature. The compressive and tensile Young's moduli of most materials are identical. Fracture of a material is caused by tensile stress that causes crack propagation. Tensile tests lend themselves well to low-temperature test methods because the use of environmental chambers necessitates longer than normal load trains. Pin connections and spherical alignment nuts can be used to take advantage of the increased length for self-alignment purposes. For most homogeneous materials, stress-strain curves obtained in tension are almost identical to those obtained in compression (Ref 6). Exceptions exist where there is disagreement between the stress-strain curves in tension and compression. This effect, termed "strength differential effect," is especially noticeable in high-strength steels (Ref 7).

There are times when compression testing is required such as when the service-condition stress is compressive or when the strength of an extremely brittle material is required. The second case is true for almost all polymers at cryogenic temperatures as they become extremely brittle, glassy materials. The fillet radius of a reduced-section tensile-test sample can create enough of a stress riser that the material fails prematurely. Stress concentrations, flaws, and submicroscopic cracks largely determine the tensile properties of brittle materials. Flaws and cracks do not play such an important role in compression tests because the stress tends to close the cracks rather than open them. The compression tests are probably a better measure of the bulk material behavior because they are not as sensitive to factors that influence brittle fracture (Ref 3). A brittle material will be nearly linear-elastic to failure, providing a well-defined ultimate compressive strength. The following table lists competing factors that influence the test method choice, many of which are generic while some are specific to conditions associated with low-temperature testing.

Tension	Compression
Advantages	
Common	Good for modulus and yield strength
Self-aligning	No grips
Well-defined gage section	No stress concentration in sample design
Good for modulus, yield, ultimate, and ductility parameters	Good for ultimate strength of brittle materials
	Easy sample installation
	Inexpensive sample cost
Disadvantages	
Sensitive to specimen design	End effects (friction/constraint)
Difficult to test brittle materials and composites where machining reduced section is not plausible	Sensitive to alignment
	Not always good for ultimate strength
	Need containment for fractured material

The temperature at which to run the test can be a simple determination such as when mechanical properties data for a material at the proposed service temperature are not available. Other cases are not so straightforward, and the temperature choice should be based on cost and the ability to provide conservative results. Sometimes, a material is to be used at a cold temperature, but testing it at room temperature will yield conservative data that are sufficient for the application. For many 4 K applications, conservative properties can be measured at 77 K in a simpler, more economical test. The degree of strengthening that will occur upon cooling from 77 to 4 K is much less than that which occurs from 295 to 77 K. When there is doubt about the applicability of data from tests at a temperature other than the service temperature, testing should be done at the service temperature. Good practice is to test above, below, and at the service temperature for a more complete understanding of the material behavior.

The relative costs and difficulty of the tests are important. Tests conducted in liquid media are simpler to perform, in general, than intermediate temperature tests that require temperature control. Below is a list of testing media and their associated temperatures (Ref 2).

Substance	Temperature, K	Bath type
Ice water	273	Slush
Isobutane	263	Liquid at BP
Carbon tetrachloride	250	Slush
Propane	231	Liquid at BP
Trichloroethylene	200	Slush
Carbon dioxide	195	Solid
Methanol	175	Slush
n-pentane	142	Slush
Iso-pentane	113	Slush
Methane	112	Liquid at BP
Oxygen	90.1	Liquid at BP
Nitrogen	77.3	Liquid at BP
Neon	27.2	Liquid at BP
Hydrogen	20.4	Liquid at BP
Helium (He_4)	4.2	Liquid at BP
Helium (He_3)	3.2	Liquid at BP

All temperatures given at 0.1 MPa (1 atm). BP, boiling point

Some of these substances are more common, cheaper, or easier to handle than others. The most commonly used substances in mechanical tests are ice water, CO_2/methanol slush, liquid-nitrogen (LN_2) cooled methanol, LN_2, and liquid helium (LHe). Obvious hazards are associated with the use of oxygen and hydrogen, and they should be avoided if possible. Safety issues concerning the use of cooled methanol, LN_2, and LHe are discussed subsequently in this chapter.

Cost of the cryogenic medium is also an issue. Since LN_2 is common and readily available, its cost is relatively low. LHe, on the other hand, is about a factor of ten times as expensive as LN_2. Liquid neon is sometimes used because it is easy to handle and its liquid boiling point temperature is relatively close to that of liquid hydrogen, but it can be 20 to 40 times as expensive as LHe. The sublimation temperature of dry ice (CO_2) is 195 K, and it can be used to cool a methanol or propanol bath with relative ease. Many of these bath cooling techniques are tried and true methods that require some practice to perfect but are usually inexpensive and simple ways to control test sample temperature. Low-temperature control can also be accomplished with electronic temperature control systems that utilize heaters and a cooling medium. Electronic temperature control systems are described in the following section.

Equipment

Low-temperature tensile tests can be performed on electromechanical or servohydraulic test machines with capacities of approximately 50 to 100 kN. The 100 kN machine is preferable for high strength materials such as steels or composites but of course larger or smaller capacities can be used as necessary. Direct tension tests usually require a simple ramp function that is possible on the more economical electromechanical (screw-drive) test machine. Computer controlled servo-hydraulic test systems are versatile and can perform a variety of tasks as well as direct tension tests.

To facilitate the low-temperature requirement, the test machine must be equipped with a temperature-controlled environmental chamber. One consideration for the suitability of the machine for low-temperature tests is the ease with which a low-temperature environmental chamber can be implemented. The physical characteristics of the test machine come into play, such as the maximum distance between crossheads and load columns.

A major factor to consider for cryogenic tests is the cryostat. "Cryostat" is a general term for an environmental chamber designed for cryogenic temperatures and can be as simple as a container (dewar) to hold a liquid cryogen. Cryostats designed for mechanical testing have the added requirement of providing structural support to react to tensile forces that are applied to the test material. Typically, a load frame is designed as an insert to a dewar. Since a dewar is a vacuum-insulated bucket to hold liquid, it is not advisable to have a hole in the bottom for pull-rod penetration because it introduces a leak potential for liquid, vacuum, and heat. The closed-bottom feature of a cryostat necessitates that the applied load and reacted load be introduced from the top. Cryostats are described further in the subsequent section "Environmental Chambers."

The simplest method to introduce the load path from the top on a servo-hydraulic machine is to use a machine that has the hydraulic actuator mounted on top of the upper crosshead. Hydraulic machines with this configuration are available, and the arrangement does not restrict normal use of the machine. Figure 2 shows a servo-hydraulic test machine equipped with a mechanical test cryostat. The screw-drive type test machine is usually accommodating and should have a movable lower crosshead with a through hole for the load train. References 8 and 9 give details of the design of cryostats for mechanical test machines.

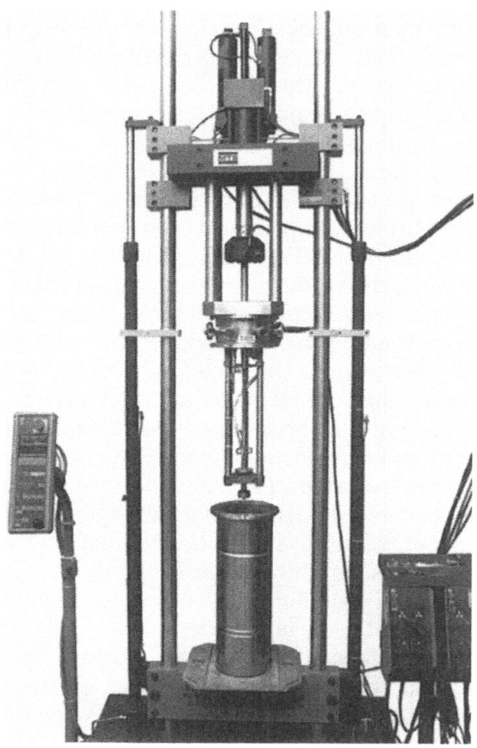

Fig. 2 A 100 kN capacity test machine equipped with cryostat for low-temperature testing

If the machine is not configured as described in the preceding paragraphs, the machine is relatively incompatible for cryogenic tests. Cryogenic tests on an incompatible machine require specially designed cryostats or an external frame system both of which are usually expensive and cumbersome alternatives. Figure 3 shows a schematic of a simple test chamber (canister) for immersion bath tests above liquid nitrogen temperature. This fixture provides an inexpensive method for conducting tests on conventional machines down to approximately 100 K.

Environmental Chambers. For low temperature tests, an environmental chamber is a thermal chamber that contains a gaseous or liquid bath media used to control the low temperature of a test. Sub-room-temperature environments are obtained with three basic chamber designs: a conventional refrigeration chamber; a thermally insulated box-container, or a cryostat designed for cryogenic temperatures with vacuum insulation; and thermal radiation shielding.

Conventional refrigeration covers the temperature range from +10 to −100 °C (+50 to −150 °F) and could be employed for tests in this range, much the same as furnaces are used on test machines to achieve elevated temperatures. Although mechanical refrigeration seems like a logical choice to cool environmental chambers, it is rarely used. This is probably because of the capital expense and the relative simplicity of other methods.

Commercial environmental chambers designed for use with test machines are available for controlling temperatures from approximately 800 K down to 80 K. Such chambers use electrical heaters for elevated temperatures and cold nitrogen gas cooling for sub-room-temperature. The cold nitrogen gas is supplied from a liquid nitrogen storage dewar. The flow of cold gas determines the cooling power and is controlled at the inlet with a variable flow valve that is regulated by the temperature controller. These systems are versatile in that a wide range of test temperature is possible with a single system. Some of the disadvantages are bulkiness, which can make setup difficult, and that the tests can be time consuming with respect to attaining equilibrated test temperatures.

As mentioned previously, cryogenic temperature tests are conducted in a cryostat. Reference 10 is an excellent historical perspective on low-temperature mechanical tests that details a number of cryostat designs, many of which use conventional machines with standard load path configurations. Pull-rod penetration through the bottom of a cryostat introduces a leak potential for liquid, vacuum, and heat and is not recommended for liquid bath-cooled tests. Modern mechanical test cryostats are typically a combination of a custom designed structural load frame fit into a commercial open-mouth bucket dewar. Some of the design details of a tensile test cryostat are shown in the schematic in Fig. 4 and photograph in Fig. 5. The design of the cryostat load frame is driven by engineering design factors such as cost, strength, stiffness, thermal efficiency, and ease of use. A good design philosophy is to produce a versatile fixture that can test a variety of specimens over a range of temperatures. The effect of lowering the temperature on the properties of a material can be evaluated by comparing the baseline room temperature properties. It is advisable to have the test apparatus capable of testing the material at both room temperature and cold temperatures. Construction materials used are austenitic stainless steels, titanium alloys, maraging steels, and FRP composites. For tensile tests, the cryostat frame reacts to the load in compression. The frame can be thermally isolated with low-thermal conductivity, FRP composite standoffs.

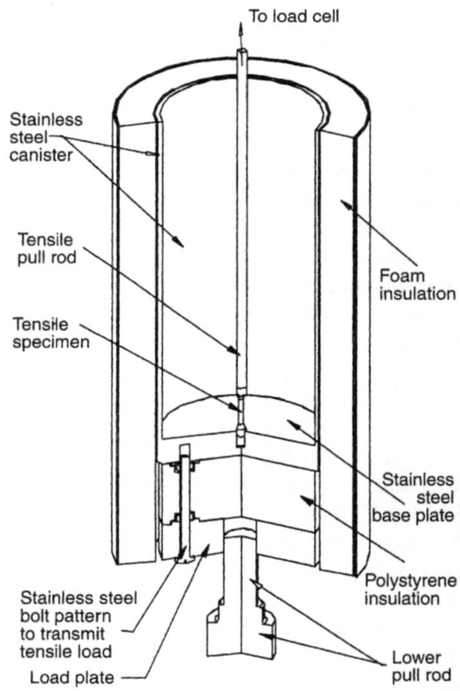

Fig. 3 Schematic of simple tensile canister from a standard-configuration machine for low-temperature testing

Cryogen Liquid Transfer Equipment. The supply and delivery of cryogenic fluids require special equipment. The equipment described here pertains to the use of the two most common cryogen test media, liquid nitrogen and liquid helium. Both liquid helium and nitrogen can be purchased from suppliers (usually welding supply distributors) in various quantities that are delivered in roll-around storage dewars. Liquid nitrogen can be transferred out of the storage dewar into the test dewar with simple or common tubing materials. Its thermal properties and inexpensive price allow its flow through uninsulated tubes. For example, butyl rubber hose can be attached to the storage dewar, and the hose will freeze as the liquid passes through. Liquid helium, on the other hand, is more difficult to handle, and it requires special vacuum-insulated transfer lines. Liquid helium transfer lines are usually flexible stainless steel lines with end fittings to match the inlet ports of the test cryostat and the supply cryostat. For both liquid nitrogen and helium, the storage tank is pressurized to enable transfer of the liquid.

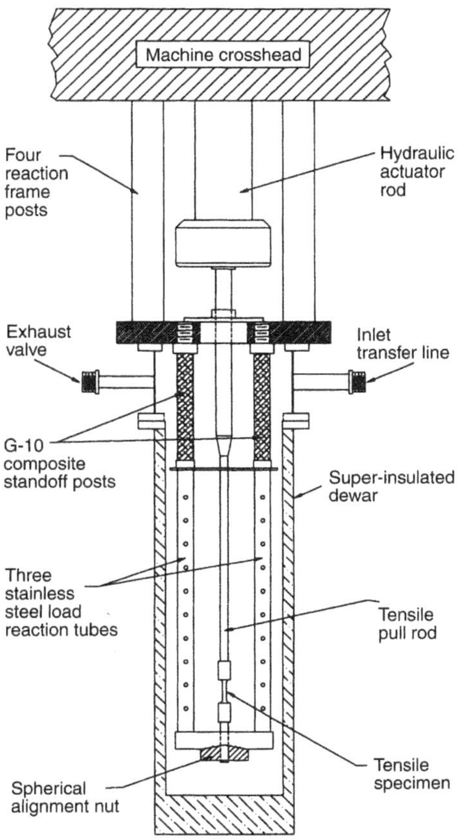

Fig. 4 Schematic of a tensile test cryostat

Instrumentation. The minimum instrumentation requirement in any tensile test is that for force measurement. Typically, forces are measured with the test machine force transducer (load cell) in the same manner as for forces measured in room temperature tests. During low-temperature tests, precautions should be taken to ensure the load cell remains at ambient room temperature.

Strain measurements may require temperature dependent calibration. Common strain measurement methods used are test machine displacement, bondable resistance strain gages, and clip-on extensometers. Also applicable to low-temperature strain measurements but less commonly used are capacitive transducer methods (Ref 11), noncontact laser extensomers, and linear variable differential transformers (LVDT) with extension rods to transmit displacements outside of the environmental chamber to the LVDT-sensing device.

Test machine displacement (stroke or crosshead movement) is a simple, low-accuracy method of estimating specimen strain. The inaccuracy comes because the displacement includes deflection of the test fixturing plus the test specimen gage section. Compensating for test fixturing compliance improves accuracy.

Bondable resistance strain gages are for sensitive measurements such as modulus and yield strength determination. The strain gage manufacturer supplies strain gage bonding procedures for use at cryogenic temperatures. The overall range of strain gages at cryogenic temperatures

Fig. 5 Tensile test cryostat. The force-reaction posts have fiber-reinforced polymer composite stand-offs.

is limited to about 2% strain. Applicable strain gages recommended by strain gage manufacturers have temperature dependent calibration data down to 77 K. Interest in their use down to 4 K has resulted in strain gage research verifying their performance to 4 K (Ref 12). A typical gage factor (GF) is 2 for NiCr alloy foil gages and it increases approximately 2 to 3% on cooling from 295 to 4 K. Thermal output strain signals are a large source of error that must be compensated for. Compensation is usually accomplished using the bridge balance of the strain circuit where zero strain can be adjusted to coincide with zero stress. If this is not possible, other steps must be taken to electrically or mathematically correct the thermal output strain.

Extensometers applicable to low-temperature tests utilize strain gages mounted to a bending beam element. The temperature sensitivity can be determined by calibrating with a precision calibration fixture that enables calibration at various temperatures. Depending on the accuracy desired, it is possible to use one or two calibration factors over a large temperature range. A typical strain-gage extensometer-calibration factor changes about ±1% over the temperature range from 295 to 4 K.

Temperature measurement is done with an assortment of temperature sensors. Reference 2 has a section devoted to temperature measurement at low temperatures. The most common method of temperature measurement is to use a thermocouple. Type E thermocouples (Chromel versus Constantan) and Type K (Chromel versus Alumel) cover a wide range of temperature and can be used at 4 K when carefully calibrated. A better choice of thermocouple, designed to have higher sensitivity at cryogenic temperatures, is a AuFe alloy versus Chromel thermocouple. Electronic temperature sensors (diodes and resistance devices) are available with readout devices that have higher precision than thermocouples. Silicon diodes, gallium-aluminum-arsenide diode, carbon glass resistor, platinum resistor, and germanium resistor are some of the more commonly used types of sensors.

Cryogenic temperature controllers that work with the types of temperature sensors named above are available. The majority of temperature controllers vary heating power and require that the test chamber environment is slightly cooler than the set-point temperature. The test engineer is responsible for the environmental chamber and cooling medium of the system. The controllers use the temperature sensors as the feedback sensor to operate a control loop and supply power for resistive heaters.

Additional Equipment Considerations. Teflon-insulated lead wires are advisable at very low temperatures because the insulation will be less likely to crack and cause problems. Electronic noise reduction can be an issue in low-temperature tests because lead wires tend to be long. Standard methods of noise reduction are shielding and grounding. Self-heating and thermocouple effects are important issues at low temperatures. Precautions should be taken to ensure that thermal effects do not mask the test data. Strain gage excitation voltages should be kept low. Reference 10 gives the parameters in terms of power density for calculating excitation voltage to be used for strain gages at 4 K.

Tensile Testing Parameters and Standards

As at room temperature, tensile tests at low temperature are for determining engineering design data as well as for studying fundamental mechanical-metallurgical behaviors of a material such as deformation and fracture processes. The usual engineering data from tensile tests are yield strength, ultimate tensile strength, elastic modulus, elongation to failure, and reduction of area. The effects of material flaws (inclusions, voids, scratches, etc.) are amplified in low-temperature testing, as materials become more brittle and sensitive to stress concentrations. Data scatter tends to increase, and the quantity of tests to characterize a material is usually greater than that used for room temperature testing. The test engineer must judge when sufficient testing has been done to provide representative data on a material.

Test fixture alignment is important at low temperatures because of necessarily long load trains. Self-alignment in tensile tests can be accomplished through the use of universal joints, spherical bearings, and pin connections. The alignment should meet specifications detailed in ASTM E 1012, "Standard Practice for Verification of Specimen Alignment Under Tensile Loading." Strain measurement should be done using an averaging technique that can reduce errors associated with misalignment or bending stress. Strain measurement equipment is detailed above in the instrumentation section.

Metals. The standard tensile test method for metals, ASTM E 8, covers the temperature range from 10 to 40 °C (50 to 100 °F) and is used as a guideline for lower temperature tests. The need for engineering data in the design of superconducting magnets has resulted in the adoption of the tensile test standard ASTM E 1450 for tests of structural alloys in liquid helium at 4.2 K.

The strain rate sensitivity of the flow stress in metals decreases as temperature is reduced. Typical strain rates in standard tensile tests are on the order of 10^{-5} s^{-1} to 10^{-2} s^{-1} and do not have a pronounced effect on the material flow stress. The strain rate becomes important in cryogenic temperature tests because of a tendency for specimen heating causing discontinuous yielding in displacement control tests. Discontinuous yielding is a subject of low-temperature research of alloys, well described in ASTM E 1450. The localized strain/heating phenomenon typically initiates after the onset of plastic strain and results in a serrated stress-strain curve. ASTM test standard E 1450 prescribes a maximum strain rate of 10^{-3} s^{-1} and notes that lower rates may be necessary. The strain required to initiate discontinuous yielding increases with decreasing strain rate. If discontinuous yielding starts before the 0.2% offset yield strength is reached, the associated load drop affects the estimation of the yield strength. It may be possible to slow the strain rate to postpone the serrated curve until after the 0.2% offset yield strength is reached and then to increase the rate, not to exceed 10^{-3} s^{-1}. Reference 13 reports research on the effect of strain rate in tensile tests at 4 K.

Test specimen sizes are preferably small for low-temperature tests. The common 12.7 mm (0.5 in.) round, ASTM-standard tensile specimen is rarely used at low temperature. Tensile specimens should be small due to size constraints placed by the environmental test chamber, which is designed for thermal efficiency. Standard capacity test machines (100 and 50 kN) favor small specimens due to high tensile strengths encountered at low temperatures. A subscale version of the 12.7 mm (0.5 in.) round that meets ASTM specifications and works well at cryogenic temperature is shown in Fig. 6(a). A 100 kN force capacity test machine can generate about 3.5 GPa stress on a 6 mm diameter gage section. Figure 6(b) shows a flat, subscale tensile specimen that is also commonly used at cryogenic temperatures.

Polymers and Fiber-Reinforced Polymer (FRP) Composites. Tensile tests of FRP composites are governed in test procedure ASTM D 3039, while polymers and low modulus (<20 GPa) composites are tested using the guidelines established in ASTM D 638. Neither have specific temperature ranges or limitations.

The problem with testing polymers at low temperatures is the tendency for polymers to be extremely brittle materials. Test temperatures below room temperature are usually well below the glass transition temperature, T_g, of the polymer. The test specimen designs in ASTM D 638 are susceptible to grip failures once the material is brittle. Traditional strain measurement techniques must be performed carefully. Clip-on extensometers must mechanically attach to the material, usually producing some sort of stress concentration that may initiate failure. Strain gages locally reinforce low-modulus material, and the associated error and correction method

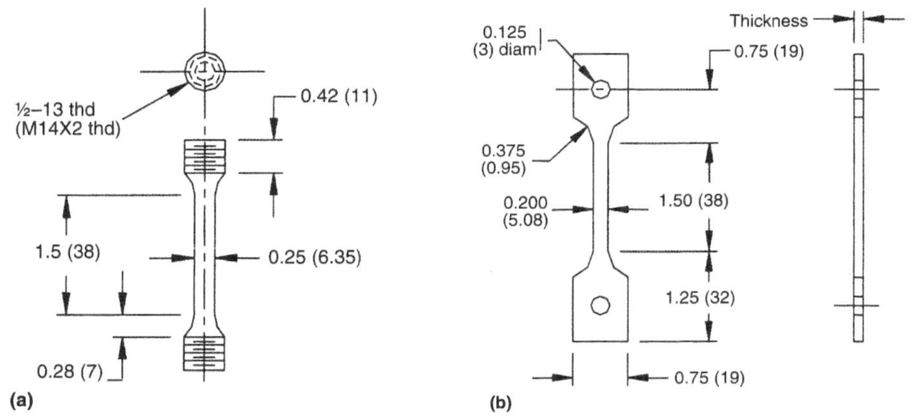

Fig. 6 Schematics of tensile specimen commonly used at low temperature. (a) Round. (b) Flat. Dimensions are in inches (millimeters). thd, threaded

is explained in Ref 14. Most tensile tests of polymers show an increase in tensile strength upon cooling from 295 to 77 K and a decrease or constant level of strength with continued cooling to 4 K (Ref 15). One would expect the strength to continue to increase with decreasing temperature. This anomalous behavior is probably an artifact of the tensile testing of an extremely brittle material.

Tensile tests of FRP composites at low temperatures are simpler than tests of neat polymers because of the more rugged sample. One challenge is the gripping of high-strength, unidirectionally reinforced composites. The convenience of hydraulic wedge grips is not usually an option in low-temperature tests. An example of a low-temperature tensile test program to characterize a unidirectionally reinforced epoxy composite from 295 to 4 K is described in Ref 16.

Temperature Control

Test temperature is controlled by a bath temperature or by controlling the temperature of a gaseous environment. The temperature of the test specimen should be maintained ± 1 K for the duration of the test. For bath-cooled tests, the temperature of the bath and the specimen should be the same. A potential source for error is the conduction path or heat sink that the load train provides, possibly causing the specimen to be warmer than the bath. For tests at temperatures below 77 K (typically 4 K tests), the test cryostat is precooled with liquid nitrogen as an economical time-saving step.

The temperature of the specimen is usually monitored for tests using gaseous environment temperature control. The temperature should be measured at the gage section and at the gripped ends to ensure that the temperature across the length of the specimen is constant. For temperatures above 80 K, the cooling can be a static method, such as a pool of liquid nitrogen in the dewar below the test fixturing. Test temperatures below the liquid nitrogen bath temperature are best accomplished through the use of cold helium gas for cooling power. This can be accomplished with a flow cryostat, where cooling power is regulated by throttling the flow of the cold gas or liquid. Between the manual regulation of the cooling medium and the regulated heater power, one can obtain constant test temperatures between 77 and 4 K. Variations of cryogenic temperature control exist such as cryogenic refrigerators, which can be applied to the temperature control of mechanical tests. Cryocoolers and other mechanical refrigeration techniques are not commonly used in mechanical tests, due to the initial capital expense and the relative simplicity of other cooling methods.

Safety

Safety in a laboratory or industrial setting is always a concern. Some of the safety issues with respect to low-temperature testing are common sense issues, while some others are not so obvious. Table 1 describes the most common safety issues and appropriate solutions for each. Safety issues associated with the use of liquid

Table 1 Safety issues associated with the use of liquid cryogens for low-temperature testing

Safety issue	Solution
Liquid cryogens can spill or splash onto the body and cause freezing of human tissue.	Personnel should wear appropriate clothing and avoid direct contact with cold parts.
Helium, nitrogen, or carbon dioxide can displace air in a confined area.	Cryogens should be used only in well-ventilated areas.
Volumetric expansion is extremely high (typically 700–1000 times) when a liquid cryogen vaporizes. Such expansion is dangerous when it occurs in closed containers or fixture components with potential for trapped gas/liquid volumes.	Cryostats must include safety pressure-relief capabilities; redundancy is necessary because mechanical relief valves may freeze or malfunction. Component parts such as tubes or threaded connections that can trap liquids or gases should be identified, and solutions (such as weep holes) included in the design. If there is potential for liquid to get into a space, provide exit relief rather than trying to seal the liquid out.
Oxygen-rich (flammable) condensation can form on chilled surfaces (surfaces that are chilled to temperatures below 90 K and then exposed to air; common on uninsulated liquid nitrogen transfer lines and inside open-mouth dewars containing liquid nitrogen or helium residue).	Inform personnel of potential fire hazard, and take appropriate precautions.
Low temperatures may cause embrittlement of the material, causing it to fail at lower-than-expected loads.	Test personnel must be aware of the potential and prepare for brittle fracture of structural components.

hydrogen or liquid oxygen are not dealt within this article and can be found in Ref 10. In addition to the issues listed in Table 1, personnel should consider general safety issues related to tension and compression testing at all temperature ranges.

ACKNOWLEDGMENT

This chapter was adapted from R.P. Walsh, Tension and Compression Testing at Low Temperatures, *Mechanical Testing and Evaluation,* Vol 8, *ASM Handbook,* ASM International, 2000, p 164–171.

REFERENCES

1. E.R. Parker, *Brittle Behavior of Engineering Structures,* John Wiley & Sons, 1957
2. R.P. Reed and A.F. Clark, Ed., *Materials at Low Temperatures,* ASM, 1983
3. L.E. Neilsen and R.F. Landel, *Mechanical Properties of Polymers and Composites,* Marcel Dekker, NY, 1994, p 249–263
4. M.B. Kasen et al., Mechanical, Electrical, and Thermal Characterization of G-10CR and G-11CR Glass-Cloth/Epoxy Laminates Between Room Temperature and 4 K, *Advances in Cryogenic Engineering,* Vol 28, 1980, p 235–244
5. R.P. Reed and M. Golda, Cryogenic Properties of Unidirectional Composites, *Cryogenics,* Vol 34 (No. 11), 1994, p 909–928
6. E.P. Popov, *Mechanics of Materials,* 2nd ed., Prentice-Hall, NJ, 1976
7. J.P. Hirth and M. Cohen, *Metalls. Trans.,* Vol 1, Jan 1970, p 3
8. G. Hartwig and F. Wuchner, Low Temperature Mechanical Testing Machine, *Rev. Sci. Instrum.,* Vol 46, 1975, p 481–485
9. R.P. Reed, A Cryostat for Tensile Test in the Temperature Range 300 to 4 K, *Advances in Cryogenic Engineering,* Vol 7, Plenum Press, NY, 1961, p 448–454
10. J.H. Lieb and R.E. Mowers, Testing of Polymers at Cryogenic Temperatures, *Testing of Polymers,* J.V. Schmitz, Ed., Vol 2, John Wiley & Sons, 1965, p 84–108
11. R.P. Reed and R.L. Durcholz, Cryostat and Strain Measurement for Tensile Tests to 1.5 K, *Advances in Cryogenic Engineering,* Vol 15, Plenum Press, NY, 1970, p 109–116
12. C. Ferrero, Stress Analysis Down to Liquid Helium Temperature, *Cryogenics,* Vol 30, March 1990, p 249–254
13. R.P. Reed and R.P. Walsh, Tensile Strain Effects in Liquid Helium, *Advances in Cryogenic Engineering,* Vol 34, Plenum Press, 1988, p 199–208
14. C.C. Perry, Strain Gage Reinforcement Effects on Low Modulus Materials, *Manual on Experimental Methods for Mechanical Testing of Composites,* R.L. Pendelton and M.E. Tuttle, Ed., Society for Experimental Mechanics, 1989, p 35–38
15. R.P. Reed and R.P. Walsh, Tensile Properties of Resins at Low Temperatures, *Advances in Cryogenic Engineering,* Vol 40, Plenum Press, NY, 1994, p 1129–1136
16. R.P. Walsh, J.D. McColskey, and R.P. Reed, Low Temperature Properties of a Unidirectionally Reinforced Epoxy Fibreglass Composite, *Cryogenics,* Vol 35 (No. 11), 1995, p 723–725

CHAPTER 15

High Strain Rate Tensile Testing

HIGH STRAIN RATE TENSILE TESTING is necessary to understand the response of materials to dynamic loading. Strain rates ranging from 100 s^{-1} to $>10^4$ s^{-1} occur in many processes or events of practical importance, such as foreign object damage, explosive forming, earthquakes, blast loading, structural impacts, terminal ballistics, and metalworking. The behavior of materials under high strain rate tensile loads may differ considerably from that observed in conventional tensile tests.

High strain rate sensitivity is primarily manifested in variations in yield and failure criteria. Yielding and failure are also affected by stress state, ratio of mean stress to deviatoric stress, stress amplitude, stress history, and temperature. Tests must be designed to simulate the most relevant load characteristics. For example, many processes involving dynamic tensile stress include compressive prestress. Strain rate sensitivity also depends on whether engineering or true strain formulations are used, because local instabilities (such as necking) are often suppressed at high rates.

Measurement of strain is a major problem in high strain rate tensile testing. In quasi-static testing, the diameter of the minimum cross section in a cylindrical specimen can be measured; such measurements are virtually impossible or highly impractical in high-rate testing. Furthermore, although strains are easily measured over a uniform gage length section in quasi-static testing, the same measurements are considerably more difficult to obtain at high strain rates. Mechanical extensometers are the primary tool used in quasi-static tests, but they are of little use at high rates of strain because of the effects of inertia.

Therefore, high-rate tests use strain gages, optical extensometers, and displacement measurements between loading fixtures to determine or infer the dynamic tensile strains in a test specimen. At very high rates of strain, strains may be measured in some experimental configurations only through wave propagation analysis. This procedure generally requires that assumptions be made about the constitutive behavior, that wave propagation analysis be carried out, and that predictions and experimental observations be compared. Unique solutions cannot be guaranteed, because some other constitutive model may conceivably provide similar results in a particular wave propagation problem.

Conventional Load Frames

Strain rate effects in tension are determined by performing conventional tensile tests at varying loading rates up to approximately 100 s^{-1}. Conventional test machines are available with increased ram velocities, as are high-speed pneumatic and hydraulic machines. The speed capability of a machine may be influenced by several factors. Speed may be a function of the load that the ram is attempting to apply, and the no-load speed may be much higher than the full-load speed. The distance traveled may also affect the speed capability. A long stroke machine may attain a given speed only after a significant amount of travel. Depending on the specimen length, considerable specimen strain could occur before final maximum velocity is obtained. Finally, the ability to control speed is a function of the response capability of a servo-controlled machine working in a closed-loop mode. Open-loop machines provide speeds that may be influenced by specimen strength and cannot easily reproduce predetermined velocities or strain rates on materials with different yield strengths or strain-hardening behaviors. Additional information on the operational characteristics of conventional tensile testing machines can be found in Chapter 4, "Tensile Testing Equipment and Strain Sensors."

Effects of Inertia and Wave Propagation. A fundamental difference between a high strain rate tension test and a quasi-static tension test is that inertia and wave propagation effects are present at high rates. It must be determined how fast a uniaxial tension test can be run to obtain valid stress-strain data. To determine this, consider a specimen of initial length L subjected to a uniform velocity v_0 at time $t = 0$, as shown in Fig. 1. This hypothesis could represent a test in a constant crosshead velocity testing machine, or a drop-weight type of test in which a large mass impacts one end of the specimen. If (x,t) denotes the displacement of any point in the x direction and assuming purely uniaxial motion—that is, neglecting radial inertia effects—the equation of motion is:

$$\frac{\partial^2 u}{\partial t^2} = c^2 \frac{\partial^2 u}{\partial x^2} \qquad \text{(Eq 1)}$$

where u is displacement and

$$c = \left(\frac{E}{\rho}\right)^{1/2} \qquad \text{(Eq 2)}$$

is the longitudinal wave velocity in a bar or rod, where E is Young's modulus, and ρ is mass density. Applying the boundary conditions of the left end fixed and the right end moving with constant velocity v_0 and assuming initial conditions of zero displacement and velocity, the solution is:

$$u(x,t) = \frac{v_0 L}{2c}\left[f\left(\tau + \frac{x}{L}\right) - f\left(\tau - \frac{x}{L}\right)\right] \qquad \text{(Eq 3)}$$

where $\tau = tc/L$ is a dimensionless time, and $\tau = 1$ represents the time it takes a wave to propagate the length of the specimen. The function $f(\tau)$ is shown in Fig. 2. Strain can be obtained from $\varepsilon = \partial u/\partial x$ and stress from $\sigma = E\varepsilon$. By introducing the dimensionless variables:

$$\xi = \frac{x}{L} \qquad \text{(Eq 4a)}$$

$$v^* = \frac{v_0}{c} \qquad \text{(Eq 4b)}$$

plots of stress and strain can be constructed as a function of time. Figure 3 illustrates strain normalized with respect to v^* against dimensionless time τ at an arbitrary position ξ along the bar. The dashed line indicates the average strain in the bar, which is normally total displacement divided by bar length.

The localized strain is measured by a strain gage with a gage length that is small compared to the length of the specimen. Figure 4 illustrates the normalized stresses at both ends of the bar. The response at the fixed end is recorded by a load cell. Figures 3 and 4 illustrate that stresses and strains accumulate from numerous waves propagating back and forth in the bar. Note that the solution to the mathematical problem has assumed an instantaneous jump in velocity at $t = 0$, whereas some finite rise time usually occurs

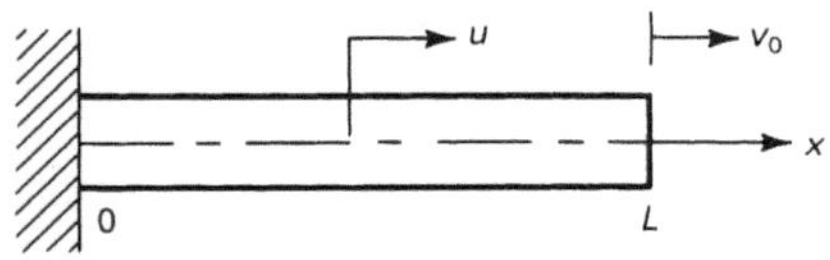

Fig. 1 Schematic of tensile test configuration. See text for details and explanation of symbols.

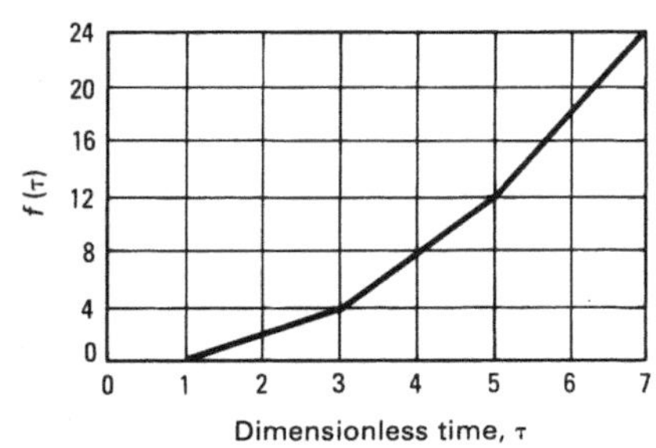

Fig. 2 Graph of the function $f(\tau)$. See text for details and explanation of symbols.

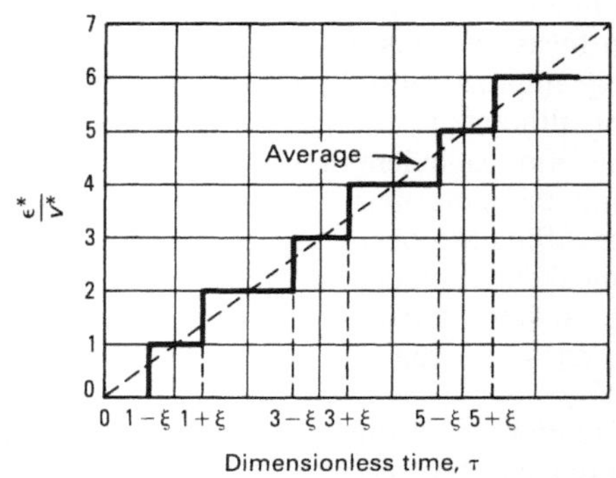

Fig. 3 Nondimensional strain profile. See text for details and explanation of symbols.

because of imperfect impact or machine response.

If many wave transits occur during a test, the use of average stresses and strains appears justified. However, if the velocity is high, then only a few wave reflections may occur before the specimen fails. In this case, individual wave propagation must be considered; average values alone cannot be considered, and the use of this test to determine dynamic stress-strain response is precluded. Note that this analysis is based on a material that is linear-elastic and assumes a zero rise time in the applied velocity. Stress waves are propagated at the elastic wave velocity. With material that has deformed into the plastic region, the plastic wave velocity is more appropriate and generally can be an order of magnitude smaller than the elastic wave velocity.

One factor in determining whether or not wave propagation effects limit the validity of a tensile test is the sample ring-up time, which is the time required for a sample to achieve a uniform state of stress. Generally, measurements are not valid for times such that $L \sim ct$. This corresponds to a situation in which strain $\varepsilon \gg v/c = \dot{\varepsilon}L/c$. Consequently, small strain measurements are difficult to obtain at very high strain rates. Another concern is that local failure may occur at the end to which the load is applied. The magnitude of the stress transient associated with the sudden application of velocity v_0 is $\sigma_m = \rho c v_0$.

The test must be designed so that $\sigma_m < Y$, the yield stress. For example, consider a bar 25 mm (1 in.) in length that is accelerated at one end to 2.5 m/s (8.2 ft/s). For many engineering materials, including steels, aluminum alloys, and titanium alloys, the elastic wave velocity is about 5000 m/s (16,400 ft/s). The maximum stress generated at the accelerated end of the bar is $\rho c v$. For a steel bar, the first stress pulse is 100 MPa (14.5 ksi), and the average strain rate is 100 s^{-1}. If a steel with a strength of 1 GPa (145 ksi) is being tested, the maximum allowable driving velocity is 25 m/s (82 ft/s). At that velocity, instantaneous failure would occur at the driven end.

Assuming that wave propagation effects may be neglected in a given test, the second aspect that must be checked is the response of the load cell. Load cell ringing is frequently encountered in high-rate tensile testing. Generally, this time period (reciprocal of the natural frequency in hertz) must be small compared to the total duration of the test. For example, if a load cell has a natural frequency of 1 kHz, its period of vibration is 10^{-3} s. This load cell could then be used only for experiments that lasted over ten times that amount, or over 10 ms.

Another condition that must be satisfied is the distance of the load cell from the end of the specimen. If a sufficient distance exists between the specimen and load cell, the finite elastic wave transit time may result in load data that are not time-coincident with strain data. To prevent phase lags from obscuring the experimental data, the wave transit time from the specimen to load cell should be negligibly small compared to the test duration. Otherwise, the load data must be corrected for the delay, and such corrections seldom are precise.

Strain Measurement. The final aspect of high-speed tensile testing is determination of strain. The most direct, reliable method uses electrical resistance strain gages. The frequency response capability of strain gages is considerably greater than the mechanical response of the combination of load train, specimen, and load cell.

Another method of measuring strain involves the use of optical extensometers, in which displacement measurements across the loading fixtures are divided by an actual or effective gage length. When using crosshead displacement measurements, caution must be exercised to ensure that these represent only specimen elongation and not machine, ram, or load train elongations. The same precautions that apply in quasi-static tests also apply in dynamic tests.

If the above precautions are observed, valid stress-strain data can be obtained up to maximum strain rates in the range of 10 to 100 s^{-1}. For higher strain rates, or for cases in which the

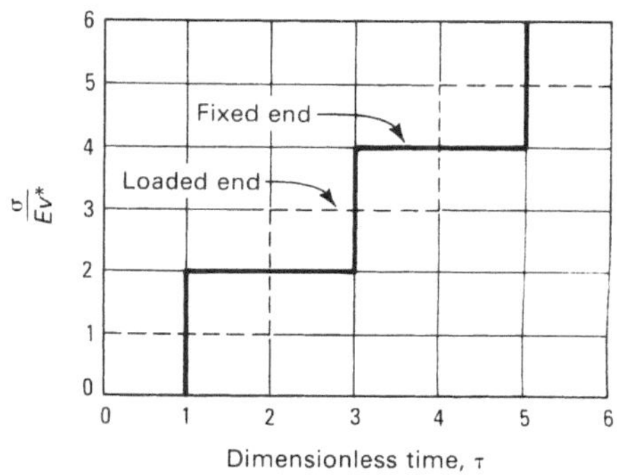

Fig. 4 Stress history at ends of bar. See text for details and explanation of symbols.

above criteria are not met, highly specialized testing techniques may have to be used. As discussed below, these include:

Applicable strain rate, s^{-1}	Testing technique
10^4	Expanding ring test
$>10^5$	Flyer plate test
$100–10^3$	Split-Hopkinson pressure bar test
$10^3–10^4$	Rotating wheel test

Expanding Ring Test

The expanding ring test is a highly sophisticated technique for subjecting metals to tensile strain rates over 10^4 s^{-1} (Ref 1, 2). Although the testing principle is simple, its performance requires specialized equipment available in only a few laboratories. The ring test can determine the high-rate stress-strain relationships, but a simplified, more widely used version can be employed to determine ultimate strain only (Ref 3, 4).

This test involves the sudden radial acceleration of a ring due to detonation of an explosive charge or electromagnetic loading. The ring rapidly becomes a free-flying body, expanding radially, and decelerating due to its own internal circumferential stresses. A thin ring must be used for the analysis to be valid; the wall thickness should be less than one tenth the ring diameter, which is typically 25 mm (1 in.). If R is the radius of the ring, ρ the density, and σ the hoop stress:

$$\sigma = -\rho R \frac{d^2R}{dt^2} \qquad \text{(Eq 5)}$$

To obtain stress-strain data, radial displacement as a function of time must be calculated. Strain is proportional to change in radius (just as engineering strain in tension is $\Delta L/L_0$); thus:

$$\varepsilon = \ln \frac{R}{R_0} \qquad \text{(Eq 6)}$$

where R_0 is the initial radius. Stress may be computed from Eq 5 by double differentiation of radial displacement data as a function of time. Ring displacement can be obtained through the use of high-speed photography, streak cameras, displacement interferometers, or other methods for measuring radius as a function of time.

It is difficult to determine stress accurately by double differentiation of displacement data. Several laboratories have used a laser velocity interferometer to measure ring velocity directly (Ref 5, 6). Thus, only a single differentiation is necessary to calculate stress, and precision is improved considerably.

Advantages of the Ring Test. The ring test has two principal advantages. The expanding ring test subjects the material to a state of dynamic uniaxial stress without the wave propagation complications that accompany other high strain rate tests. Also, the maximum strain rate available in the ring test is higher than in any other common tension tests involving large plastic strains.

Limitations of the Ring Test. Strain rate in the expanding ring test is not usually constant. The strain rate is computed from $(dR/dt)/R$, and both of these terms vary continually. Strain rate is usually greatest at the start of ring deceleration, when strain is smallest. Values in excess of 10^4 s^{-1} are readily obtained. If the ring does not rupture, the strain rate falls to zero at the end of the test.

Ring specimens also experience a compressive preload in the radial direction that often exceeds the yield stress during the acceleration phase. Because load history is known to affect the subsequent stress-strain behavior of many materials, data obtained from expanding ring tests do not always agree with results from other tests at slightly lower strain rates.

The difficulties, expense, and limitations of the expanding ring test preclude its use as a standard test technique for generating high strain rate stress-strain data in tension. Only a few laboratories are capable of performing this test. However, if subjecting a material to high strain rates in tension without determining stress-strain data is of primary interest, the expanding ring test is much easier to conduct. A number of investigators have used this test to determine strain to failure under dynamic loading (Ref 3, 4). Here, the accurate determination of radial displacement versus time is not as critical, because stresses are not calculated. Less precise displacement data provide reasonably accurate determinations of strain rate. The ambiguity arising from possible strain rate history effects still exists when the expanding ring test is used in this simpler manner.

The expanding cylinder test, a variation of the ring test, provides a dynamic stress state equivalent to that produced in a quasi-static tensile test on a wide sheet versus a thin strip of material. A difficulty encountered in this type of test is the need for an impulse to be generated simultaneously in time along the axis of the cylinder. Because explosive detonation along a wire, for example, propagates at a finite wave speed, uniform deformation along the length of the axis cannot be ensured. Dimensions, detonation wave speeds, and synchronization of multiple detonation all must be considered carefully to ensure that the cylinder is deformed as uniformly as possible and that axial stress waves are not generated (Ref 7).

Flyer Plate and Short Duration Pulse Loading

Traditionally, flat plate impact tests have been used to obtain high strain rate yield data, shock wave response data, and equation of state data for materials undergoing uniaxial strain. Uniaxial strain refers to a three-dimensional state of stress in which deformation or strain occurs in only one direction—the direction of loading. The uniaxial strain condition persists for only a short period of time, until stress waves originating at lateral boundaries reach the specimen interior. In a typical experiment, this time period is on the order of several to tens of microseconds. Uniaxial strain is defined mathematically as:

$$u_x \neq 0,\ u_y = u_z = 0 \quad \text{(Eq 7)}$$

where x is the direction of loading; u_x is the displacement in that direction; and y and z are orthogonal directions in a plane normal to x. The strains are obtained from the displacement derivatives, thus:

$$\varepsilon_x \neq 0,\ \varepsilon_y = \varepsilon_z = 0 \quad \text{(Eq 8)}$$

The flat plate impact test is performed by launching a flat flyer plate against a second stationary target plate. Compressed gas guns, propellant guns, magnetic accelerators, and explosives have all been used to launch the flyer plate (Ref 8). Extreme precision must be achieved to eliminate relative tilt at the instant of impact. A typical experimental setup using a gas gun is shown in Fig. 5. The flyer plate is carried in the gas gun in a plastic sabot. Velocity of the flyer is determined from the transit time between the shorting pin in the gun barrel and time-of-arrival pins in the target. The target is supported by a spall ring that suppresses late-time radial tensile waves.

The stress waves along the axis normal to the impact plane are shown in Fig. 6. A flyer plate of thickness d, moving left to right, strikes an initially stationary target of thickness T; the impact occurs at the origin, O, of the (x,t) coordinates. Elastic-plastic behavior is assumed in Fig. 6. Elastic waves propagate at approximately c_L, the longitudinal elastic sound speed. Plastic waves propagate at approximately $\sqrt{(B/\rho)}$, where B is the bulk modulus. The arrivals of the elastic and plastic waves at the target rear surface are denoted as E and P.

Propagation speeds are always relative to the material into which the wave is moving. Strain occurs only at the wave fronts. The amplitude of the E wave in Fig. 6 is known as the Hugoniot elastic limit (HEL) and is simply related to the uniaxial yield stress, Y, as:

$$\sigma_{\text{HEL}} = \left[\frac{B + 4\mu/3}{2\mu}\right] Y \quad \text{(Eq 9)}$$

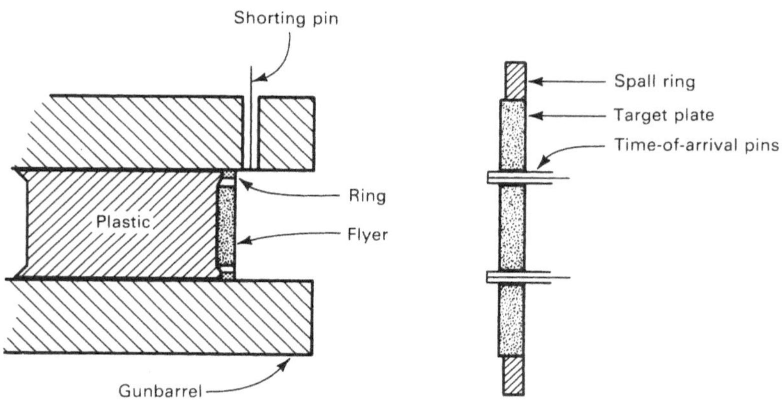

Fig. 5 Schematic of gas-gun-launched flyer plate impact test setup

where μ is the shear modulus. The final state of the shocked material is characterized by a stress and particle velocity. The functional relationship between these two variables depends on the material and is known as the Hugoniot. The state behind the P wave in Fig. 6 lies on the Hugoniot (see Ref 9 for a discussion of Hugoniots). If the flyer and target plates are composed of the same material, the particle velocity behind the P wave is one half the impact velocity.

Reference 10 discusses determination of particle velocity when the flyer and target are composed of different materials. If the Hugoniot of the target is known, then the stress can be calculated from the particle velocity. Hugoniots for most engineering metals can be found in Ref 11.

Compressive waves reflect from a free surface as tensile (rarefaction) waves, which begin to arrive at the target rear surface at point R in Fig. 6. The tensile (rarefaction) waves may interact and cause spall failure. This causes material separation in the target, which is indicated at point SP in Fig. 6. The sudden relaxation of tensile stress generates a shock wave that arrives at the free surface at point S.

Spall is a form of tensile failure under an extremely high strain rate and a nearly spherical stress tensor. Spall usually is characterized by the spall stress, σ_{spall}, defined as the highest tensile stress that exists in the material prior to rupture. When designing spall experiments, the flyer plate diameter, a, must be large enough so that the phenomena of interest occur within a time $a/2c_L$ after the impact.

Flat plate impact tests normally are used to measure σ_{HEL} and spall strength. For example, consider the characterization of a steel by this technique. The value of σ_{HEL} for steel is usually between 5 and 15 kilobar (kbar), a useful unit for analyzing shock experiments; 1 kbar = 0.1 GPa. When density is expressed as g/cm^3 × 10 and velocity is given in km/s (or, equivalently, mm/μs), stress is given in kbar.

To measure the Hugoniot elastic limit, the impact velocity must be sufficient for the peak stress to exceed σ_{HEL}. Peak stress is given by:

$$\sigma = \rho U u \qquad \text{(Eq 10)}$$

where U is shock propagation speed, and u is particle velocity. Peak particle velocity is half the impact velocity, u_0, for a symmetric impact. For steel-on-steel impacts, Eq 10 becomes approximately $\sigma = 200\ u_0$. For $\sigma > \sigma_{HEL} = 15$ kbar, $u_0 > 75$ m/s (245 ft/s) is required. This presents no problem when a gas gun is used.

Experiments with $u_0 < 100$ m/s (<330 ft/s) are often difficult because of impact tilt, which becomes more critical at low velocities. Also, impact velocity must not be so high that the velocity of the P wave (Fig. 6) exceeds c_L. That limit for steels usually is greater than 1 km/s (0.6 mile/s). The limit for other materials can be found by consulting the tables in Ref 11.

Given an appropriate impact velocity, to determine σ_{HEL} one of the following measurements must be made. The peak particle velocity behind the E wave can be measured. This can be accomplished at the free surface with capacitor gages, sloping mirrors, or a velocity interferometer. The velocity behind the wave is half the free surface velocity. The stress is related to the free surface velocity by Eq 10 with $u = c_L$.

Direct measurement of σ_{HEL} can be obtained by embedded piezoresistive gages. Manganin and carbon gages frequently are used for this purpose. This technique requires sectioning the target or using a backing plate and correcting for partial transmission of the wave transmitted through the interface. Magnetic particle velocity gages can be used for nonconducting targets such as plastics and rocks, but they are not suitable for metals.

Spall stress can be determined by two methods. The simplest, in terms of analysis, interpre-

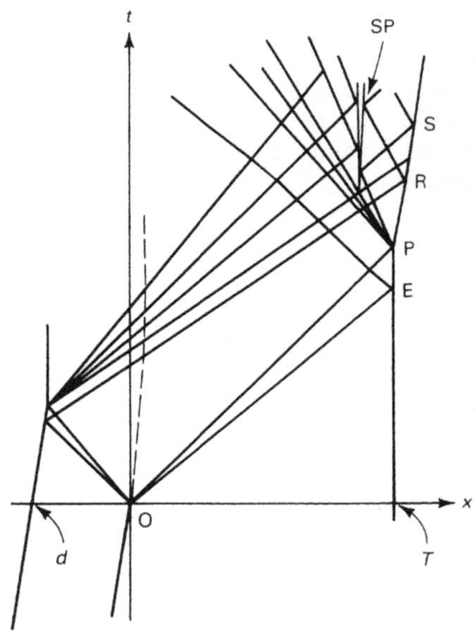

Fig. 6 Lagrangian diagram showing stress waves in flyer plate experiment. See text for details and explanation of symbols.

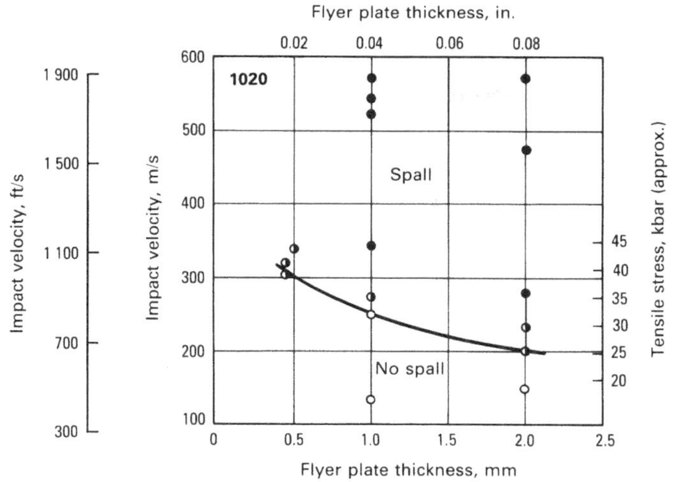

Fig. 7 Spall data for low-carbon 1020 steel

tation, and experimental technique, is to vary systematically the flyer plate thickness, d, and impact velocity, u_0, to determine the critical values at which rupture occurs. As the flyer plate thickness is increased, the duration of the compressive and tensile load increases; the load duration is approximately $2d/c_L$.

Eventually, for flyer plate thicknesses exceeding about 5 mm (0.2 in.), the spall stress reaches a load duration limit. In many metals, the limiting spall strength is several times the value of σ_{HEL}. Figure 7 illustrates typical spall stress data for low-carbon steel. The data illustrate that for pulse durations longer than a few microseconds, the greatest tensile stress that the material can sustain without rupture is 25 kbar.

Interpretation of experiments using thinner flyer plates is more complex, because a computer code must be used to calculate the stress history on the spall plane. Finite difference codes (Ref 9) or method of characteristics codes (Ref 12) can be used. Finite difference codes are more accurate and more widely applicable than method of characteristics codes, but the user must be specially trained in this subject.

Another approach to spall characterization is to initiate impact above the spall threshold and to deduce the material behavior from the free surface velocity, Δv_s, data. Figure 8 illustrates a typical free surface velocity history with spalling, E, P, R, and S refer to the same arrivals as explained in text for Fig. 6. The spall stress is given approximately by $\rho c_L \Delta v_s/2$. However, a more exact determination requires code analysis.

The Split-Hopkinson Pressure Bar Technique (Ref 13)

The split-Hopkinson pressure bar (SHPB) technique is named for Bertram Hopkinson who, in 1914, used the induced-wave propagation in a long elastic metallic bar to measure the pressures produced during dynamic events. Through the use of momentum traps of differing lengths, Hopkinson studied the shape and evolution of stress pulses as they propagated down long metallic rods as a function of time. Based on this pioneering work, the experimental apparatus using elastic stress-wave propagation in long rods to study dynamic processes in materials was named the Hopkinson pressure bar. Later work used two Hopkinson pressure bars in series, with the sample sandwiched in between, to measure the dynamic stress-strain response of materials.

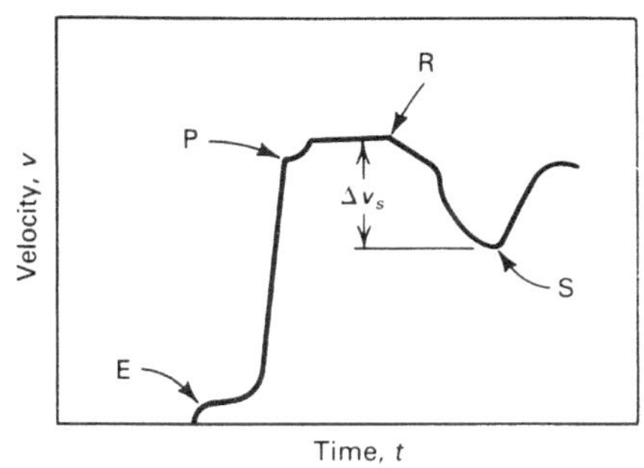

Fig. 8 Free surface velocity data when spall occurs. See text for details and explanation of symbols.

This technique was referred to as the split-Hopkinson pressure bar. Although the original split-Hopkinson pressure bar apparatus was developed to measure compressive mechanical behavior of materials, alternate Hopkinson bar schemes were later developed for loading the samples in uniaxial tension and torsion. Information on split-Hopkinson pressure bar testing in compression and torsion can be found in Ref 13–15.

Tensile Loading Techniques (Ref 13)

The principles and the data analysis for the tensile split-Hopkinson pressure bar are similar to those for the compression SHPB (Ref 13, 14). The primary differences are the methods of generating a tensile-loading pulse, specimen geometry, and the method of attaching the specimen to the two bars (incident and transmitted). Three separate general types of tension split-Hopkinson pressure bar design have been developed (Ref 13). All three loading techniques use measures of the tensile pulses in the input and transmitter bars, as in the compressive SHPB, to study the dynamic tensile response of a material.

Method 1. In the first method, developed by Lindholm and Yeakley (Ref 16), the incident bar is solid, while the transmitted bar is a hollow tube of the same cross-sectional area as the input bar. A complex "top-hat" type of sample geometry, as shown in Fig. 9, is machined from the material of interest. The specimen essentially comprises four parallel tensile bars of equal cross-sectional area. Although specimen machining is somewhat complex in this method, the actual SHPB test is conducted in the identical manner because compressive testing and the data analysis is identical to that outlined previously. The advantage of this tensile loading method is that, given a suitable hollow transmitted bar matched to the incident bar, tensile Hopkinson bar tests can be conducted using a standard compressive SHPB loading setup.

Method 2. The second type of tensile split-Hopkinson bar test, and the most commonly implemented mode of loading, involves direct tensile loading of the incident bar to subject a sample in a uniaxial tensile stress state. This loading mode can be accomplished using a standard type of axisymmetric circular tension specimen threaded directly into the ends of the incident and transmitted pressure bars, a dumbbell-shaped sample loaded through flanges attached to the incident and transmitted bars, or a flat tensile sample loaded using a small compression grip assembly designed into the ends of the incident and transmitted bars. A tensile pulse in each instance is generated in the incident bar either by loading the end of the incident bar through direct impact of a mass with a flange on the end of the incident bar or by releasing a tensile pulse stored in the incident bar using a clamping fixture. Figure 10 shows a schematic of a tensile split-Hopkinson bar setup using the hollow-striker-bar loading method. In this loading method, a long tensile pulse, similarly stable as in a compressive bar, can be imparted using a hollow striker tube accelerated along the incident bar from a compressed gas breech or from a falling weight in a vertically configured tensile bar. In the second variation, tensile wave loading in the incident bar is generated through the release of a tensile load that is initially stored in a section of the incident bar.

Method 3. The third type of Hopkinson bar loading in tension also uses a circular specimen threaded into the ends of the two pressure bars but uses the reflection of the compression pulse at the free end of the transmitted bar to load the sample in tension and a circular collar to protect

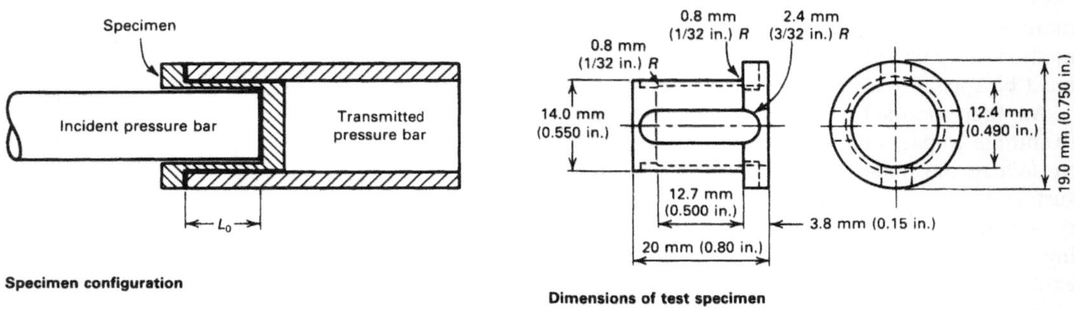

Fig. 9 "Top-hat" tensile split-Hopkinson bar sample design. Source: Ref 16

the specimen from the initial compressive pulse. After the specimen has been screwed into the incident and transmitted bars, a split shoulder or collar is placed over the specimen, and it is screwed in until the pressure bars fit tightly against the shoulder. The shoulder is made of the same material as the pressure bars, has the same outer diameter, and has an inner diameter that just clears the specimen. The ratio of the cross-sectional area of the shoulder to that of the pressure bars is typically 3 to 4, while the ratio of the area of the shoulder to the net cross-sectional area of the specimen is typically 12 to 1. When the striker bar impacts the incident bar, a compressive pulse travels down the incident bar until it reaches the specimen. The amplitude of the pulse, which is a function of the striker velocity and length, is twice the elastic wave transit time in the striker bar.

In this loading method, the compression pulse travels through the composite cross section of the loading collar and specimen in an essentially undisturbed manner. The relatively loose fit of the threaded joint of the specimen into the bars and the large area ratio of the collar to the specimen ensure that no compression beyond the elastic limit is transmitted through the specimen. Ideally, the entire compression pulse passes through the supporting circular collar as if the specimen were not present, although in practice it is operationally difficult to prevent prestraining of the specimen to some degree. The compression pulse continues to propagate until it reaches the free end of the transmitted bar where it reflects and propagates back as a tensile pulse. Upon reaching the specimen, the tensile pulse is partially transmitted through the specimen and partially reflected back into the bar, which is now acting as the incident bar. Because the shoulder, which carried the entire compressive pulse around the specimen, is not rigidly connected to the pressure bars, it will not support any tensile load. Tight fitting of the collar against the two pressure bars is critical in transmitting the compression pulse down the bars without significant wave dispersion or prestraining of the sample. Similarly, the fit of the threaded tensile specimen against the bars is essential to achieve smooth and rapid loading of the specimen as the tensile pulse arrives. Failure to remove all play from the threaded joint results in uneven loading of the specimen and spurious wave reflections because of the open gaps in the loading thread area.

Data Analysis (Ref 13)

As mentioned earlier in this section, data analysis for a tensile split-Hopkinson pressure bar test is essentially identical to that of compression Hopkinson bar analysis detailed in Ref 13 and 14. The additional complications encountered in tensile Hopkinson techniques are related to the following:

- Modification of the pressure bar ends to accommodate gripping of complex samples, which alter wave propagation in the sample and bars
- Potential need for additional diagnostics to calculate true stress

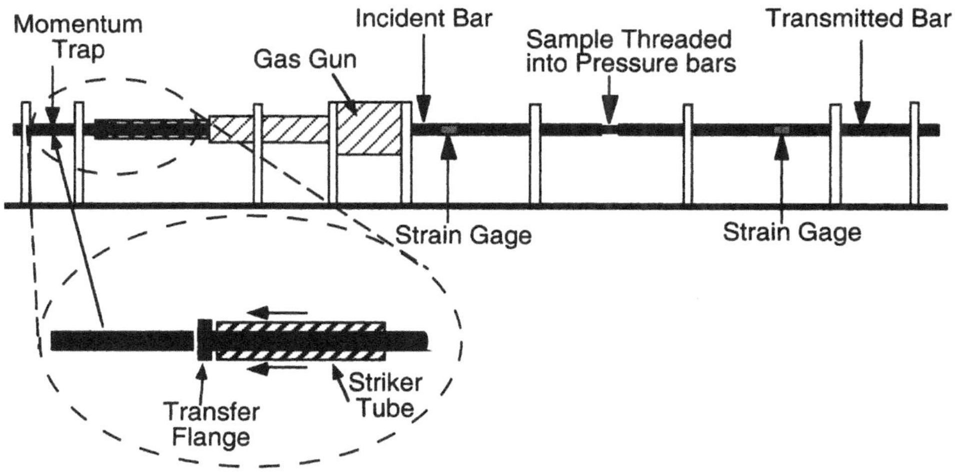

Fig. 10 Schematic of a tensile split-Hopkinson pressure bar test setup

- Increased need to accurately incorporate inertial effects into data reduction to extract quantitative material constitutive behavior
- More complicated stress pulse generation systems required for tensile and torsion bars

Alteration of the bar ends to accommodate threaded or clamped samples leads to complex boundary conditions at the bar specimen interface and, therefore, introduces uncertainties in the wave mechanics description of the test.

When complex sample geometries are used, signals measured in the pressure bars record the structural response of the entire sample, not just the gage section, where plastic deformation is assumed to be occurring. When plastic strain occurs in the sections adjacent to the sample's uniform gage area, accurate determination of the stress-strain response of the material is more complicated. In these cases, additional diagnostics, such as high-speed photography, are mandatory to quantify the loaded section of the deforming sample. In the tensile bar case, an additional requirement is that exact quantification of the deforming sample cross-sectional area as a function of strain is necessary to achieve true-stress data. Contrary to a compressive SHPB test, in which a right-circular cylindrical sample is most often utilized, the tensile SHPB test uses a cylindrical specimen with an attached shoulder and additional gripping, including threads. Because the split-Hopkinson bar data analysis only provides data on the relative displacement between the ends of the incident and transmitter bars, an effective gage length generally must be used. This is equivalent to determining strain in a tensile test through cross-head displacement measurement. The use of strain gages on test samples to determine an effective gage length is strongly recommended. This calibration is accomplished easily at low strain rates, preferably in a conventional test machine in which the cross-head displacement is monitored separately.

As with any uniaxial tensile test, once localized necking occurs, it is no longer possible to simply convert load-displacement data to stress-strain data. This lack of valid stress-state analysis is related to both the lack of uniform plastic deformation in the sample and the attendant volumetric sample expansion, which damage-evolution processes represent. The range of application of the Hopkinson bar test can be extended by high-speed photography of necking specimens, although an accurate measure of the deforming volume of the sample as necking proceeds is difficult at best, given a lack of knowledge of the damage processes evolving within the sample. An analysis that allows estimation of effective stress and strain from the profile of the necking specimen can be obtained. Using the apparatus described in method 3, photographs can be made with a suitable high-speed camera system through windows provided in the collar.

The final complexity inherent to the tension Hopkinson loading configurations has to do with the increased sample dimensions required. Valid dynamic characterization of many material product forms, such as thin sheet materials and small-section bar stock, may be significantly complicated or completely impractical using either tensile or torsion Hopkinson bars because of an inability to fabricate test samples. Techniques have been developed, however, to address these concerns in the case of testing sheet materials in the Hopkinson bar.

Rotating Wheel Test (Ref 14)

Another method for tensile testing at high strain rates consists of a rotating wheel with claws or noses that quickly stroke a yoke containing test pieces. An early test machine was developed by Mann in 1936 (Ref 17), and in 1944 Fehr et al. (Ref 18) reached strain rates of nearly 10^3 s^{-1} with some bearable ringing. In the 1960s, Schopper produced and sold about 100 "rotating wheel machines," which also had a 200 kg (440 lb) wheel with a releasable claw and a specimen within a yoke fixed in front of the wheel (Fig. 11). By careful adjustments, velocities of about 40 m/s (130 ft/s) were reached without any bending moments. However, the overall frequency response of the fixture (despite the 50 kHz quartz transducer for force-time recording) was only 2.5 kHz, which is too low for high-rate testing.

The essential improvement has been to introduce load-measuring gages as close to the gage length as possible. This is realized by measuring elastic strains and converting to stresses by the elastic modulus. To assure low barriers for reflections of stress wave propagation, the strain gages for load measurement are positioned at cones of 8° on the smallest possible diameter. With this technique, stress-strain records are possible up to velocities of 30 m/s, which corresponds to strain rates of $\dot{\varepsilon} = 2500\ s^{-1}$ with a gage length of 10.5 mm (0.41 in.) and a diameter of 3.5 mm (0.14 in.) ($L/D = 3$). Even with high strain hardening and highly deformable materials, there is practically no influence of the tested materials on the history of velocity or strain rate like in a Hopkinson bar setup because the energy content of the rotating wheel exceeds the fracture energy of the specimen more than 10 times (for striking velocities $v > 10$ m/s, or 33 ft/s).

A similar setup, but based on a moving mass of a few kilograms guided along straight bars, has also been developed by Stelly and Dormeval at CEA, France with good results (Ref 19). Higher strain rates of $\dot{\varepsilon} = 10^4\ s^{-1}$, for example, can be reached with short gage length. To pro-

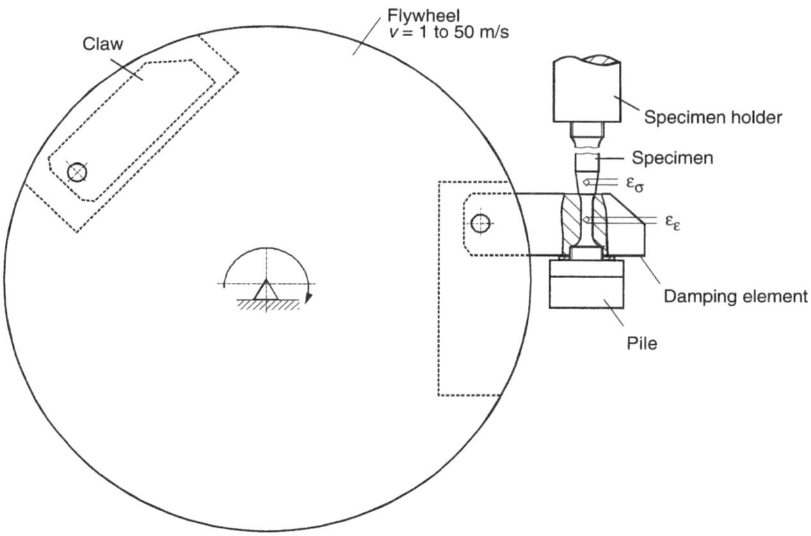

Fig. 11 Principle of high-rate tensile testing with flywheel setup

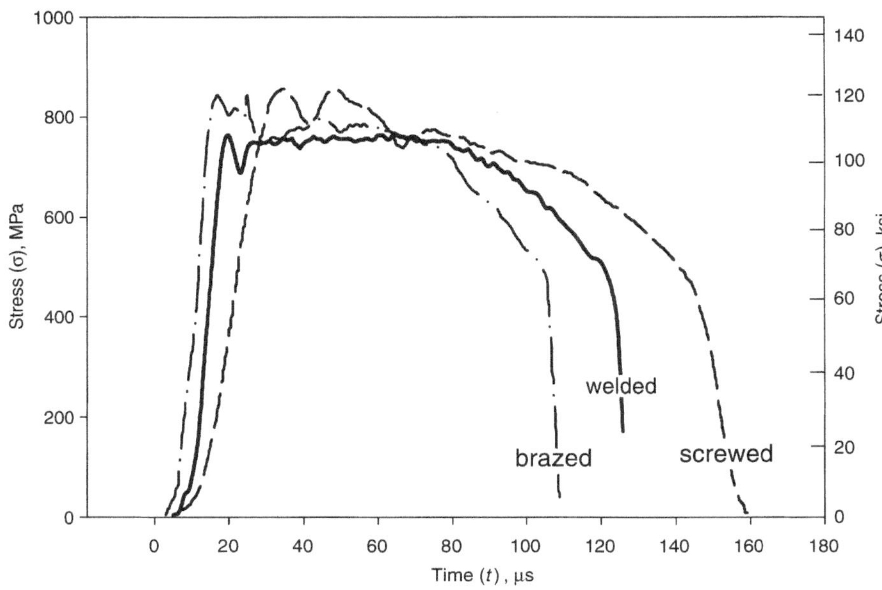

Fig. 12 Influence of joining method on stress-time curves for high strain rate tension test specimens

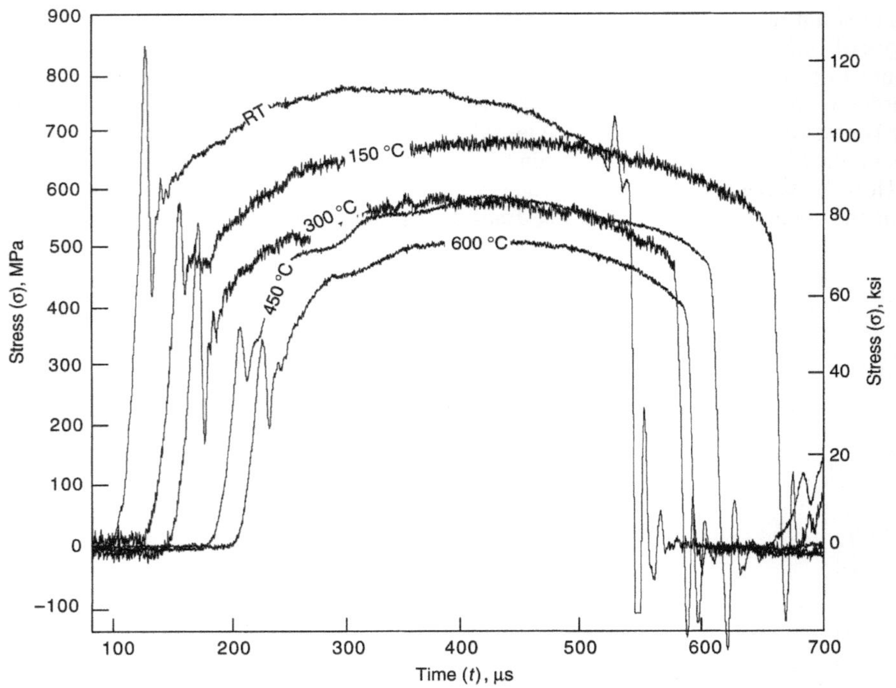

Fig. 13 Stress-time diagrams from high strain rate tensile testing of carbon steel (0.45% C) between room temperature and 600 °C (1100 °F)

vide easy strain recording for the complete loading up to fracture, electro-optical cameras or noncontact laser interferometers can be used. The advantage of this instrumentation is that the strain is measured, not calculated under certain assumptions.

In order to avoid more reflections of the stress wave from the upper end of the specimen, the length beside the gage length is extended to $2s > v_c \cdot t$ (where s is the rod length between gage length and fixture, v_c is the sound velocity, and t is the time to fracture at the used striking velocity). In the case of short lengths, a wave transmitter bar is connected to the specimen. This procedure requires the evaluation of impedance transfer between the sample and bar. Figure 12 shows stress-time diagrams of screwed, brazed, and welded joints tested under high strain rate conditions at about $\dot{\varepsilon} = 1000\ s^{-1}$. These results reveal that screwing and brazing are insufficient methods to obtain stress-time diagrams of good quality. Therefore, in this test setup welded joints are most suitable to perform tensile tests at high and very high strain rates.

This procedure was successfully applied for high-rate, high-temperature tests (Fig. 13). Notice the occurrence of the upper and lower yielding and the following nearly undisturbed stress-time records. The wave transmitter bar was used for a stress measurement because the strain gages at the gage length were unsuitable at test temperature. Temperatures up to 600 °C (1100 °F) are reached with heated air; higher temperatures should be possible using small infrared ovens (Ref 20, 21) or induction heating.

ACKNOWLEDGMENTS

This chapter was adapted from T. Nicholas and S.J. Bless, High Strain Rate Tension Testing, *Mechanical Testing,* Vol 8, 9th ed., *Metals Handbook,* American Society for Metals, 1985, p 208–214. Information was also taken from Ref 13 and 14.

REFERENCES

1. F.I. Niordson, A Unit for Testing Materials at High Strain Rates, *Exp. Mech.,* Vol 5, 1965, p 29–32

2. C.R. Hoggatt and R.F. Recht, Stress-Strain Data Obtained at High Strain Rates Using an Expanding Ring, *Exp. Mech.,* Vol 9, 1969, p 441–448
3. D.E. Grady and D.A. Benson, Fragmentation of Metal Rings by Electromagnetic Loading, *Exp. Mech.,* Vol 28, 1983, p 393–400
4. A.M. Rajendran and I.M. Fyfe, Inertia Effects on the Ductile Failure of Thin Rings, *J. Appl. Mech.,* Vol 104, 1982, p 31–36
5. L.M. Barker and R.E. Hollenback, Laser Interferometer for Measuring High Velocities of Any Reflecting Surface. *J. Appl. Phys.,* Vol 43, 1972, p 4669–4674
6. R.H. Warnes *et al.,* An Improved Technique for Determining Dynamic Material Properties Using the Expanding Ring, in *Shock Waves and High-Strain-Rate Phenomena in Metals,* M.A. Meyers and L.E. Murr, Ed., Plenum Press, New York, 1981
7. D. Bauer and S.J. Bless, Strain Rate Effects on Ultimate Strain of Copper, in *Shock Waves in Condensed Matter,* North Holland, Amsterdam, 1983
8. G.R. Fowles, Experimental Technique and Instrumentation, in *Dynamic Response of Materials to Intense Impulse Loading,* P.C. Chou and A.K. Hopkins, Ed., Air Force Materials Laboratory, Wright-Patterson AFB, OH, 1973
9. J.A. Zukas, T. Nicholas, H.F. Swift, L.B. Greszczuk, and D.R. Curran, *Impact Dynamics,* John Wiley & Sons, New York, 1982
10. R.G. McQueen, S.P. Marsh, J.W. Taylor, J.N. Fritz, and W.J. Carter. The Equation of State of Solids from Shock Wave Studies, in *High Velocity Impact Phenomena,* R. Kinslow, Ed., Academic Press, New York, 1970
11. S.P. Marsh, *LASL Shock Hugoniot Data,* University of California Press, Berkeley, 1980
12. L.M. Barker and E.G. Young, "SWAP-9: An Improved Stress Wave Analyzing Program," Sandia National Laboratories Report No. SLA-74-0009. Albuquerque, NM, 1974
13. G.T. Gray III, Classic Split-Hopkinson Pressure Bar Testing, *Mechanical Testing and Evaluation,* Vol 8, *ASM Handbook,* ASM International, 2000, p 462–476
14. High Strain Rate Tension and Compression Tests, *Mechanical Testing and Evaluation,* Vol 8, *ASM Handbook,* ASM International, 2000, p 429–446
15. A. Gilat, Torsional Kolsky Bar Testing, *Mechanical Testing and Evaluation,* Vol 8, *ASM Handbook,* ASM International, 2000, p 505–515
16. U.S. Lindholm and L.M. Yeakley, High Strain Rate Testing: Tension and Compression, *Exp. Mech.,* Vol 8, 1968, p 1–9
17. H.C. Mann, High Velocity Tension Impact Tests, *Proc. ASTM,* Vol 36 (part 2), 1936, p 85–109
18. R.O. Fehr, E. Parker, and D.J. DeMichael, Measurement of Dynamic Stress and Strain in Tensile Test Specimens, *J. Appl. Mech. (Trans. ASME),* Vol 6A, 1944, p 65–71
19. R. Dormeval and M. Stelly, Influence of Grain Size and Strain Rate of the Mechanical Behavior of High-Purity Polycrystalline Copper, Second Conf. on Mechanical Properties at High Rates of Strain, 1979 (Oxford), Institute of Physics, London, Serial No. 47, p 154–165
20. C.E. Frantz, P.S. Follansbee, and W.E. Wright, New Experimental Techniques with the Split Hopkinson Pressure Bar, *High Energy Rate Forming,* Berman and Schroeder, Ed., American Society of Mechanical Engineers, 1984, p 229
21. A.M. Lennon and K.T. Ramesh, A Technique for Measurement of the Dynamic Behaviour of Materials at High Temperatures, *Int. J. Plast.,* Vol 14 (No. 12), 1998, p 1279–1292

Glossary of Terms

A

accuracy. (1) The agreement or correspondence between an experimentally determined value and an accepted reference value for the material undergoing testing. The reference value may be established by an accepted standard (such as those established by ASTM), or, in some cases, the average value obtained by applying the test method to all the sampling units in a lot or batch of the material may be used. (2) The extent to which the result of a calculation or the reading of an instrument approaches the true value of the calculated or measure quantity. Compare with *precision.*

alligator skin. See preferred term *orange peel.*

anisotropy. The characteristic of exhibiting different values of a property in different directions with respect to a fixed reference system in the material. See also *planar anisotropy.*

average linear strain. See *engineering strain.*

axial strain. See *uniaxial strain.*

B

Bauschinger effect. The phenomenon by which plastic deformation increases yield strength in the direction of plastic flow and decreases it in other directions.

biaxial stress. See *principal stress (normal).*

breaking load. The maximum load (or force) applied to a test specimen or structural member loaded to rupture.

breaking stress. The stress at failure. Also known as rupture stress. See also *fracture stress.*

brittle fracture. Separation of a solid accompanied by little or no macroscopic plastic deformation. Typically, brittle fracture occurs by rapid crack propagation with less expenditure of energy than for *ductile fracture.*

bulk modulus. See *bulk modulus of elasticity.*

bulk modulus of elasticity (K). The measure of resistance to change in volume; the ratio of hydrostatic stress to the corresponding unit change in volume. This elastic constant can be expressed by the equation:

$$K = \frac{\sigma_m}{\Delta} = \frac{-p}{\Delta} = \frac{1}{\beta}$$

where K is bulk modulus of elasticity, σ_m is hydrostatic or mean stress tensor, p is hydrostatic pressure, and β is the coefficient of compressibility. Also known as bulk modulus, compression modulus, hydrostatic modulus, and volumetric modulus of elasticity.

C

chord modulus. The slope of the chord drawn between any two specific points on a stress-strain curve. Compare with *secant modulus.* See also *modulus of elasticity.*

coefficient of thermal expansion. (1) Change in unit of length (or volume) accompanying a unit change of temperature, at a specified temperature. (2) The linear or volume expansion of a given material per degree rise of temperature, expressed at an arbitrary base temperature or as a more complicated equation applicable to a wide range.

conventional strain. See *engineering strain.*

conventional stress. See *engineering stress.*

crack-growth rate. Rate of propagation of a crack through a material due to statically or dynamically applied load.

crazing. Region of ultrafine cracks, which may extend in a network on or under the surface of a resin or plastic material. May appear as a white band. Often found in a filament-wound pressure vessel or bottle. In many plastics, craze growth precedes crack growth, often generating additional strength because crazes are load bearing.

cross linking. With thermosetting and certain thermoplastic polymers, the setting up of chemical links between the molecular chains. When extensive, as in most thermosetting resins, cross linking makes an infusible supermolecule of all the chains. In rubbers, the cross linking is just enough to join all molecules into a network.

cup fracture (cup-and-cone fracture). A mixed-mode fracture, often seen in tensile-test specimens of a ductile material, where the central portion undergoes plane-strain fracture and the surrounding region undergoes plane-stress fracture. One of the mating fracture surfaces looks like a miniature cup; it has a central depressed flat-face region surrounded by a *shear lip.* The other fracture surface looks like a miniature truncated cone.

cupping test. A mechanical test used to determine the ductility and drawing properties of sheet metal. It consists of measuring the maximum depth of bulge that can be formed before fracture. The test is commonly carried out by drawing the test piece into a circular die by means of a punch with a hemispherical end. See also *Erichsen cup test, Olsen cup test,* and *Swift cup test.*

D

deformation. A change in the form of a body due to stress, thermal change, change in moisture, or other causes. Measured in units of length.

dimpled rupture. A fractographic term describing ductile fracture that occurs through the formation and coalescence of microvoids along the fracture path. The fracture surface of such a ductile fracture appears dimpled when observed at high magnification and usually is most clearly resolved when viewed in a scanning electron microscope. See also *ductile fracture.*

discontinuous yielding. The nonuniform plastic flow of a metal exhibiting a *yield point* in which plastic deformation is inhomogeneously distributed along its length. Under some circumstances, it may occur in metals not exhibiting a distinct yield point, either at the onset of or during plastic flow.

ductile fracture. Fracture characterized by tearing of metal accompanied by appreciable gross *plastic deformation* and expenditure of considerable energy. Contrast with *brittle fracture.*

ductility. The ability of a material to deform plastically before fracturing. Measured by elongation or reduction in area in a tensile test, by height of cupping in a cupping test, or by the radius or angle of bend in a bend test.

dynamic modulus. The ratio of stress to strain under cyclic conditions (calculated from data obtained from either free or forced vibration tests, in shear, compression, or tension).

dynamic strain aging. A behavior in metals in which solute atoms are sufficiently mobile to move toward and interact with dislocations. This results in strengthening over a specific range of elevated temperature and strain rate.

E

effective yield strength. An assumed value of uniaxial yield strength that represents the influence of plastic yielding on fracture-test parameters.

elastic calibration device. A device for use in verifying the load readings of a testing machine consisting of an elastic member(s) to which loads may be applied, combined with a mechanism or device for indicating the magnitude (or a quantity proportional to the magnitude) of deformation under load.

elastic constants. The factors of proportionality that relate elastic displacement of a material to applied forces. See also *modulus of elasticity, bulk modulus of elasticity,* and *Poisson's ratio.*

elastic deformation. A change in dimensions directly proportional to and in phase with an increase or decrease in applied force.

elastic energy. The amount of energy required to deform a material within its elastic range of behavior, neglecting small heat losses due to internal friction. The energy absorbed by a specimen per unit volume of material contained within the gage length being tested. It is determined by measuring the area under the stress-strain curve up to a specified elastic strain. See also *modulus of resilience* and *strain energy.*

elastic limit. The maximum stress which a material is capable of sustaining without any permanent strain (deformation) remaining on complete release of the stress. Compare with *proportional limit.*

elastic recovery. Amount the dimension of a stressed elastic material returns to its original (unstressed) dimension on release of an applied load.

elastic strain. See *elastic deformation.*

elastic strain energy. The energy expended by the action of external forces in deforming a body elastically. Essentially, all the work performed during elastic deformation is stored as elastic energy, and this energy is recovered upon release of the applied force.

elasticity. The property of a material by virtue of which deformation caused by stress disappears on removal of the stress. A perfectly elastic body completely recovers its original shape and dimensions after release of stress.

elongation. A term used in mechanical testing to describe the amount of extension of a test piece when stressed. See also *elongation, percent.*

elongation, percent. The extension of a uniform section of a specimen expressed as percentage of the original gage length:

$$\text{Elongation, \%} = \frac{L_x - L_o}{L_o} = 100$$

where L_o is original gage length, and L_x is final gage length.

engineering strain (*e*). A term sometimes used for average linear strain or conventional strain in order to differentiate it from *true strain.* In tensile testing it is calculated by dividing the change in the gage length by the original gage length.

engineering stress (*s*). A term sometimes used for conventional stress in order to differentiate it from *true stress.* In tensile testing, it is calculated by dividing the breaking load applied to the specimen by the original cross-sectional area of the specimen.

Erichsen cup test. A *cupping test* used for assessing the ductility of sheet metal. The method consists of forcing a conical or hemispherical-ended plunger into the specimen and measuring the depth of the impression at fracture. Compare with *Olsen cup test* and *Swift cup test.*

extensometer. An instrument for measuring changes in length over a given gage length caused by application or removal of a force. Commonly used in tensile testing of metal specimens.

F

flexibility. The quality or state of a material that allows it to be flexed or bent repeatedly without undergoing rupture. See also *flexure.*

flexural modulus. Within the elastic limit, the ratio of the applied stress on a test specimen in flexure to the corresponding strain in the outermost fiber of the specimen. Flexural modulus is the measure of relative stiffness.

flexure. A term used in the study of strength of materials to indicate the property of a body, usually a rod or beam, to bend without fracture. See also *flexibility.*

formability. The ease with which a metal can be shaped through *plastic deformation.* The evaluation of the formability of a metal involves measurement of strength and ductility, as well as the amount of deformation required to cause fracture. *Workability* is used interchangeably with formability; however, formability refers to the shaping of sheet metal, while workability refers to shaping materials by bulk deformation (i.e., forging or rolling).

forming limit diagram (FLD). A diagram on which the major strains at the onset of necking in sheet metal are plotted vertically and the corresponding minor strains are plotted horizontally. The onset-of-failure line divides all possible strain combinations into two zones: the "safe" zone, in which failure during forming is not expected, and the "failure" zone, in which failure during forming is expected.

fracture stress. The true normal stress on the minimum cross-sectional area at the beginning of fracture. This term usually applies to tensile tests of unnotched specimens.

G

gage length. The original length of the portion of a specimen over which strain, change of length, or other characteristics are determined.

H

Hall-Petch relationship. A general relationship for metals that shows that the yield strength is linearly related to the reciprocal of the square root of the grain diameter.

Hartmann lines. See *Lüders lines.*

Hooke's law. Observation that, in the elastic region of solid material, stress is directly proportional to strain and can be expressed as:

$$\frac{\text{Stress}}{\text{Strain}} = \frac{\sigma}{\varepsilon} = \text{constant} = E$$

where E is the modulus of elasticity, or Young's modulus. The constant relationship between stress and strain applies only below the proportional limit. See also *modulus of elasticity.*

hysteresis. The phenomenon of permanently absorbed or lost energy that occurs during any cycle of loading or unloading when a material is subjected to repeated loading.

L

limiting dome height (LDH) test. A mechanical test, usually performed unlubricated on sheet metal, that simulates the fracture conditions in a practical press-forming operation. The results are dependent on the sheet thickness.

linear (tensile or compressive) strain. The change per unit length due to force in an original linear dimension. An increase in length is considered positive.

load. For testing machines, a force applied to a test piece that is measured in units such as pound-force, newton, or kilogram-force.

longitudinal direction. The principal direction of flow in a worked metal. See also *normal direction* and *transverse direction.*

Lüders lines. Elongated surface markings or depressions, often visible with the unaided eye, that form along the length of sheet metal or a tension specimen at an angle of approximately 45° to the loading axis. Caused by localized plastic deformation, they result from discontinuous (inhomogeneous) yielding. Also known as Lüders bands, Hartmann lines, Piobert lines, or stretcher strains.

M

macrostrain. The mean strain over any finite gage length of measurement large in comparison with interatomic distances. Macrostrain can be measured by several methods, including electrical-resistance strain gages and mechanical or optical extensometers. Elastic macrostrain can be measured by x-ray diffraction. Compare with *microstrain.*

maximum load (P_{max}). (1) The load having the highest algebraic value in the load cycle. Tensile loads are considered positive and compressive loads negative. (2) Used to determine the strength of a structural member; the load that can be borne before failure is apparent.

maximum stress (S_{max}). The stress having the highest algebraic value in the stress cycle, tensile stress being considered positive and compressive stress negative. The *nominal stress* is used most commonly.

mechanical hysteresis. Energy absorbed in a complete cycle of loading and unloading within the elastic limit and represented by the closed loop of the stress-strain curves for loading and unloading. Sometimes called *elastic hysteresis.*

mechanical metallurgy. The science and technology dealing with the behavior of metals when subjected to applied forces.

microstrain. The strain over a gage length comparable to interatomic distances. These are the strains being averaged by the *macrostrain* measurement. Microstrain is not measurable by currently existing techniques. Variance of the microstrain distribution can, however, be measured by x-ray diffraction.

modulus of elasticity (E). The measure of rigidity or stiffness of a metal; the ratio of stress, below the proportional limit, to the corresponding strain. In terms of the *stress-strain diagram,* the modulus of elasticity is the slope of the stress-strain curve in the range of linear proportionality of stress to strain. Also known as *Young's modulus.* For materials that do not conform to *Hooke's law* throughout the elastic range, the slope of either the tangent to the stress-strain curve at the origin or at low stress, the secant drawn from the origin to any specified point on the stress-strain curve, or the chord connecting any two specific points on the stress-strain curve is usually taken to be the modulus of elasticity. In these cases, the modulus is referred to as the *tangent modulus, secant modulus,* or *chord modulus,* respectively.

modulus of resilience. The amount of energy stored in a material when loaded to its elastic limit. It is determined by measuring the area under the stress-strain curve up to the *elastic limit.* See also *elastic energy, resilience,* and *strain energy.*

***m*-value.** See *strain-rate sensitivity.*

N

necking. (1) Reducing the cross-sectional area of metal in a localized area by stretching. (2) Reducing the diameter of a portion of the length of a cylindrical shell or tube.

nominal stress. The stress at a point calculated on the net cross section by simple elasticity theory without taking into account the effect on the stress produced by stress raisers such as holes, grooves, fillets, etc.

normal direction. Direction perpendicular to the plane of working in a worked metal. See also *longitudinal direction* and *transverse direction.*

normal stress. The stress component perpendicular to a plane on which forces act.

notch brittleness. Susceptibility of a material to *brittle fracture* at points of stress concentration. For example, in a notch tensile test, the material is said to be notch brittle if the *notch strength* is less than the tensile strength of an unnotched specimen. Otherwise, it is said to be notch ductile.

notch depth. The distance from the surface of a notched test specimen to the bottom of the notch. In a cylindrical test specimen, the percentage of the original cross-sectional area removed by machining an annular groove.

notch ductility. The percentage *reduction in area* after complete separation of the metal in a tensile test of a notched specimen.

notch strength. The maximum load on a notched tensile-test specimen divided by the minimum cross-sectional area (the area at the root of the notch). Also known as notch tensile strength.

***n*-value.** See *strain hardening exponent.*

O

offset. The distance along the strain coordinate between the initial portion of a stress-strain curve and a line parallel to the initial portion that intersects the stress-strain curve at a value of stress (commonly 0.2%) that is used as a measure of the *yield strength.* Used for materials that have no obvious *yield point.*

offset modulus. The ratio of the offset yield stress to the extension at the offset point (plastics).

offset yield strength. The stress at which the strain exceeds by a specified amount (the *offset*) an extension of the initial proportional portion of the stress-strain curve. Expressed in force per unit area.

Olsen cup test. A *cupping test* in which a piece of sheet metal, restrained except at the center, is deformed by a standard steel ball until fracture occurs. The height of the cup at time of fracture is a measure of the ductility. Compare with *Erichsen cup test* and *Swift cup test.*

orange peel. A surface roughening in the form of a pebble-grained pattern where a metal of unusually coarse grain is stressed beyond its elastic limit. Also known as pebbles and alligator skin.

P

permanent set. The deformation or strain remaining in a previously stressed body after release of load.

Piobert lines. See *Lüders lines.*

planar anisotropy. A variation in physical and/or mechanical properties with respect to direction within the plane of material in sheet form. See also *plastic strain ratio.*

plastic deformation. The permanent (inelastic) distortion of a material under applied stress that strains the material beyond its *elastic limit.*

plastic instability. The stage of deformation in a tensile test where the plastic flow becomes nonuniform and *necking* begins.

plasticity. The property that enables a material to undergo permanent deformation without rupture.

plastic strain. Dimensional change that does not disappear when the initiating stress is removed. Usually accompanied by some *elastic deformation.*

plastic strain ratio (*r*-value). The ratio of the true width strain to the true thickness strain in a tensile test, $r = \varepsilon_w/\varepsilon_t$. Because of the difficulty in making precise measurement of thickness strain in sheet material, it is more convenient to express r in terms of initial and final length and width dimensions. It can be shown that

$$r = (\ln W_o W_f) - (\ln L_f W_f / L_o W_o)$$

where L_o and W_o are initial length and width of gage section, respectively; and L_f and W_f are final length and width, respectively.

Poisson's ratio (ν). The absolute value of the ratio of transverse (lateral) strain to the corresponding axial strain resulting from uniformly distributed axial stress below the *proportional limit* of the material.

precision. The closeness of agreement between the results of individual replicated measurements or tests. The standard deviation of the error of measurement may be used as a measure of "imprecision."

principal stress (normal). The maximum or minimum value of the *normal stress* at a point in a plane considered with respect to all possible orientations of the considered plane. On such principal planes the shear stress is zero.

There are three principal stresses on three mutually perpendicular planes. The state of stress at a point may be: (1) uniaxial, a state of stress in which two of the three principal stresses are zero; (2) biaxial, a state of stress in which only one of the three principal stresses is zero; or (3) triaxial, a state of stress in which none of the principal stresses is zero. Multiaxial stress refers to either biaxial or triaxial stress.

proof stress. (1) The stress that will cause a specified small *permanent set* in a material. (2) A specified stress to be applied to a member or structure to indicate its ability to withstand service loads.

proportional limit. The greatest stress a material is capable of developing without a deviation from straight-line proportionality between stress and strain. Compare with *elastic limit.* See also *Hooke's law.*

R

reduction in area (RA). The difference between the original cross-sectional area of a tensile specimen and the smallest area at or after fracture as specified for the material undergoing testing. Also known as reduction of area.

residual stress. Stresses that remain within a body as the result of thermal or mechanical treatment or both.

resilience. The ability of a material to absorb energy when deformed elastically and return to its original shape on release of load. See also *modulus of resilience.*

rosette. Strain gages arranged to indicate, at a single position, strain in three different directions.

rupture stress. The stress at failure. Also known as *breaking stress.* See also *fracture stress.*

S

sample. (1) One or more units of product (or a relatively small quantity of a bulk material) that are withdrawn from a lot or process stream and that are tested or inspected to provide information about the properties, dimensions, or other quality characteristics of the lot or process stream. Not be confused with *specimen.* (2) A portion of a material intended to be representative of the whole.

secant modulus. The slope of the secant drawn from the origin to any specified point on a stress-strain curve. Compare with *chord modulus.* See also *modulus of elasticity.*

shear lip. A narrow, slanting ridge along the edge of a fracture surface. The term sometimes also denotes a narrow, often crescent-shaped, fibrous region at the edge of a fracture that is otherwise of the cleavage type, even though this fibrous region is in the same plane as the rest of the fracture surface.

specimen. A test object, often of standard dimensions or configuration, that is used for destructive or nondestructive testing. One or more specimens may be cut from each unit of a *sample.*

stiffness. (1) The ability of a metal or shape to resist elastic deflection. (2) The rate of stress increase with respect to the rate of increase in strain induced in the metal or shape; the greater the stress required to produce a given strain, the stiffer the material is said to be.

strain. The unit of change in the size or shape of a body due to force. Also known as nominal strain. See also *engineering strain, linear strain,* and *true strain.*

strain aging. The changes in ductility, hardness, yield point, and tensile strength that occur when a metal or alloy that has been cold worked is stored for some time. In steel, strain aging is characterized by a loss of ductility and a corresponding increase in hardness, yield point, and tensile strength.

strain energy. A measure of the energy absorption characteristics of a material determined by measuring the area under the stress-strain diagram. Also known as deformation energy. See also *elastic energy, resilience,* and *toughness.*

strain gage. A device for measuring small amounts of strain produced during tensile and similar tests on metal. A coil of fine wire is mounted on a piece of paper, plastic, or similar carrier matrix (backing material), which is rectangular in shape and usually about 25 mm (1.0 in.) long. This is glued to a portion of metal under test. As the coil extends with the specimen, its electrical resistance increases in direct proportion. This is known as bonded resistance-strain gage. Other types of gages measure the actual deformation. Mechanical, optical, or electronic devices are sometimes used to magnify the strain for easier reading. See also *rosette.*

strain hardening. An increase in hardness and strength caused by plastic deformation at temperatures below the recrystallization range. Also known as work hardening.

strain-hardening coefficient. See *strain-hardening exponent.*

strain-hardening exponent (*n* value). The value *n* in the relationship $\sigma = K\varepsilon^n$, where σ is the *true stress,* ε is the *true strain,* and *K,* the strength coefficient, is equal to the true stress at a true strain of 1.0. The strain hardening exponent is equal to the slope of the true stress/true strain curve up to maximum load, when plotted on log-log coordinates. The *n*-value relates to the ability of a sheet of material to be stretched in metalworking operations. The higher the *n*-value, the better the formability (stretchability). Also known as the strain-hardening coefficient.

strain rate. The time rate of straining for the usual tensile test. Strain as measured directly on the specimen gage length is used for determining strain rate. Because strain is dimensionless, the units of strain rate are reciprocal time.

strain-rate sensitivity (*m* value). The increase in *stress* (σ) needed to cause a certain increase in plastic-strain rate ($\dot{\varepsilon}$) at a given level of *plastic strain* (ε) and a given temperature (*T*).

$$\text{Strain-rate sensitivity} = m = \left(\frac{\Delta \log \sigma}{\Delta \log \dot{\varepsilon}}\right)_{\varepsilon T}$$

strength. The maximum nominal stress a material can sustain. Always qualified by the type of stress (tensile, compressive, or shear).

stress. The intensity of the internally distributed forces or components of forces that resist a change in the volume or shape of a material that is or has been subjected to external forces. Stress is expressed in force per unit area and is calculated on the basis of the original dimensions of the cross section of the specimen. Stress can be either direct (tension or compression) or shear. See also *engineering stress, nominal stress, normal stress, residual stress,* and *true stress.*

stress raisers. Changes in contour or discontinuities in structure that cause local increases in stress.

stress ratio (*A* or *R*). The algebraic ratio of two specified stress values in a stress cycle. Two commonly used stress ratios are the ratio of the alternating stress amplitude to the mean stress, $A = S_a/S_m$, and the ratio of the minimum stress to the maximum stress, $R = S_{min}/S_{max}$.

stress-strain curve. See *stress-strain diagram.*

stress-strain diagram. A graph in which corresponding values of stress and strain are plotted against each other. Values of stress are usually plotted vertically (ordinates or *y*-axis) and values of strain horizontally (abscissas or *x*-axis). Also known as deformation curve and stress-strain curve.

stretcher strains. See *Lüders lines.*

Swift cup test. A simulative *cupping test* in which circular blanks of various diameters are clamped in a die ring and deep drawn into cups by a flat-bottomed cylindrical punch. Compare with *Erichsen cup test* and *Olsen cup test.*

T

tangent modulus. The slope of the stress-strain curve at any specified stress or strain. See also *modulus of elasticity.*

tensile strength. In tensile testing, the ratio of maximum load to original cross-sectional area. Also known as ultimate strength. Compare with *yield strength.*

tensile stress. A stress that causes two parts of an elastic body, on either side of a typical stress plane, to pull apart.

tensile testing. A method of determining the behavior of materials subjected to uniaxial loading, which tends to stretch the metal. A longitudinal specimen of known length and diameter is gripped at both ends and stretched at a slow, controlled rate until rupture occurs. Also known as tension testing.

tension. The force or load that produces elongation.

tension testing. See *tensile testing.*

testing machine (load-measuring type). A mechanical device for applying a load (force) to a specimen.

total elongation. A total amount of permanent extension of a test piece broken in a tensile test. See also *elongation, percent.*

total-extension-under-load yield strength. See *yield strength.*

toughness. The ability of a metal to absorb energy and deform plastically before fracturing.

transverse direction. Literally, the "across" direction, usually signifying a direction or plane perpendicular to the direction of working. In rolled plate or sheet, the direction across the width is often called long transverse, and the direction through the thickness, short transverse.

triaxial stress. See *principal stress (normal).*

true strain. (1) The ratio of the change in dimension, resulting from a given load increment, to the magnitude of the dimension immediately prior to applying the load increment. (2) In a body subjected to axial force, the natural logarithm of the ratio of the gage length at the moment of observation to the original gage length. Also known as natural strain. Compare with *engineering strain.*

true stress. The value obtained by dividing the load applied to a member at a given instant by the cross-sectional area over which it acts. Compare with *engineering stress.*

U

ultimate strength. The maximum stress (tensile, compressive, or shear) a material can sustain without fracture, determined by dividing maximum load by the original cross-sectional area of the specimen. Also known as nominal strength or maximum strength.

uniaxial strain. Increase (or decrease) in length resulting from a stress acting parallel to the longitudinal axis of the specimen.

uniaxial stress. See *principal stress (normal).*

uniform elongation. The elongation at maximum load and immediately preceding the onset of *necking* in a tension test.

uniform strain. The strain occurring prior to the beginning of localization of strain *(necking);* the strain to maximum load in the tension test.

V

viscoelasticity. A property involving a combination of elastic and viscous behavior. A material having this property is considered to combine the features of a perfectly elastic solid and a perfect fluid. A phenomenon of time-dependent, in addition to elastic, deformation (or recovery) in response to load.

volumetric modulus of elasticity. See *bulk modulus of elasticity.*

W

workability. See *formability.*

work hardening. See *strain hardening.*

wrinkling. A wavy condition obtained in deep drawing of sheet metal, in the area of the metal between the edge of the flange and the draw radius. Wrinkling may also occur in other forming operations when unbalanced compressive forces are set up.

Y

yielding. Evidence of plastic deformation in structural materials.

yield point. The first stress in a material, usually less than the maximum attainable stress, at which an increase in strain occurs without an increase in stress. Only certain metals—those which exhibit a localized, heterogeneous type of transition from elastic to plastic deformation—produce a yield point. If there is a decrease in stress after yielding, a distinction may be made between upper and lower yield points. The load at which a sudden drop in the flow curve occurs is called the upper yield point. The constant load shown on the flow curve is the lower yield point.

yield-point elongation. During discontinuous yielding, the amount of strain measured from the onset of yielding to the beginning of strain hardening.

yield strength. The stress at which a material exhibits a specified deviation from proportionality of stress and strain. An offset of 0.2% is used for many metals. Compare with *tensile strength.*

yield stress. The stress level of highly ductile materials, such as structural steels, at which large strains take place without further increase in stress.

Young's modulus. A term used synonymously with *modulus of elasticity.* The ratio of tensile or compressive stresses to the resulting strain.

SELECTED REFERENCES

- *Compilation of ASTM Standard Definitions,* 8th ed., ASTM, 1994
- H.E. Davis, G.E. Troxell, and G.F.W. Hauck, *The Testing of Engineering Materials,* 4th ed., McGraw Hill, 1982
- J.R. Davis, Ed., *ASM Materials Engineering Dictionary,* ASM International, 1992
- G.E. Dieter, Mechanical Metallurgy, 2nd ed., McGraw Hill, New York, 1976
- *Glossary of Metallurgical Terms and Engineering Tables,* American Society for Metals, 1979
- D.N. Lapedes, Ed., *Dictionary of Scientific and Technical Terms,* 2nd ed., McGraw Hill, 1974
- A.D. Merriman, *A Dictionary of Metallurgy,* Pitman Publishing, London, 1958
- "Metal Test Methods and Analytical Procedures," *Annual Book of ASTM Standards,* Vol 03.01 and 03.02, ASTM, 1984
- J.G. Tweeddale, *Mechanical Properties of Metals,* American Elsevier, 1964

Reference Tables

Table 1 Room-temperature tensile yield strength comparisons of metals and plastics

Material	Tensile yield strength			
	High		Low	
	MPa	ksi	MPa	ksi
Cobalt and its alloys	1999	290	179	26
Low-alloy hardening steels; wrought, quenched and tempered	1986	288	524	76
Stainless steels, standard martensitic grades; wrought, heat treated	1896	275	414	60
Rhenium	1862	270	...	...
Ultrahigh strength steels; wrought, heat treated	1862	270	1172	170
Stainless steels, age hardenable; wrought, aged	1634	237	724	105
Nickel and its alloys	1586	230	69	10
Stainless steels, specialty grades; wrought, 60% cold worked	1558	226	703	102
Tungsten	1517	220	...	...
Molybdenum and its alloys	1448	210	565	82
Titanium and its alloys	1317	191	186	27
Carbon steels, wrought; normalized, quenched and tempered	1296	188	400	58
Low-alloy carburizing steels; wrought, quenched and tempered	1227	178	427	62
Nickel-base superalloys	1186	172	276	40
Alloy steels, cast; quenched and tempered	1172	170	772	112
Stainless steels; cast	1138	165	214	31
Tantalum and its alloys	1089	168	331	48
Steel P/M parts; heat treated	1062	154	517	75
Ductile (nodular) irons, cast	1034	150	276	40
Copper casting alloys(a)	965	140	62	9
Stainless steels, standard austenitic grades; wrought, cold worked	965	140	517	75
Niobium and its alloys	931	135	241	35
Iron-base superalloys; cast, wrought	924	134	276	40
Cobalt-base superalloys, wrought	800	116	241	35
Bronzes, wrought(a)	786	114	97	14
Heat treated low-alloy constructional steels; wrought, mill heat treated	758	110	621	90
High-copper alloys, wrought(a)	758	110	62	9
Stainless steels, standard martensitic grades; wrought, annealed	724	105	172	25
Cobalt-base superalloys, cast	689	100	517	75
Heat treated carbon constructional steels; wrought, mill heat treated	690	100	290	42
Hafnium	662	96	221	32
Brasses, wrought(a)	638	92.5	69	10
Aluminum alloys, 7000 series	627	91	97	14
Alloy steels, cast; normalized and tempered	627	91	262	38
Copper-nickel-zinc, wrought(a)	620	90	124	18
Copper nickels, wrought(a)	586	85	90	13
Malleable irons, pearlitic grades; cast	552	80	310	45
High-strength low-alloy steels; wrought, as-rolled	552	80	290	42
Stainless steels, specialty grades; wrought, annealed	552	80	186	27
Stainless steels, standard ferritic grades; wrought, cold worked	552	80	310	45
Carbon steels, wrought; carburized, quenched and tempered	531	77	317	46
Carbon steel, cast; quenched and tempered	517	75	...	...
Stainless steel (410)P/M parts; heat treated	517	75	...	...
Steel P/M parts; as-sintered	517	75	207	30
Coppers, wrought(a)	496	72	69	10
Aluminum alloys, 2000 series	455	66	69	10
Ductile (nodular) austenitic irons, cast	448	65	193	28
Zinc foundry alloys	441	64	207	30

(continued)

At 0.2% offset for metals, unless otherwise noted; tensile strength at yield for plastics, per ASTM D 638. P/M, powder metallurgy; ABS, acrylonitrile-butadiene-styrene; PVC, polyvinyl chloride. (a) At 0.5% offset. Adapted from Guide to Engineering Materials, *Advanced Materials and Processes,* Dec 1999

Table 1 (continued)

Material	Tensile yield strength			
	High		Low	
	MPa	ksi	MPa	ksi
Zinc alloys, wrought	421	61	159	23
Stainless steels, standard ferritic grades; wrought, annealed	414	60	241	35
Aluminum alloys, 5000 series	407	59	41	6
Aluminum alloys, 6000 series	379	55	48	7
Aluminum casting alloys	379	55	55	8
Carbon steels, cast; normalized and tempered	379	55	331	48
Stainless steels, standard austenitic grades; wrought, annealed	379	55	207	30
Stainless steel P/M parts, as sintered	372	54	276	40
Rare earths	365	53	66	9.5
Zirconium and its alloys	365	53	103	15
Depleted uranium	345	50	241	35
Aluminum alloys, 4000 series	317	46	. . .	. . .
Thorium	310	45	179	26
Magnesium alloys, wrought	303	44	90	13
Silver	303	44	55	8
Carbon steels, cast; normalized	290	42	262	38
Beryllium and its alloys	276	40	34	5
Aluminum alloys, 3000 series	248	36	41	6
Carbon steel, cast; annealed	241	35	. . .	. . .
Malleable ferritic cast irons	241	35	221	32
Palladium	207	30	34	5
Gold	207	30	. . .	. . .
Magnesium alloys, cast	207	30	83	12
Polyimides, reinforced	193	28	34	5
Platinum	186	27	14	2
Iron P/M parts; as-sintered	179	26	76	11
Aluminum alloys, 1000 series	165	24	28	4
Polyphenylene sulfide, 40% glass reinforced	145	21	. . .	. . .
Polysulfone, 30–40% glass reinforced	131	19	117	17
Acetal, copolymer, 25% glass reinforced	128	18.5	. . .	. . .
Styrene acrylonitrile, 30% glass reinforced	124	18	. . .	. . .
Phenylene oxide based resins, 20–30% glass reinforced	117	17	100	14.5
Polyamide-imide	117	17	92	13.3
Polystyrene, 30% glass reinforced	97	14	. . .	. . .
Zinc die-casting alloys	96	14	. . .	. . .
Polyimides, unreinforced	90	13	52	7.5
Nylons, general purpose	87	12.6	49	7.1
Polyethersulfone	84	12.2	. . .	. . .
Polyphenylene sulfide, unreinforced	76	11	. . .	. . .
Polysulfone, unreinforced	70	10.2	. . .	. . .
Acetal, homopolymer, unreinforced	69	10	. . .	. . .
Nylon, mineral reinforced	69	10	62	9
Polypropylene, glass reinforced	69	10	41	6
Polystyrene, general purpose	69	10	34	5.0
Phenylene oxide based resins, unreinforced	66	9.6	54	7.8
Acetal, copolymer, unreinforced	61	8.8	. . .	. . .
ABS/polycarbonate	55	8.0	. . .	. . .
Lead and its alloys	55	8	11	1.6
Polyarylsulfone	55	8	. . .	. . .
ABS/polysulfone (polyarylether)	52	7.5	. . .	. . .
Acrylic/PVC	48	7.0	45	6.5
Tin and its alloys	45	6.6	7	1.3
ABS/PVC, rigid	41	6.0	. . .	. . .
Polystyrene, impact grades	41	6.0	19	2.8
Polypropylene, general purpose	36	5.2	33	4.8
ABS/polyurethane	31	4.5	26	3.7
Polypropylene, high impact	30	4.3	19	2.8

At 0.2% offset for metals, unless otherwise noted; tensile strength at yield for plastics, per ASTM D 638. P/M, powder metallurgy; ABS, acrylonitrile-butadiene-styrene; PVC, polyvinyl chloride. (a) At 0.5% offset. Adapted from Guide to Engineering Materials, *Advanced Materials and Processes*, Dec 1999

Table 2 Room-temperature tensile modulus of elasticity comparisons of various materials

Material	Tensile modulus: High, GPa	High, 10^6 psi	Low, GPa	Low, 10^6 psi
Silicon carbide	655	95	90	13
Tungsten carbide-base cermets	650	94.3	425	61.6
Tungsten carbide	648	94	448	65
Osmium	551	80	...	...
Iridium	545	79	...	...
Titanium, zirconium, hafnium borides	503	73	490	71
Ruthenium	469	68	...	...
Rhenium	469	68	...	...
Boron carbide	448	65	290	42
Boron	441	64	...	...
Tungsten	406	59.0	...	...
Beryllia	399	58	270	39
Titanium carbide-base cermets	393	57	290	42
Rhodium	379	55	...	...
Titanium carbide	379	55	248	36
Molybdenum and its alloys	365	53	317	46
Tantalum carbide	365	53	...	...
Magnesia	345	50	241	35
Alumina ceramic	345	50	207	30
Niobium carbide	338	49	...	...
Beryllium carbide	317	46	207	30
Chromium	289	42	...	...
Beryllium and its alloys	289	42.0	186	27.0
Graphite-epoxy composites	276	40	134	20
Cobalt-base superalloys	248	36.0	199	29.0
Zirconia	241	35	158	23
Nickel and its alloys	234	34.0	131	19.0
Cobalt and its alloys	231	33.6	207	30.0
Nickel-base superalloys	231	33.5	126	18.3
Iron-base superalloys; cast and wrought	214	31	193	28
Silicon nitride	214	31	62	9
Alloy steels; cast	207	30	200	29
Boron-epoxy composites	207	30	...	...
Carbon steels; cast	207	30	...	...
Carbon steel, carburizing grades; wrought	207	30	200	29
Carbon steels, hardening grades; wrought	207	30	200	29
Depleted uranium	207	30	138	20
Stainless steels, age hardenable; wrought	207	30	193	28
Stainless steels, specialty grades; wrought	207	30	186	27
Ultrahigh strength steels; wrought	207	30	186	27
Stainless steels; cast	200	29	165	24
Stainless steels, standard austenitic grades; wrought	200	29	193	28
Stainless steels, standard ferritic grades; wrought	200	29	...	...
Stainless steels, standard martensitic grades; wrought	200	29	...	...
Boron-aluminum composites	193	28	...	...
Malleable irons, pearlitic grades; cast	193	28	179	26
Tantalum and its alloys	186	27.0	144	21.0
Ductile (nodular) irons; cast	172	25	152	22
Malleable ferritic cast irons	172	25	...	...
Platinum	172	25	...	...
Gray irons; cast	165	24	66	9.6
Copper nickels, wrought	151	22.0	124	18.0
Mullite	145	21	...	...
Zircon	145	21	...	...
Ductile (nodular) austenitic irons; cast	138	20	90	13
Hafnium	138	20	...	...
Copper casting alloys	133	19.3	76	11.0
Vanadium	131	19	124	18
High-copper alloys, wrought	131	19.0	117	17.0
Coppers, wrought	129	18.7	117	17.0
Titanium and its alloys	127	18.5	76	11.0
Copper-nickel-zinc; wrought	124	18.0	124	18.0
Palladium	124	18.0	...	...
Brasses; wrought	124	18.0	103	15.0
Bronzes; wrought	120	17.5	110	16.0
Polycrystalline glass	119	17.3	86	12.5

(continued)

PET, polyethylene terephthalate; ECTFE, ethylene tetrafluoroethylene; ETCFE, ethylene chlorotrifluoroethylene; PVC, polyvinyl chloride; PVF, polyvinyl formal; FEP, fluorinated ethylene propylene; PTFE, polytetrafluoroethylene

Table 2 (continued)

Material	Tensile modulus			
	High		Low	
	GPa	10^6 psi	GPa	10^6 psi
Niobium and its alloys	110	16.0	79	11.5
Silicon	107	15.5	...	...
Zirconium and its alloys	96	14.0	95	13.8
Zinc alloys; wrought	96	14.0	43	6.2
Rare earths	84	12.2	15	2.2
Gold	82	12.0	...	...
Aluminum alloys, 4000 series	79	11.4	...	...
Silver	76	11.0	...	...
Boron nitride	76	11	48	7
Aluminum alloys, 2000 series	74	10.8	70	10.2
Silica	72	10.5	...	...
Aluminum alloys, 7000 series	72	10.4	71	10.3
Aluminum alloys, 5000 series	71	10.3	69	10.0
Thorium	71	10.3	...	...
Aluminum alloys, 1000 series	69	10.0	69	10.0
Aluminum alloys, 3000 series	69	10.0	69	10.0
Aluminum alloys, 6000 series	69	10.0	69	10.0
Thorium	69	10.0	...	...
Tin and its alloys	53	7.7	41	6.0
Cordierite	48	7	...	...
Magnesium alloys; wrought	45	6.5	41	6.0
Magnesium alloys; cast	45	6.5	45	6.5
Polyesters, thermoset, pultrusions, general purpose	41	6.0	16	2.3
Epoxy, glass laminates	40	5.8	23	3.3
Glass fiber-epoxy composites	34	5	...	...
Bismuth	32	4.6	...	...
Polyimides; glass reinforced	31	4.5	...	...
Carbon graphite	28	4.0	4	0.6
Graphite, pyrolytic	28	4.0	...	...
Phenolics; reinforced	23	3.3	2.4	0.35
Alkyds	20	2.9	13	1.9
Graphite; recrystallized	19	2.7	5.5	0.8
Hickory (shag bark)	15	2.2	...	...
Locust (black)	14	2.1	...	...
Polyester, thermoplastic, PET; 45 and 30% glass reinforced	14	2.1	9	1.3
Birch (yellow)	13	2.0	...	...
Douglas fir (coat type)	13	2.0	...	...
Lead and its alloys	13	2.0	...	...
Pine (long needle, ponderosa)	13	2.0	9	1.3
Polyesters, thermoset, reinforced moldings	13	2.0	8.3	1.2
Ash (white)	12	1.8	...	...
Graphite, general purpose	12	1.8	3.4	0.5
Maple (sugar)	12	1.8	...	...
Oak (red, white)	12	1.8	...	...
Styrene acrylonitrile; 30% glass reinforced	12	1.8	...	...
Beech	11	1.7	...	...
Carbon and graphite, fibrous reinforced	12	1.8	2	0.3
Graphite, premium	11.7	1.7	4.8	0.7
Walnut (black)	11.7	1.7	...	...
Polycarbonate, 40% glass reinforced	11.7	1.7	5.9	0.86
Spruce (sitka)	11.0	1.6	...	...
Poplar (yellow)	11.0	1.6	...	...
Carbon, petroleum coke base	11.0	1.6	15.8	2.3
Indium	10.8	1.57	...	...
Basswood	10.3	1.5	...	...
Elm (rock)	10.3	1.5	...	...
Polysulfone, 30–40% glass reinforced	10.3	1.5	7.6	1.1
Cypress (Southern bald)	9.6	1.4	...	...
Nylons; 30% glass reinforced	9.7	1.4	6.9	1.0
Polyester, thermoplastic, PBT; 40 and 15% glass reinforced	9.7	1.4	5.5	0.8
Cedar (Port Orford)	8.9	1.3	...	...
Cottonwood (black)	8.9	1.3	...	...
Phenylene oxide based resins; 20–30% glass reinforced	9.0	1.3	6.4	0.93
Redwood (virgin)	8.9	1.3	...	...
Acetal, copolymer; 25% glass reinforced	8.6	1.25	...	...
Carbon, anthracite coal base	8.2	1.2	4.1	0.6

(continued)

PET, polyethylene terephthalate; ECTFE, ethylene tetrafluoroethylene; ETCFE, ethylene chlorotrifluoroethylene; PVC, polyvinyl chloride; PVF, polyvinyl formal; FEP, fluorinated ethylene propylene; PTFE, polytetrafluoroethylene

Table 2 (continued)

Material	Tensile modulus, High, GPa	Tensile modulus, High, 10^6 psi	Tensile modulus, Low, GPa	Tensile modulus, Low, 10^6 psi
Diallyl phthalates, reinforced	8.3	1.2	4.1	0.6
Fir (balsam)	8.3	1.2	...	...
Hemlock (Eastern, Western)	8.3	1.2	10.3	1.5
Pine (Eastern white)	8.3	1.2	...	...
Polybutadienes	8.3	1.2	2.8	0.4
Polystyrene, 30% glass reinforced	8.3	1.2	...	...
Polyphenylene sulfide, 40% glass reinforced	7.7	1.12	...	...
Fluorocarbon, ETFE and ECTFE; glass reinforced	7.6	1.1	...	...
Melamines, cellulose electrical	7.6	1.1	6.9	1.0
Cedar (Eastern red)	6.2	0.9	...	...
Polyimides, unreinforced	4.8	0.70	3.1	0.45
Polyesters, thermoset, cast, rigid	4.5	0.65	1.0	0.15
Acetal, homopolymer; unreinforced	3.6	0.52	...	...
Acrylics, cast, general purpose	3.4	0.50	2.4	0.35
Acrylics, moldings	3.4	0.50	1.6	0.23
Nylon, mineral reinforced	3.4	0.5	...	...
Polystyrene, general purpose	3.4	0.50	3.2	0.46
Styrene acrylonitrile; unreinforced	3.4	0.50	2.8	0.40
Nylons; general purpose	3.3	0.48	1.9	0.28
Polyphenylene sulfide, unreinforced	3.3	0.48	...	...
Polystyrene, impact grades	3.2	0.47	1.0	0.15
Epoxies, cast	3.1	0.45	0.3	0.05
Polycarbonate, unreinforced	3.1	0.45	2.3	0.34
ABS	2.9	0.42	2.0	0.29
Acetal, copolymer; unreinforced	2.8	0.41	...	...
Phenylene oxide based resins; unreinforced	2.6	0.38	2.5	0.36
ABS/polycarbonate	2.6	0.37	...	...
Acrylic/PVC	2.6	0.37	2.3	0.34
Polyaryl sulfone	2.6	0.37	...	...
Polysulfone; unreinforced	2.5	0.36	...	...
Polyether sulfone	2.4	0.35	...	...
ABS/PVC, rigid	2.3	0.33	...	...
ABS/polysulfone (polyaryl ether)	2.2	0.32	...	...
Allyl diglycol carbonate	2.1	0.30	...	...
Fluorocarbon, PTFCE	2.1	0.30	1.3	0.19
Fluorocarbon, ETFE and ECTFE; unreinforced	1.7	0.24	...	...
ABS/polyurethane	1.5	0.22	1.1	0.16
Polypropylene, general purpose	1.5	0.22	1.1	0.16
Polymethylpentene	1.4	0.21	...	...
Fluorocarbon, PVF	1.4	0.2	1.2	0.17
Vinylidene chloride copolymer; oriented	1.38	0.20	...	...
Polypropylene, high impact	0.9	0.13	...	...
Polyethylene, high molecular weight	0.69	0.1	...	...
Fluorocarbon, FEP	0.5	0.07	0.3	0.05
Fluorocarbon, PTFE	0.5	0.07	0.3	0.04
Vinylidene chloride copolymer; unoriented	0.48	0.07	...	...
Polybutylene, homopolymer	0.25	0.036	0.23	0.034
Polybutylene, copolymer	0.23	0.034	0.08	0.012
Polyacrylate, unfilled	0.20	0.29	...	...
Polyethylenes, low density	0.19	0.027	0.14	0.020
PVC, PVC-acetate, nonrigid	0.021	0.003	0.0027	0.0004

PET, polyethylene terephthalate; ECTFE, ethylene tetrafluoroethylene; ETCFE, ethylene chlorotrifluoroethylene; PVC, polyvinyl chloride; PVF, polyvinyl formal; FEP, fluorinated ethylene propylene; PTFE, polytetrafluoroethylene

Index

E

F

G

H

I

J

L

M

T

U

V

W

Y

Z

FSC
www.fsc.org
MIX
Papier | Fördert gute Waldnutzung
FSC® C083411

Kolibri 360 GmbH
Ferdinand-Jühlke-Straße 7
99095 Erfurt, Deutschland
produktsicherheit@kolibri360.de